PROCESSING AND PROPERTIES OF LIGHTWEIGHT CELLULAR METALS AND STRUCTURES

Edited by

A.K. Ghosh
T.H. Sanders
T.D. Claar

Global Symposium on
Materials Processing & Manufacturing

PROCESSING AND PROPERTIES OF LIGHTWEIGHT CELLULAR METALS AND STRUCTURES

Proceedings of a symposium sponsored by
the Materials Processing and Manufacturing Division (MPMD) of
TMS (The Minerals, Metals & Materials Society).
Held during the 2002 TMS Annual Meeting in Seattle, Washington
February 17-21, 2002

Edited by

A.K. Ghosh
T.H. Sanders
T.D. Claar

A Publication of

TMS

A Publication of The Minerals, Metals & Materials Society
184 Thorn Hill Road
Warrendale, Pennsylvania 15086-7528
(724) 776-9000

Visit the TMS web site at
http://www.tms.org

The Minerals, Metals & Materials Society is not responsible for statements or opinions and is absolved of liability due to misuse of information contained in this publication.

Printed in the United States of America
Library of Congress Catalog Number 200109
ISBN Number 0-87339-527-1

Authorization to photocopy items for internal or personal use, or the internal or personal use of specific clients, is granted by The Minerals, Metals & Materials Society for users registered with the Copyright Clearance Center (CCC) Transactional Reporting Service, provided that the base fee of $7.00 per copy is paid directly to Copyright Clearance Center, 27 Congress Street, Salem, Massachusetts 01970. For those organizations that have been granted a photocopy license by Copyright Clearance Center, a separate system of payment has been arranged.

© 2002

If you are interested in purchasing a copy of this book, or if you would like to receive the latest TMS publications catalog, please telephone 1-800-759-4867 (U.S. only) or 724-776-9000, EXT. 270.

PREFACE

Cellular metals and designed porous structures are attractive lightweight structural materials with potential applications in transportation, electronics, biomedical, defense, building and energy-based industry. Such materials as metal foams, sponge-like metals as well as periodic geometry core structures, e.g. trusscore, honeycomb, hollow sphere, triangular core, etc. are being increasingly considered a solution to problems of lightweight construction, heat exchangers, passive safety, sound dampening, filtering, thermal stability and other purposes. Other cellular structures are designed and built per mathematical design methodology to yield uncommon properties. While both closed and open cell metal foams have been of interest, the open cell materials appear to have many possible applications in addition to their possible use as load carrying structures. A number of companies are involved in the manufacturing such materials and evaluating their properties, and the possible application of cellular metals in a variety of functions. Modeling of thermal effects, deformation and other characteristics have also made steady progress, and design criteria for their use is evolving. While widespread use of the technologies for cellular metals is still in the future, the momentum in the basic technologies has been rising in the United States, Europe and Japan in the past 10 years. Much of the effort in the U.S.A. has been funded through government sponsors such as DARPA and ONR to launch the first wave of developments in this field. However, the field still remains at an early developmental stage, waiting for major breakthroughs in performance, lower cost and knowledge to integrate into design. This is thus a key point in time to capture the status of this emerging enabling technology.

This book underscores the enthusiasm in the outstanding potential in this unique group of materials and structures, and we hope that it will stimulate yet greater interest in the scientific community, among the research sponsors and industry. The authors of the papers here have expounded on a variety of key areas of development, such as, new design concepts and functional properties, new processing technologies, microstructure and property characterization, reinforcement of foam metals, modeling of properties, component fabrication, and applications. The papers have been organized into following broad categories: Multifunctional Properties and Processing, Novel and Low-Cost Processes and Principles, Oxide-Reduction and P/M Processes, and Mechanical and Service Properties.

Publication of these proceedings has been made possible as part of the Third Global symposium of TMS, held during the TMS Annual Conference in Seattle, Washington, February 17-21, 2002. The editors gratefully acknowledge the moral encouragement of Dr. Steve Fishman of the U.S. Office of Naval Research who has supported much of the research and development activities documented here, and a financial grant from ONR to defray the cost of publication. We also acknowledge the help and cooperation of the TMS staff, its publication department, and the able workmanship of McKay Press of Midland, Michigan.

Amit Ghosh	Tom Sanders	Dennis Claar
University of Michigan	Georgia Institute of Technology	Fraunhofer USA
Ann Arbor, Michigan	Atlanta, Georgia	Newark, Delaware

TABLE OF CONTENTS

Preface v

Section I: Multifunctional Properties and Processing

Multifunctionality of Closed-Cell Aluminum Foams 3
    T.D. Claar, V. Irick, J. Adkins. and K. Kremer

The Stiffness and Weldability of an Ultra-Light Steel Sandwich Sheet Material
with a Fibrous Metal Core 15
    A.E. Markaki, S.A. Westgate and T.W. Clyne

Reinforced Alulight for Structural Use 25
    F. Simancik, W. Rajner, and R. Laag

Constructed Cellular Metals 35
    D.J. Sypeck

Cellular Structure by Direct Materials Deposition 47
    E. Stiles and J. Mazumder

Section II: Novel and Low-Cost Processes and Principles

Processing of Controlled Porosity Titanium-Based Materials 61
    D. Kupp, D. Claar, K. Flemmig, U. Waag and H. Goehler

Novel Lightweight Open Cell Metal Foam Process and Resulting Properties.............. 73
    A.K. Ghosh and X. Tong

Processing of In-718 Lattice Block Castings 85
    M.G. Hebsur

Advances in the Melt Route Production of Closed Cell Aluminium Foams using
Gas-Generating Agents.. 97
    V. Gergely, D.C. Curran, and T.W. Clyne

Rheological Changes of Cellular Structure in Foaming Metal 107
    Z. Song and S.R. Nutt

Role of Wetting Behavior in the Stabilization of Liquid Metal Foams by Ceramic Particles: An Experimental Simulation Study 115
    T. Gao and Y.Q. Sun

Section III: Oxide-Reduction and P/M Processes

Multifunctional Metallic Honeycombs by Thermal Chemical Processing 127
    J.K. Cochran, K.J. Lee, D. McDowell, and T. Sanders

Metal Honeycomb from Oxide Paste: Maraging Steel and Super Invar Structure and Properties 137
    J.L. Clark, J.K. Cochran, T.H. Sanders, and K.J. Lee

Reduction of Iron and Chromium Sesquioxide Powder Mixtures for Metal Honeycomb Structures 147
    J.H. Nadler, T.H. Sanders, Jr., and J.K. Cochran

Copper Alloys from Oxide Reduction for High Conductivity Applications 157
    B.C. Church, J.K. Cochran, and T.H. Sanders, Jr.

Modeling Powder Extrusion Pastes for Forming Lightweight Multifunctional Structures 167
    K.M. Hurysz, R. Oh, J.K. Cochran, T.H. Sanders, Jr., and K.J. Lee

Effect of Processing Variables on Solid-State Foaming of Titanium by Superplasticity 177
    N.G. Davis and D.C. Dunand

An investigation of the Effects of Sintering Duration and Powder Sizes on the Porosity and Compression Strength of Porous Ti-6Al-4V 189
    R. Cirincione, R. Anderson, J. Zhou, D. Mumm, and W.O. Soboyejo

Section IV: Mechanical and Service Properties

An Investigation of Compressive Deformation Mechanisms in Open Cell Metallic Foams 201
    J. Zhou, P. Shrotriya, and W.O. Soboyejo

Fabrication and Creep Studies on Open Cell Nickel Structures 211
    R.K. Oruganti and A.K. Ghosh

Properties of Maraging Steel Honeycomb under Compressive Quasistatic and
Dynamic Loading 223
    A.M. Hayes, D.L. McDowell, and J.K. Cochran, Jr.

Mechanical Testing of In718 Lattice Block Structures 233
    D.L. Krause, J.D. Whittenberger, P.T. Kantzos, and M.G. Hebsur

The Collapse and Energy Absorption of Egg-Box Panels 243
    M. Zupan, N.A. Fleck, M.F. Ashby

Mechanical Properties and In-Situ Diffraction Strain Measurements in Aluminum-Mullite
Microsphere Syntactic Foams Produced by Liquid Metal Infiltration 251
    D.K. Balch and D.C. Dunand

Vibration Properties of Closed-Cell Aluminum Foam 261
    H.F. von Bremen, V.S. Sokolinsky, S. Rajaram, H. Shen, J.A. Lavoie, Y. Zhou, and S.R. Nutt

3D Microtomography (μCT) of Cellular Metals using an up to 320kV X-ray Tube 271
    B. Illerhaus, E. Jasinien, A. Kottar, and J. Goebels

Section V: Additional Abstracts

Perspectives On High-Volume Automotive Applications Of Lightweight
Cellular Metals And Structures..283
    R. Jahn and A. M. Sherman

Mechanical Behavior Of Periodic Hollow Sphere Foam..284
    W. S. Sanders and L. J. Gibson

Mechanical Behavior Of A Closed-Cell Aluminum Foam...285
    C. M. Cady, G. T. Gray, C. P. Trujillo and T. Mukai

Tailored Component Fabrication Using Stabilized Aluminum Foam...........................286
    G. Mills and S. Nichol

Processing And Properties Of Steel Foam Materials And Structures...........................287
    T. D. Claar, V. Irick, K. Kremer and J. Adkins

Development Of Vibration Damping Techniques In Metallic-Intermetallic
Laminate (MIL) Composites..288
    A.Rohatgi, K. S. Vecchio and J. B. Kosmatka

Author Index..289

Subject Index...291

SECTION I

MULTIFUNCTIONAL PROPERTIES AND PROCESSING

MULTIFUNCTIONALITY OF CLOSED-CELL ALUMINUM FOAMS

D. Claar, V. Irick, J. Adkins, and K. Kremer
Fraunhofer USA - Delaware
Center for Manufacturing and Advanced Materials
501 Wyoming Road
Newark, DE 19716

Abstract

Closed-cell aluminum foams are an emerging class of ultra-lightweight materials that can be tailored to exhibit a broad range of properties. The powder metallurgy-based foaming process is capable of fabricating foams with porosity levels of 40 to 90 vol%. Various geometries are possible, including 3-D molded shapes, foam-filled tubes, and aluminum foam sandwich (AFS) panels. A significant characteristic of aluminum foams is their ability to absorb energy during compressive deformation. Energy is absorbed during crushing at a nearly constant "plateau" stress to approximately 60% strain, before the stress increases significantly. Aluminum foam inserts inside metal tubes increased the specific energy absorption by 35-40% compared to the hollow tubes, based on axial compression tests. Foam inserts also increased the bending stiffness of hollow tubes by 40-45%. Foamed aluminum has demonstrated superior fire resistance compared to solid aluminum plates. During exposure to a gas flame at 850-870°C, a 6061 Al plate experienced melt-through, while an Al-Si alloy foam plate was essentially unaffected. Aluminum foams also improve the performance of lightweight armor, providing improved projectile defeat, lower dynamic deflection, and enhanced ballistic shock wave attenuation. AFS panels have very high specific stiffness, allowing significant weight reduction. Applications that take advantage of these multifunctional features of Al foams will be presented.

Processing and Properties of Lightweight Cellular Metals and Structures
Edited by Amit Ghosh, Tom Sanders and Dennis Claar
TMS, 2002

Introduction

Metal foams with high levels of controlled porosity are an emerging class of ultra-lightweight materials that are receiving increased attention for a broad range of applications (1-3). The Fraunhofer Institute for Applied Materials Research in Bremen, Germany, has developed a powder metallurgy process (4-6) for producing ultra-lightweight aluminum foam materials and structures that are attractive for a broad range of transportation applications, including commercial and military vehicles, lightweight armor systems, crash energy absorption, and other uses. Further work on process scale-up and applications is being conducted at the Fraunhofer USA Center - Delaware (FhUSA-DE), which is part of Fraunhofer USA, a wholly owned subsidiary of Fraunhofer Gesellschaft.

The Fraunhofer USA Center - Delaware is working with designers, manufacturers and end-users to evaluate and demonstrate the advantages of metal foams in advanced vehicles, ships, railroad equipment, and aerospace structures. This paper provides an overview of the potential multifunctional advantages that metal foam technology can offer for the flexible design and enhanced performance in advanced systems applications.

The Fraunhofer powder metallurgy-based foaming agent process for producing ultra-lightweight aluminum foams is shown in Figure. 1.

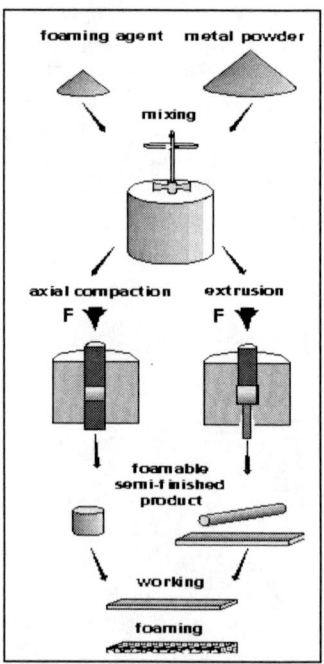

Figure 1: Schematic diagram of Fraunhofer powder metallurgy-based foaming agent process for producing ultra-lightweight aluminum foams.

A compacted powder mixture of metal + foaming agent is heated to a temperature near the metal melting point. The foaming agent, which is titanium hydride for the production of Al foam, decomposes and forms a gas that is trapped inside the compacted powder body. Gas bubbles create voids within the expanding body of semi-solid metal and are retained during solidification.

This process results in a lightweight structure with a high degree of closed-cell porosity, which can be controlled over the range of approximately 40 to 90 vol %. The microstructure of closed cell aluminum foam with a porosity of about 80 vol % is shown in Figure. 2.

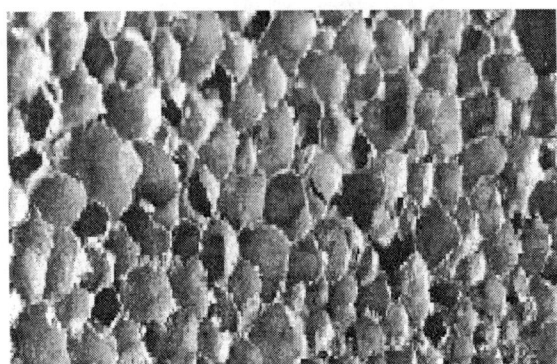

Figure 2: Microstructure of aluminum foam produced by the powder metallurgy process.

Aluminum foams can be produced in several different configurations:

Complex 3-dimensional metal foam parts can be molded to net-shape, as shown in Figure. 3. The foamable preform of consolidated metal powder + foaming agent is placed inside the cavity of a forming tool. This mold assembly is then placed inside a furnace and heated to near the melting point of the metal alloy powder. The action of the foaming agent causes significant volume expansion resulting from the evolution of gas bubbles, which form the closed-cell pores. The molded foam part expands to fill the entire mold cavity, resulting in a 3-D shape. Figure 3 shows an aluminum foam part produced in a mold to replicate the shape of a barbell weight. This part has a bulk density of approximately 0.5 g/cm^3, which corresponds to about 80 vol % porosity. A thin skin of dense aluminum forms on the outer surface of the part. This outer skin of metal is very useful in joining metal foams to other materials to fabricate engineering structures.

Sandwich panels of metal foam cores inside metal face sheets can also be produced. Aluminum foam sandwich panels, such as that shown in Figure 4, are being used by the specialty German automaker Karmann GmbH. Karmann has shown the high specific stiffness and weight reduction advantages of aluminum foam sandwich panels in convertible body structures (7,8). These foam sandwich panels are formed by roll cladding aluminum face sheets onto an extruded billet of aluminum + foaming agent. The resulting preform sheet, consisting of the densified powder core sandwiched between aluminum face sheets, is heated to activate the foaming agent. This results in expansion of the foam core thickness by about 400%, yielding 80 vol% porosity. Sandwich panels having complex contoured surfaces, as shown in

Figure 4, are produced by press-forming the semi-finished panel into the desired shape before the foaming step. Aluminum foam sandwich panels up to about 3 ft x 5 ft x 0.5 in thick are being produced for prototype vehicles.

Metal tubes or hollow profiles filled with metal foam have been produced out of aluminum, as exhibited in Figure 5. These Al foam-filled tubes were fabricated by inserting consolidated powder preforms into the tubes and heating to initiate the foaming reaction, thus filling the inside of the tubes with lightweight foam. In this case, the Al foams are metallurgically bonded to the inner surface of the tubes, providing an excellent interfacial bond. Foam-filled tubes provide significant stiffening, resistance to buckling, and ability to absorb impact energy.

Alternatively, foam cores that have been formed separately can be inserted inside hollow tubes. This type of foam core may be either force fit into the tube or adhesively bonded to the tube ID surface. The foam can also be selectively located at critical high-load regions of the tubes. Tubular shapes with aluminum foam cores can also be hydroformed prior to the foaming step to create complex structures for specific applications.

Figure 3: Three-dimensional aluminum foam part molded to net shape in tooling. Thin skin on the outer surface covers the inner foam core and provides a means of joining Al foam to other materials.

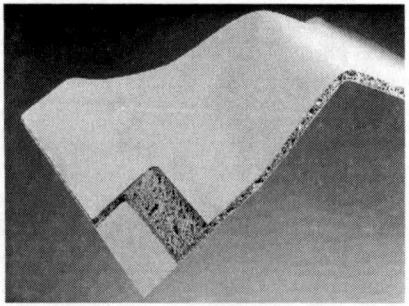

Figure 4: Aluminum foam sandwich panel press formed into contoured surface before foaming. Cut-away section shows the Al foam core inside thin Al outer face sheets.

Figure 5: Aluminum foams of Al-6Si-4Cu in-situ bonded metallurgically inside 6061 Al tubes during the foaming step.

The P/M foaming agent process allows a high degree of closed porosity to be introduced into metal foams such as aluminum. Furthermore, the porosity can be varied over the broad range of approximately 40 to 90 vol %, allowing the properties of Al foams to be tailored to meet the multifunctional requirements of specific applications. Typical characteristics of Al foams include:

- Ultra-lightweight materials with high degree of homogeneous closed-cell porosity
- Foam microstructures tailorable over the range 40 to 90% porosity (1.6 to 0.3 g/cm^3)
- High stiffness-to-weight and strength-to-weight ratio
- Ability to absorb energy from impact, crash, and explosive blasts via compressive deformation

  mechanisms to high levels of strain (up to 80% compression are possible)
- Fire resistance and thermal insulating properties
- Vibration damping and sound absorption
- Capability of forming into various geometries, including molded foam shapes, sandwich panels

  with aluminum foam core, and aluminum foam-filled tubes
    - Readily recycled

Applications of Closed-Cell Aluminum Foams

<u>Vehicle Crash Energy Management</u>

Aluminum foams are being evaluated by Fraunhofer for crash energy management in automotive and high speed rail systems (9). Because of their high porosity levels and energy absorption mechanisms during compressive crushing, Al foams can absorb high levels of kinetic energy via plastic deformation. Axial compression tests were conducted on tubes of 6063 Al and 304 stainless steel – both empty tubes and tubes filled with 80% porous Al foam inserts. The tubular test specimens were 3 x 3 in. cross-section x 12 in. long. Figure 6 shows the axial compression load-deformation curves for the various test specimens of Al foam, empty 304 SS tube, and 304 SS tubes with Al foam inserts either force fit or adhesive bonded into the tubes. These test results indicated that metal alloy tubes filled with Al foam inserts can absorb more energy per unit mass than comparable empty tubes: 33–38% higher specific energy absorption for Al foam-filled Type 304 stainless steel tubes and 28–32% greater for Al foam-filled 6063 Al tubes. Figure 7 is a cross-section through an Al-foamed filled 304 SS tube subjected to about 60% axial compression. This photo shows the densification of the Al foam and interaction between the Al foam insert and fold convolutions in the outer tube wall, which led to the increased energy absorption capabilities observed with Al foams. In addition to improving axial crush energy absorption, Al foam inserts inside 6063 Al and 304 SS tubes increased their load bearing capabilities by 40 – 45% when subjected to 3-point transverse bending loads.

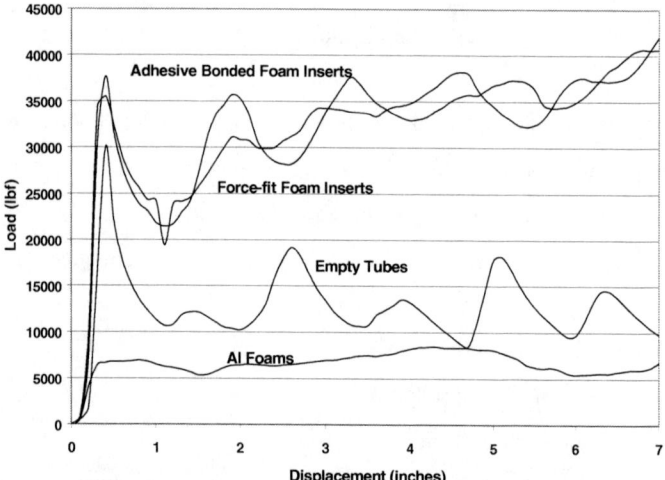

Figure 6: Load vs. deformation curves for axial crushing tests of Al Foams, empty 304 SS tubes, and Al foam-filled 304 SS tubes. Each curve represents average of four tests per sample type.

The Al foam inserts increased the specific energy absorption during axial compression by 33 to 38% compared to the empty 304 SS tubes.

Protection Against Ballistic Threats

The Army Research Office and the Army Research Laboratory have sponsored programs at Fraunhofer USA – Delaware to evaluate the effectiveness of aluminum foams in lightweight

Figure 7: Al foam-filled 304 SS tubes axially crushed to approximately 60% strain. Foam-filled tubes absorbed 38% greater energy per unit weight than empty 304 SS tubes.

hybrid armor systems. Hybrid armor layups with selective placement of 80% porosity Al foams have been tested ballistically using 20 mm Fragment Simulating Projectiles (FSP) at velocities of 2,750 ft.sec. By comparing the performance of the baseline hybrid armor developed for the Army's Composite Armored Vehicle vs. hybrid armor layups containing thin layers of Al foam, the metal foams were found to provide the following advantages in armor performance (10):

- The unique microstructure of closed cell Al foams provides the capability of delaying and attenuating stress wave propagation through the armor layup, acting as a stress wave filter,
- Plastic deformation of Al foams significantly reduces dynamic backface deflection and leads to less volumetric delamination of the fiber-reinforced polymer composite backing plate,
- Al foams provide improved ceramic armor fracture behavior, overall ballistic performance, and reduced areal density of the hybrid armor system,
- Less damage to the over cover layer with Al foam inserts is expected to lead to improved multi-hit capabilities.

Figure 8 shows a cross-section through an Al foam insert from a 20 mm FSP ballistic test, showing the plastic compressive deformation and shear bands in the deformed foam. This Al foam layer totally attenuated and absorbed the elastic shock wave associated with the ballistic event, as evidenced by the absence of any signal from the stress gage in the instrumented test and complete protection of a fragile Al_2O_3 ceramic tile located behind the Al foam insert.

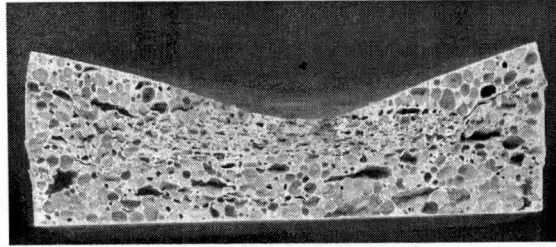

Figure 8: Plastic deformation of Al foam insert from hybrid armor lay-up tested against 20 mm FSP threat at 2,750 ft/sec. Al foam totally absorbed elastic shock wave and protected Al_2O_3 ceramic tile behind against fracture.

Protection Against Explosive Blast Threats

Based on the attenuation and absorption of elastic shock waves in the ballistic tests described above, it is expected that Al foams will be very effective in blast mitigation, such as protection against military mines and explosive blasts resulting from acts of terrorism. Several preliminary blast tests were performed on panels of Al foam with 80% porosity (18 x 18 x 2 in. size) sandwiched between front and back steel face sheets. A cannister containing 2 lb of PETN explosive was detonated at various stand-off distances from the blast test panel (see blast test set-up and photograph of blast event in Figure 9 below). At 0.5 meter stand-off, the Al foam panel was densified about 50% in the center by the blast but remained intact, and the steel backplate was not deformed. At 0.1 meter stand-off, the Al foam panel was densified about 80% in local regions and fractured into several pieces. In both cases, the foam protected the steel sheets against perforation, as shown in Figure 10.

Figure 9: Set-up for blast testing of aluminum foam panels using 2 lb PETN explosive charge in cannister (left).
Photo of actual blast event (right).

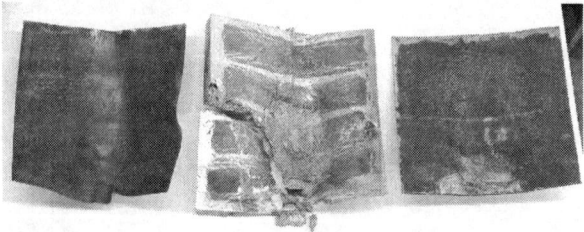

Figure 10: Aluminum foam (center) deformed by blast loading at 0.1 meter stand-off. Foam absorbed blast energy and prevented perforation of front (left) and back (right) steel plates.

Fire Resistance of Aluminum Foam Panels

The fire resistance of aluminum foam panels was tested according to the principles of Harmathy's Rules of Fire Endurance (11), which is often used in testing materials for building construction. This test procedure specifies that a representative sample of the construction material be exposed on one side to the atmosphere of a specially built test furnace in which the temperature is controlled to follow a prescribed curve representing, in an idealized way, the temperature in a burning room. The other side of the sample remains at all times in contact with the cooler ambient atmosphere.

The first fire test was conducted on a single piece of solid aluminum 6061 plate while the second test was on two pieces of Al-7wt % Si foam panels joined together and having approximately the same thermal mass as the plate. The dimensions were 18 x 36 x 0.5 in. for the aluminum plate and 18 x 36 x 2 in. for the aluminum foam panel. The test apparatus consisted of a concrete test oven chamber, 35 x 16 x 16 in. open on one side, fired with a propane torch. The test sample was affixed to the open side of the test chamber using clamps. Three thermocouples were installed, one in the test chamber near the exposed side of the test sample, and two affixed to the top, unexposed side of the sample. The temperature data were recorded in a data logging device. After the test sample and thermocouples were positioned in place, the propane torch was ignited and manually regulated to a starting flame temperature of about 1400°F in the test chamber.

The flame temperature increased gradually from about 1400°F to 1580-1600°F during the course of the 90 minute test runs. Temperatures measured with thermocouples at various locations were recorded every minute during the tests. Figure 11 shows a piece of solid aluminum plate while Figure 6 shows two pieces of joined aluminum foam panels after being fire tested at temperatures between 1400 and 1600°F for 90 minutes. It can be seen in Figure 11 that significant melting of the localized area (left side of the plate) occurred in the aluminum plate. The observed melting of the 6061 Al plate, which melts over the temperature range of 1080 to 1206°F, was consistent with the measured plate temperature exceeding 1200°F. The photograph in Figure 12 shows that Al foam panels - which had essentially the same thermal mass as the solid alloy plate and was subjected to the same fire testing cycle - did not exhibit any significant melting. The melting temperature of Al-7 wt% Si alloy ranges from 1035 to 1135°F. The lack of significant melting of Al foam is attributed to the lower thermal conductivity of high-porosity aluminum foam, compared to bulk aluminum plates. Thus, it was expected that the temperature of the outer surface (unexposed to flame) of aluminum foam would experience a less rapid temperature rise than the bulk aluminum plate. This effect was in fact observed. The measured temperatures of the two thermocouples affixed to the outer, unexposed surfaces of aluminum foam panels were lower than those in the aluminum plate (the difference was measured to be about 100°F). In summary, under similar heating conditions, aluminum foam made from a low melting temperature Al-Si alloy did

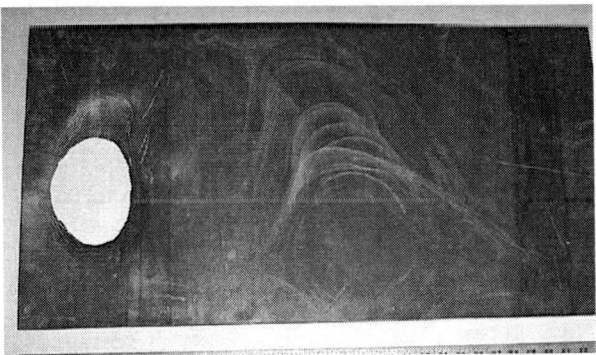

Figure 11: Photograph of solid 6061 aluminum plate after fire testing, showing area of localized melting through the plate.

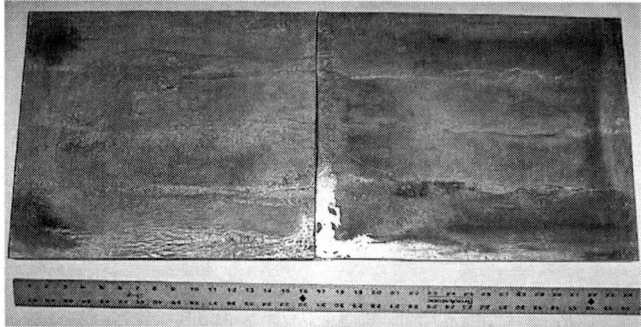

Figure 12: Photograph of the inner surfaces of Al-7Si alloy foam panels after fire testing. Two pieces of Al foam were segmented together to form the 18 x 36 x 2 inch thick test panel.

not melt, while the aluminum plate made from a high melting point Al 6061alloy exhibited melting entirely through the 0.5 inch thick plate at a location down-stream of the propane flame. Aluminum foam also seemed to spread the heat uniformly without localized melting as was observed for the dense aluminum plate. These results indicate that aluminum foams exhibit very good fire resistance and endurance, although more testing is required to quantify these findings.

Conclusions

Ultra-lightweight aluminum foams produced by the powder metallurgy process demonstrate numerous attractive physical, mechanical, and thermal properties. Foam properties can be varied by controlling the porosity level to achieve a broad range of properties tailored to specific applications. Foam characteristics that are especially attractive include lightweight; very high specific stiffness; energy absorption in compression for vehicle crash energy management, enhancement of ballistic armor performance, and explosive blast mitigation; fire resistance; and thermal and sound insulation. Furthermore, it is possible to design structures that take advantage of more than one foam characteristic simultaneously, providing true multifunctionality.

These mutifunctional features of metal foam suggest numerous potential applications of aluminum foam materials and structures:

- Automotive components, such as fire walls, floor panels, and crash energy absorbers
- Energy absorbing crashworthiness structures in high speed rail and aerospace vehicles
- Foam inserts for shock wave attenuation and overall improved performance of lightweight
- hybrid armor systems
- Ballistic armor and blast containment components in military and VIP vehicles
- Anti-terrorism blast mitigation structures
- Fire and flame-resistant structures
- Vibration damping and sound attenuation applications

Based on the promising performance of aluminum foams in a variety of industrial and military applications, Fraunhofer USA is pursuing manufacturing scale-up and commercialization of aluminum foam materials and structures with strategic business partners.

Acknowledgments

The authors wish to thank the following government agencies for partial funding of portions of the work reported herein. The Army Research Office, the Army Research Laboratory, the Office of Naval Research, and the National Academy of Sciences/Transportation Research Board High Speed Rail IDEA Program Office.

References

1. T.D. Claar, C.-J. Yu, J.D. Adkins, and H.H. Eifert, "Properties and Applications of Aluminum Foams: Two Case Studies," Cellular Metals and Metal Foaming Technology, ed. J. Banhart, M, Ashby, and N. Fleck (Verlag MIT; Bremen, Germany, 2001), 37-42.
2. M.F. Ashby et al., Metal Foams: A Design Guide, (Butterworth Heinemann; Boston, MA, 2000.
3. T.D. Claar, C.-J. Yu, I. Hall, J. Banhart, J. Baumeister, and W. Seeliger, "Ultra-lightweight Aluminum Foam Materials for Automotive Applications," Society of Automotive Engineers Paper 2000-01-0335, (2000).
4. J. Baumeister and H. Schrader, "Method for Manufacturing Foamable Metal Bodies", U.S. Patent No. 5,151,246.
5. J. Banhart, J. Baumeister, M. Weber, *Powder Metallurgical Technology for the Production of Metal Foam,* Proc. European Conf. on Advanced PM Materials, Birmingham, (Oct. 1995) 201.
6. J. Banhart and J. Baumeister, "Production Methods for Metallic Foams", Proceedings of 1998 MRS Metal Foam Symposium, 521 (1998), 121-132.
7. W. Seeliger, "Complex Shaped Aluminum Foam Sandwich Panels for Automotive Applications," Fraunhofer USA Metal Foam Symposium, ed. J. Banhart and H. Eifert, (Verlag MIT; Bremen, Germany, Oct. 7-8, 1997), 79-89.
8. W. Seeliger, "New Aspects of Getting Aluminum Foam Sandwich Parts into Volume Production," Cellular Metals and Metal Foaming Technology, ed. J. Banhart, M, Ashby, and N. Fleck (Verlag MIT; Bremen, Germany, 2001), 5-16.
9. C.-J. Yu, T.D. Claar, J. Adkins, and H.H. Eifert, "Metal Foams for Improved Crash Energy Absorption in Rail Passenger Equipment," Proc. of 2001 Int. Conf. on Powder Metallurgy and Particulate Materials, New Orleans (2001).
10. B. Gama, T. Bogetti, B. Fink, C. Yu, D. Claar, H. Eifert, and J. Gillespie, Jr., "Aluminum Foam Integral Armor: A New Dimension in Armor Design," Composite Structures 52 (2001) 381-395.
11. T.Z. Harmathy, Fire Safety Design: Concrete Design and Construction, (Addison Wesley, London 1993).

THE STIFFNESS AND WELDABILITY OF AN ULTRA-LIGHT STEEL SANDWICH SHEET MATERIAL WITH A FIBROUS METAL CORE

A.E. Markaki[†], S.A. Westgate[§] and T.W. Clyne[†]

[†]Department of Materials Science & Metallurgy
University of Cambridge, Pembroke Street
Cambridge CB2 3QZ, UK

[§]TWI
Granta Park, Great Abington
Cambridge CB1 6AL, UK

Abstract

A sandwich material, based on a pair of thin stainless steel faceplates separated by a core incorporating stainless steel fibres, has recently been developed. This material has the potential to exhibit an attractive combination of properties, while retaining formability and general handling characteristics similar to those of conventional steel sheet. Three different core structures have been investigated: (a) transversely-aligned fibres bonded to the faceplates by adhesive, (b) a sintered fibre mat bonded to the faceplates by adhesive and (c) a sintered fibre mat brazed to the face plates. The beam stiffnesses of these three structures have been measured and compared with theoretical predictions. Through-thickness electrical resistances were measured and compared with predictions from simple analytical models. These data were correlated with the results of spot welding tests. It is shown that structure (c) is readily weldable, whereas there are certain difficulties with the adhesively-bonded sheets.

Processing and Properties of Lightweight Cellular Metals and Structures
Edited by Amit Ghosh, Tom Sanders and Dennis Claar
TMS, 2002

Introduction

It has long been recognized that sandwich structures, composed of stiff outer layers held apart by a low density core, offer the potential for very high specific stiffness and other attractive mechanical properties. Most such structures are created by an assembly step of some sort, before or during component manufacture. This limits the flexibility of the production process and is relatively expensive. Nevertheless, there is considerable current interest in producing sandwich structures of different types, many of them based on metallic faceplates and having metal-containing cores - often made of metallic foams [1-5] or some more regular structure such as a truss assembly[6]. A novel type of composite sandwich steel sheet has recently been proposed[7, 8] based on a pair of thin (~ 200 µm) stainless steel faceplates, with a core structure incorporating stainless steel fibers. Certain variants of this structure have also been manufactured. The overall thickness of the sheet, and certain features of the core structure, are such that the material can in many ways be handled and processed as a conventional monolithic metallic sheet. Moreover, not only does it have a very high specific stiffness, but it is also expected to exhibit other potentially attractive characteristics, such as good thermal insulation and high acoustic and vibration damping capacity.

A further aim is to ensure that the sheets can be joined by conventional resistance spot welding techniques. This is considered to be highly desirable, for example, for use in the automobile industry[9]. There has been extensive work on simulation of heat and current flow during spot welding[10-13], on applied force characteristics[14] and on welding of thin monolithic metallic sheets[15]. However, there is little or no information available on welding of metallic sandwich structures, although there has been some work[16] on welding of vibration damping steel. This consists of two mild steel sheets typically 0.3 to 1 mm thick, separated by a thin layer of adhesive (~ 20 to 500 µm). In the present paper, a preliminary investigation is presented into the beam stiffness of these sheets and into their welding characteristics. Experimental data are correlated with predictions from some simple analytical treatments of relevant characteristics.

Experimental Procedure

Material Production

Three materials have been studied. These are termed (a) Hybrid Stainless Steel Assembly (HSSA), (b) Cambridge Bonded Steel Sheets (CAMBOSS) and Cambridge Brazed Steel Sheets (CAMBRASS). The first of these is made by a flocking process[7], in which short (~ 1 mm) 316L stainless steel fibres, of about 25 µm diameter, are aligned approximately normal to the plane of the faceplates (200 µm thick 316L stainless steel sheets) - see Fig.1.

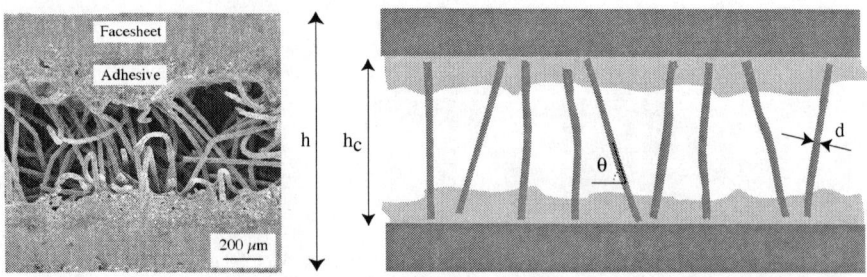

$h=1.2$ mm, $h_c=0.8$ mm, $f \approx 8\%$, $d \approx 25$ µm, $\bar{\theta} \approx 75°$

Fig.1: SEM micrograph and schematic depiction of the structure of HSSA sheet.

The other two materials were obtained by bonding to the faceplates a pre-manufactured mat of the same fibres. This was done either by adhesive bonding, using a two-component epoxy adhesive (Araldite® 420A/B), or by brazing, using a Ni-14Cr-4Fe-2.8B-3.3Si-0.6C braze alloy and a brazing temperature of 1000°C for a period of about 5 minutes. For these materials, the cores were denser than the HSSA, the fibres occupying about 19% of the volume of the core, compared with about 8% in the HSSA. The structure is depicted in Fig.2.

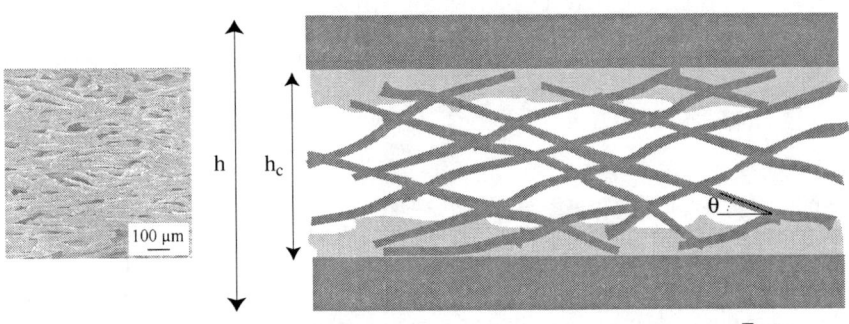

h=1.3 mm, h_c=0.9 mm, d≈25 μm, f≈19%, $\bar{\theta}$≈10°

Fig.2: SEM micrograph and schematic depiction of the structure of CAMBOSS sheet.

Stiffness and Density Measurement

The beam stiffness was measured by cantilever bending, using a Lasermike scanning laser extensometer to monitor the beam deflection and correcting for the possibility of errors arising from extraneous beam torsion. Areal density measurements were made using a Sartorius microbalance with a precision of ±10 μg.

Electrical Conductivity Measurement

The specimen was in the form of a small rectangular coupon (10 x 14 mm). An AC circuit was used for the measurements. A fixed current (1 A in this case) was passed through the specimen via flat ended probes on the faceplates. This current was modulated at a constant frequency of 1 kHz. The potential drop across the specimen, from which the resistivity of the core was deduced, was then amplified and measured. In making this measurement, the potential drop across other resistances in the sensing circuit must be eliminated. This drop generates an offset in the mean voltage of the AC signal. This DC offset was removed by passing the signal through an AC amplifier, after which it was rectified by a demodulator and then passed through an integrator to remove noise and finally displayed as a DC voltage (few mV to ~ 1 V for these specimens). Full details of the technique are given elsewhere[17].

Resistance Welding Experiments

The resistance welding trials were conduced on a single phase AC machine. Cu/Cr/Zr electrodes, of 16 mm diameter, were used to ISO 5821 type E design, having a 6 mm radius dome tip. An electrode force of 1.5 kN was chosen for the initial trials. In cases where the matrix would not allow current to flow at the low voltages used in resistance welding (typically about 5 V), a clamp tool was used to provide a current shunt path, in parallel with the electrodes, linking the outer sheet surfaces. Welding current was measured with a commercial meter and also, together with the voltage across the electrodes, on a transient recorder to illustrate the waveforms.

Microstructural Examination

Scanning electron micrographs through sections of the sandwich sheet material and the fibrous metal core were obtained using a Jeol-5800 scanning electron microscope. Spot weld sections were infiltrated in vacuum by a low viscosity resin (Epofix, supplied by Struers). After vacuum impregnation, the specimens were ground with a series of SiC papers and afterwards further polished with 6, 1 and 1/4 µm diamond paste. Macrographs of the welded specimens were acquired using a Nikon Coolpix 950 digital camera attached to a Zeiss optical microscope. To reveal the grain structure, a standard microetchant for stainless steel grades was used (5 g $CuCl_2$, 40 ml HCL, 30 ml H_2O, 25 ml ethanol).

Beam Stiffness

Fig. 3 shows predicted and measured beam stiffnesses, plotted against the areal density. Fig.3(a) compares plots for solid steel, titanium and aluminium sheets with that for a sandwich composed of 200 µm steel faceplates and an isotropic epoxy core (E = 2 GPa, density = 1.56 Mg m^{-3}). This demonstrates that such a structure can be simultaneously stiffer, lighter and thinner than a monolithic aluminium sheet. The comparison with experimental results shown in Fig.3(b) indicates that all 3 sheets are less stiff than expected. The HSSA and CAMBOSS sheets have only about 50% of the predicted stiffness, while the figure for CAMBRASS is about 70%. This is attributed to the low resistance to compression and shear exhibited by these cores. Deeper study of this would require evaluation of the 5 elastic constants (assuming in-plane isotropy) of the core. It may, however, be noted that considerable improvements in beam stiffness are expected[8] by arranging for the fibres to lie at angles of approximately 45° and by increasing the fibre diameter.

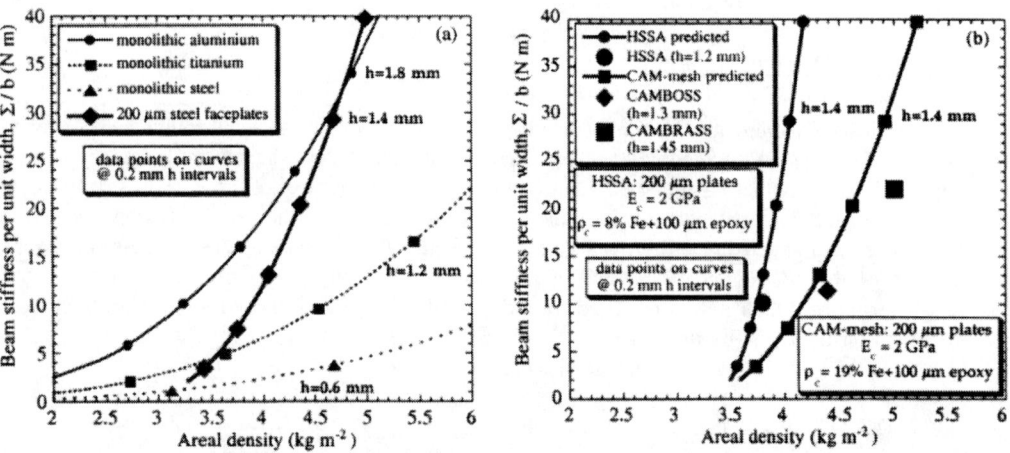

Fig.3: Beam stiffness as a function of areal density, showing (a) benefits of a sandwich structure (isotropic core with E = 2 GPa, ρ = 1.56 Mg m^{-3}) and (b) comparison between theory and experiment for the three cases.

Through-thickness Electrical Resistance

Simple analytical models have been developed (Fig.4), giving the resistivity of the core as a function of geometrical variables, for the two fibre arrangements. For the transverse fibre core, a square array is assumed, with side of length L. The fibre volume fraction is thus given by

$$f = \frac{\pi\, d^2}{4\, L^2} \tag{1}$$

The resistance of a fibre column, R_{col}, is equal to the sum of the resistance offered by the fibre itself and the contact resistance at each interface with the faceplates, R_*.

$$R_{col} = R_{fib} + 2R_* = \frac{4\, \rho_{fib}\, h_c}{\pi\, d^2} + 2R_* \tag{2}$$

where h is the separation of the faceplates (~ fibre length), d is the fibre diameter and ρ_{fib} is the fibre material resistivity. Since there is, in effect, one fibre column per square array, the apparent resistivity of the core, ρ_{core}, is given by

$$\rho_{core} = \frac{\left(\dfrac{4\, \rho_{fib}\, h_c}{\pi\, d^2} + 2R_*\right) L^2}{h_c} \tag{3}$$

After substitution for L in terms of f (eqn.(1)), this leads to an expression for the core resistivity in terms of known dimensions, the resistivity of the fibre material and the contact resistance.

$$\rho_{core} = \frac{1}{f}\left(\frac{R_*\, \pi\, d^2}{2\, h_c} + \rho_{fib}\right) \tag{4}$$

It can be seen that the thickness of the core comes into this expression.

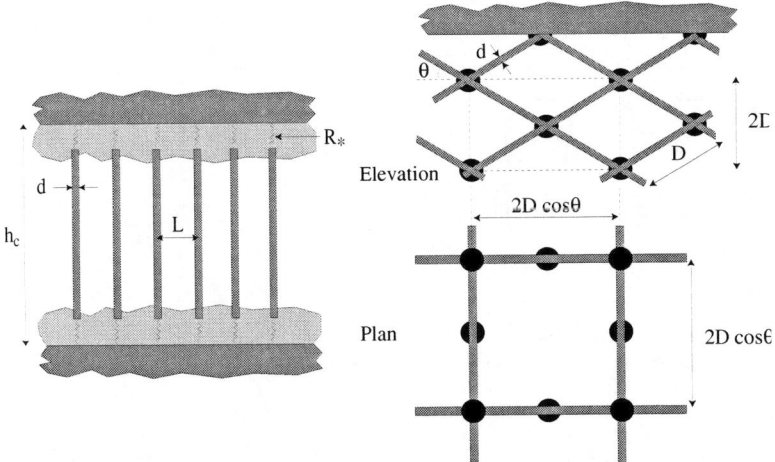

Fig.4: Schematic showing modelled fibre distributions and electrical resistances for the transverse fibre (left) and the sintered mat (right) structures.

For the brazed mat core, it is assumed that there is no contact resistance at the interface between fibres and faceplates, so the treatment just concerns a small representative volume of core material. A tetragonal unit cell is identified, of side $2D\cos\theta$ and height $2D\sin\theta$, where D is the length of a segment of fibre having one end at a cell corner and the other at the mid-point of a vertical face (Fig.4). Each cell thus contains 16 fibre segments, all of which are shared between 2 cells. The fibre volume fraction is thus given by

$$f = \frac{8\left[D\left(\frac{\pi d^2}{4}\right)\right]}{(2D\cos\theta)^2(2D\sin\theta)} = \frac{\pi d^2}{4D^2\cos^2\theta\sin\theta} \tag{5}$$

The resistance to current flow presented by the unit cell, R_{cell}, can be expressed in terms of the resistance of a segment of fibre R_{seg}. Since, in effect, current passes through two sets of 4 parallel segments as it progresses down the height of a unit cell, it follows that

$$\frac{1}{R_{cell}} = \frac{4}{2R_{seg}} = \frac{2}{\dfrac{D\,\rho_{fib}}{\left(\dfrac{\pi d^2}{4}\right)}} = \frac{\pi d^2}{2D\,\rho_{fib}} \tag{6}$$

The resistance offered by the unit cell can also be expressed in terms of the resistivity of the core

$$R_{cell} = \frac{\rho_{core}\,2D\sin\theta}{(2D\cos\theta)^2} = \frac{\rho_{core}\,\sin\theta}{2D\cos^2\theta} \tag{7}$$

Combining eqns. (5), (6) and (7) leads to

$$\rho_{core} = \frac{\rho_{fib}}{f\sin^2\theta} \tag{8}$$

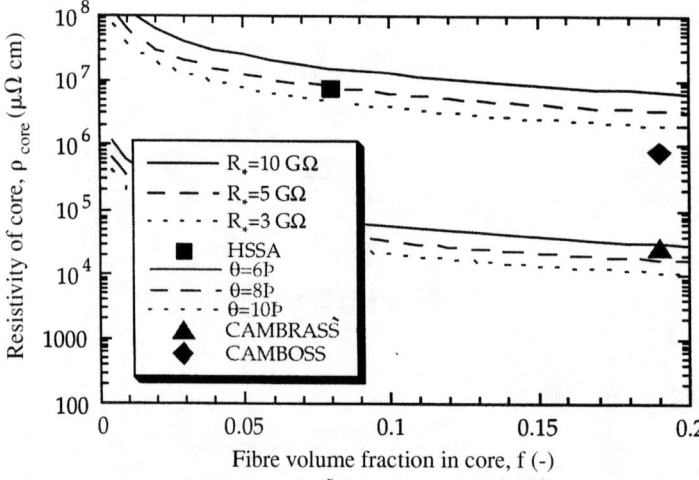

Fig.5: Core resistivity as a function of fibre content. The three points correspond to experimental measurements, while the curves are predictions, obtained using eqn.(4), with three values of R_* and $h_c = 0.8$ mm, for the transverse fibre (HSSA), and eqn.(8) with three values of θ, for the sintered mat (CAMBOSS and CAMBRASS). A value of 60 $\mu\Omega$ cm was used for the resistivity of the stainless steel fibres.

It can be seen (Fig.5) that the average resistance between fibre end and faceplate in the transverse fibre core is relatively high (few GΩ), while the resistivity of the sintered mat core is consistent with the simple geometrical model shown, with the fibres inclined at about 7° to the horizontal - which is approximately correct. The resistance of the brazed contact to the faceplates (CAMBRASS) is thus apparently negligible. The resistance of the CAMBOSS material is considerably higher than the predicted levels for a sintered mat, which is attributed to poor electrical contact with the faceplates. However, it can be seen that this is considerably better than in the HSSA material. This is probably because pressure was applied while the adhesive was setting, bringing the fibres into better electrical contact with the faceplates than is possible with the HSSA procedure.

Welding Characteristics

The HSSA sheet could not be welded directly, as no significant current flowed through the material with the electrode force and voltage used. This is consistent with the high measured electrical resistivity of the core. By using a shunt, however, it was found to be possible to create a weld. Initially, sufficient current flowed through the faceplates and across the shunt to cause heating of the core between the electrodes, leading to softening, consolidation and hence sufficient reduction in core resistance to allow a substantial direct current to flow and melting to occur. However, during these initial trials the breakthrough of the core was inconsistent and the faceplates were susceptible to local burn-through by the shunt current. In the example shown in Fig.6, breakthrough and current increase occurred only in the last of the 10 cycles of weld time. Even then, the poor shape of the final half cycles of current indicates intermittent current flow. In some cases, depending on the position of the shunt, no breakthrough occurred at all.

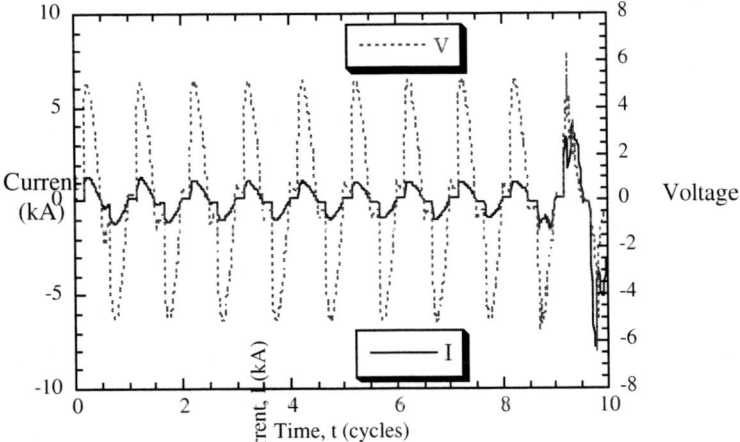

Fig.6: Voltage-time and current-time plots obtained during welding together of two HSSA sheets, with a weld time of 10 cycles (0.2 s).

Even when a weld was created between HSSA sheets, it was invariably of poor quality. This can be seen in the micrograph shown in Fig.7, where it is clear that the pressure has resulted in much of the faceplate material being melted and squeezed out laterally. There has also been vaporisation of the adhesive, leading to blow-holes, and cracking of the faceplates. Melt expulsion of this type is often problematic, particularly with thin metal sheets in composite materials such as vibration damping steels[16]. Such a weld would be mechanically very weak.

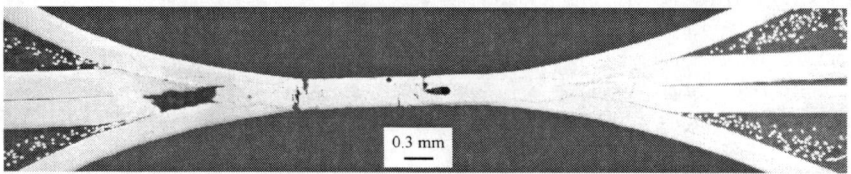

Fig.7: Optical micrograph of a polished transverse section from a pair of HSSA sheets after resistance welding.

The CAMBRASS sheet, on the other hand, was readily weldable. The plot in Fig.8 shows typical voltage and current characteristics. The current rises quickly to the set value and substantial heat is generated from the start in the sheets between the electrodes. Sections through a corresponding weld are shown in Fig.9. It can be seen that the weld is of good quality, with some lateral flow of melted fibres, but the inner faceplates retaining their integrity and the outer faceplates remaining unmelted.

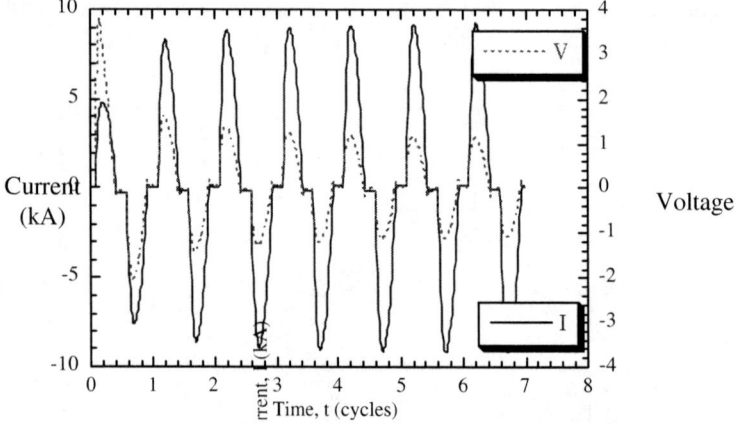

Fig.8: Voltage-time and current-time plots obtained during welding together of two CAMBRASS sheets, with a weld time of 7 cycles (0.14 s).

Conclusions

1. The beam stiffnesses for HSSA, CAMBRASS and CAMBOSS sheets are appreciably lower than predicted. This is attributed to the fact that the core is not stiff enough to maintain the

separation of the faceplates and to prevent excessive lateral displacement (shearing) between them when the sheet is subjected to a bending moment. Fibres with larger

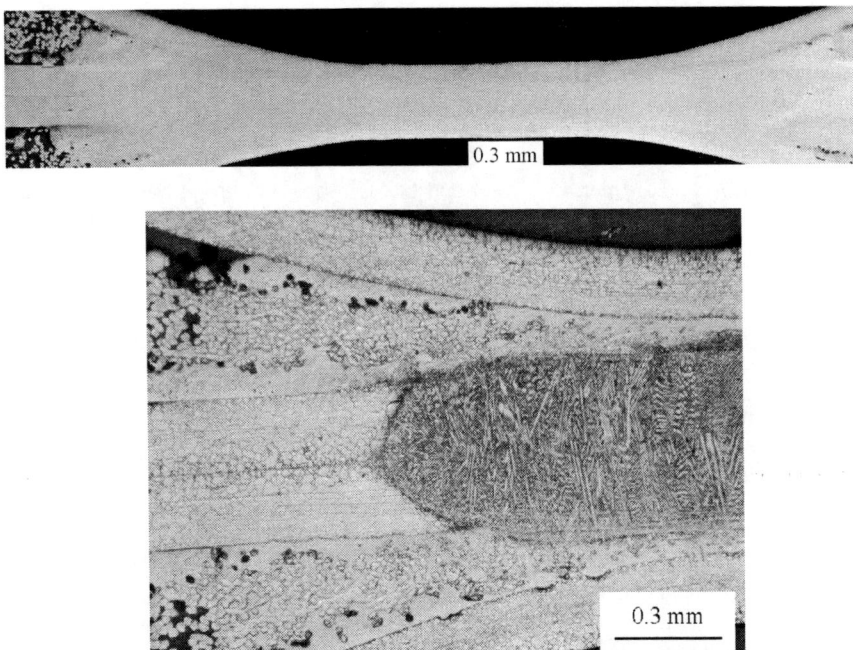

Fig.9: Optical micrographs of transverse sections from a pair of HSSA sheets after resistance welding, showing an as-polished complete section (above) and a higher magnification view of an etched sample (below), in which the fused zone is clearly visible.

diameter (~ 100–300 µm) and orientation at intermediate angles (~ 30-60•) will enhance the core stiffness.

2. Through-thickness electrical resistivities have been measured for the cores of HSSA (~10 Ω cm), CAMBOSS (~1 Ω cm) and CAMBRASS (~0.01 Ω cm). The experimentally measured resistivities compared well with predictions using simple geometrical models. The high resistivity of the HSSA core is attributed to poor electrical contact between the fibre ends and the faceplates.

3. The HSSA adhesively-bonded material could not be successfully welded in these initial trials. A shunt was required to achieve any melting but electrical contact through the core was variable. In addition, weld flaws and faceplate damage occurred during welding. The

CAMBRASS brazed material, on the other hand, exhibited good welding behaviour, with good current flow from the start of the welding period.

Acknowledgements

Support for AEM via the recently-established Cambridge-MIT institute (CMI) is gratefully acknowledged. Also, the authors are grateful to R.N-G Gustafsson (Volvo Technological Development Corporation, Sweden) and J. Karlsson (HSSA AB, Sweden) for useful discussions and for providing the HSSA material.

References

1. T.M. McCormack, R. Miller, O. Kesler and L.J. Gibson, "Failure of Sandwich Beams with Metallic Foam Cores", Int. J. of Solids and Structures, 38 (28-29) (2001), 4901-4920.
2. C. Chen, A.M. Harte and N.A. Fleck, "The Plastic Collapse of Sandwich Beams with a Metallic Foam Core", Int. J. Mech. Sci., 43 (6) (2001), 1483-1506.
3. A.M. Harte, N.A. Fleck and M.F. Ashby, "Sanwich Panel Design using Aluminium Alloy Foam", Adv. Engng. Mater., 2 (4) (2000), 219-222.
4. J.K. Paik, A.K. Thayamballi and G.S. Kim, "The Strength Characteristics of Aluminium Honeycomb Sandwich Panels", Thin-Walled Structures, 35 (3) (1999), 205-231.
5. G. De-Matteis and R. Landolfo, "Structural Behaviour of Sandwich Panel Shear Walls: An Experimental Analysis", Materials and Structures, 32 (219) (1999), 331-341.
6. T.S. Lok and Q.H. Chen, "Elastic Stiffness Properties and Behavior of Truss-Core Sandwich Panel", J. Structural Engineering - ASCE, 126 (5) (2000), 552-559.
7. R. Gustavsson, "Formable Sandwich Construction Material and Use of the Material as Construction Material in Vehicles, Refrigerators, Boats etc", patent WO 98/01295, 15th Jan. 1998, AB Volvo, International.
8. A.E. Markaki and T.W. Clyne, "Ultra Light Stainless Steel Sheet Material", patent filed 31 Oct. 2001, Cambridge University.
9. M. Jou, "Experimental investigation of resistance spot welding for sheet metals used in automotive industry", JSME Int. J. Series C-Mech. Systems Machine Elements and Manufacturing, 44 (2) (2001), 544-552.
10. S.S. Babu, M.L. Santella, Z. Feng, B.W. Riemer and J.W. Cohron, "Empirical Model of Effects of Pressure and Temperature on Electrical Contact Resistance of Metals", Science and Technology of Welding and Joining, 6 (3) (2001), 126-132.
11. S.C. Wang and P.S. Wei, "Modeling Dynamic Electrical Resistance during Resistance Spot Welding", J. Heat Transfer - Trans. ASME, 123 (3) (2001), 576-585.
12. J.A. Khan, L.J. Xu, Y.J. Chao and K. Broach, "Numerical Simulation of Resistance Spot Welding Process", Numerical Heat Transfer Part A-Applications, 37 (5) (2000), 425-446.
13. U. Dilthey, H.C. Bohlmann, U. Reisgen, W. Sudnik, W. Erofeew and R. Kudinow, "Modelling and Numerical Simulation of Resistance Spot Welding with Experimental Verification" (Paper presented at the 9th International Conference on the Joining of Materials, Helsingor, Denmark, 16-19 May 1999), 38-43.
14. H. Tang, W. Hou, S.J. Hu and H. Zhang, "Force Characteristics of Resistance Spot Welding of Steels", Welding Journal, 79 (7) (2000), 175S-183S.
15. Y. Zhou, P. Gorman, W. Tan and K.J. Ely, "Weldability of Thin Sheet Metals during Small-Scale Resistance Spot Welding using an Alternating-Current Power Supply", J. Electronic Materials, 29 (9) (2000), 1090-1099.
16. H. Oberle, C. Commaret, R. Magnaud, C. Minier and G. Pradere, "Optimizing Resistance Spot Welding Parameters for Vibration Damping Steel Sheets", Welding Journal, 77 (1) (1998), 8S-13S.
17. A.F. Whitehouse, C.M. Warwick and T.W. Clyne, "The Electrical Resistivity of Copper Reinforced with Short Carbon Fibres", J. Mat. Sci., 26 (1991), 6176-6182.

REINFORCED ALULIGHT FOR STRUCTURAL USE

František Simancik, Walter Rajner* and Rainhard Laag**

Institute of Materials & Machine Mechanics, Račianska 75, SK-831 02 Bratislava, Slovakia
*Alulight Int. GmbH, Ranshofen, Austria
**Alulight Deutschland GmbH, Wasseralfingen, Germany

Abstract

Alulight aluminum foam parts prepared by PM-techniques are very promising for lightweight structures because of their excellent stiffness-to-weight ratio. However their structural use is considerably limited by the fact that they are not able to withstand higher tensile stresses, which provide sufficient conditions for rapid propagation of surface cracks in highly porous structure. This essential drawback can be overcome via strengthening of tensile loaded surface of the foamed part with various wires or fibers woven into webs, similar to the case of reinforced concrete. According to a novel foaming technique, the reinforcements are placed in the foaming mould together with foamable precursor and during foam expansion they are infiltrated with molten cell-wall material. The main advantages of this method are the simplicity and the possibility to reinforce the foamed part selectively and anisotropically according to the applied load. The required mechanical properties of the structure can thus be achieved at minimum weight and manufacturing costs.

Introduction

Metallic foams are relatively unknown structural materials, however with enormous future potential for applications where lightweight combined with high stiffness (see Fig.1), and acceptable manufacturing costs are of prime interest. The metallic foams are crushable, offering high impact energy absorption capacity at low stresses, they are stable at elevated temperatures, they possess superior fire resistance and do not evolve toxic fumes in a fire. They are also efficient in shielding of electromagnetic waves and in vibration and noise damping [1]. Although metallic foams appeared as early as 1940's [2], they have remained almost unused by the designers up to now. Due to the development of new process technologies, which enable the manufacturing at reasonable costs [3], this type of material becomes again very attractive and provides a strong impetus for rapid increase of R&D activities in this field.

The metallic foams generally are generally formed by nucleation and subsequent growth of gas pores in a liquid or semi-liquid metal. The distribution of pores is therefore fairly random. The pores are initially round and closed, although some inevitable cracks or other defects originate in the cell walls due to shrinkage of solidifying cell wall metal and/or due to pressure reduction inside the pores during cooling [4].

Alulight is a net shape aluminum foam made by a powder metallurgical (PM) route [5], which comprises foaming of the precursor prepared by compacting of powdered metal or alloy, wherein the pore forming gas is developed during melting of this precursor from admixed foaming agent. The powder must be thoroughly compacted in order to seal the particles of foaming agent. This avoids the premature release of the gas during heating. The formation of the pores starts during the melting of the precursor in the mould in which the precursor is distributed and then heated (together with the mould) to the melting point of used alloy. The foam fills the mould and is formed into the shape of the mold cavity. After melting and foaming, the mould is rapidly cooled to prevent collapse of the foamed structure. There are almost no constraints concerning complexity of the outer shape and geometry. The possibility to use the simple-form precursor e.g. extruded rods, for foaming of components with various shapes and sizes [6] significantly reduces the production costs usually accompanying the PM manufacturing techniques.

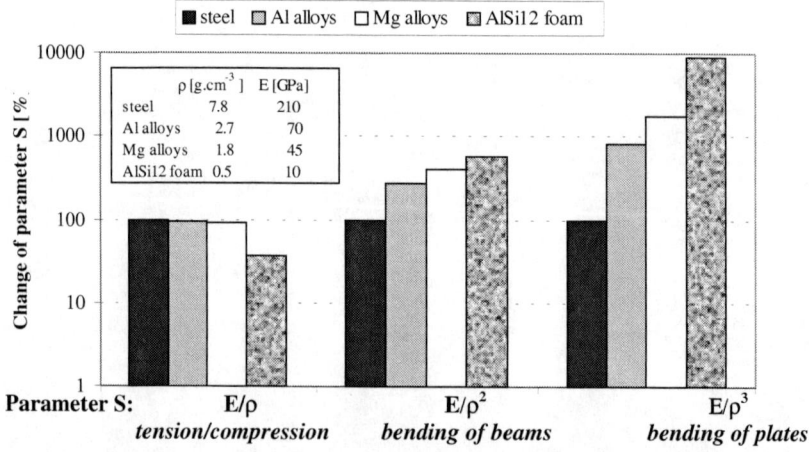

Figure 1: Parameter S defining stiffness-to-weight ratio for various materials and loading conditions (E - elasticity modulus, ρ - density, normalized relative to that of steel).

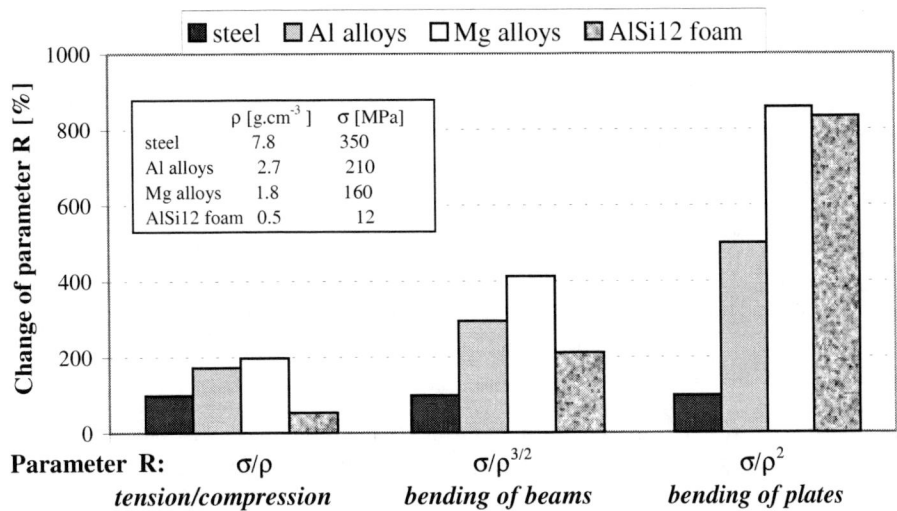

Figure 2: Parameter R defining strength-to-weight ratio for various materials and loading conditions (σ – yield strength, ρ - density normalized relative to that of steel).

Alulight foamed parts are always covered with dense skin, which significantly improves their appearance and the properties (e.g. the bending stiffness). However this skin possesses variable thickness and always contains microcracks or even holes. These inevitable defects can initiate premature fracture of the foam, especially if it is loaded in tension, when the conditions for crack displacement in highly porous structure are favorable. Therefore the tensile strength of foamed parts is often insufficient and exhibits too high a variability to allow their efficient use as the load bearing structural elements (see Fig.2), although their stiffness to weight ratio is excellent.

The growth of cracks in cell walls can be prohibited by strengthening of the foam with some kind of reinforcing elements, similar to the case of reinforced concrete (see Fig.3). If the reinforcements have higher elasticity modulus and sufficient yield strength, the tensile stresses are transferred from the foam onto these elements and existing cracks in cell walls become inactive.

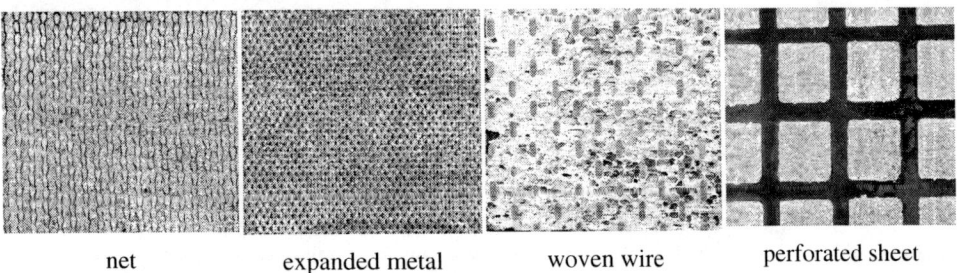

net · expanded metal · woven wire · perforated sheet

Figure 3: Aluminum foam AlSi12 reinforced in the surface with stainless steel in various forms.

Manufacturing of Reinforced Foams

The PM process developed for manufacturing of Alulight foam offers also simple and cost effective possibility for making reinforced foams or "metal foam matrix composites (MFMC)". According to a novel foaming technique [7], the reinforcements are inserted into the foaming mould together with foamable precursor and properly positioned. In the course of foam expansion the reinforcements are moved to the mould surface, where they are infiltrated with molten cell-wall material. The liquid foam alloy reacts with reinforcements, forming metallurgical bond. The quality of this bond depends on the chemical composition of both materials (foam and reinforcement) and can be controlled by contact time and proper surface treatment of the reinforcement. If appropriate, the mutual reaction can be prohibited by coatings providing diffusion barriers. Fig. 4 shows $Al_{12}Fe_3Si$ interfacial phase formed between reinforcement made of expanded stainless steel and AlSi12 foam. This type of metallurgical bonding provides a certain formability of the foam and results in a significant improvement of the mechanical properties and the thermal stability in comparison with glued or brazed sandwiches. In distinction to typical metal matrix composites, in this case the interfacial layer does not represent "the weakest link"; its properties are usually better than the properties of highly porous and brittle AlSi12 foam matrix.

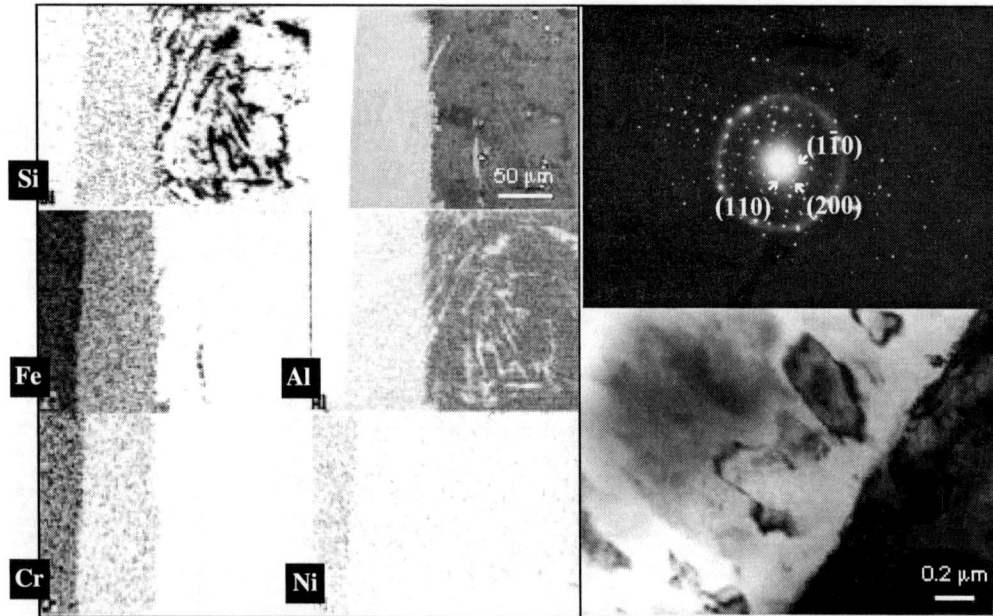

Figure 4: Interfacial reaction zone $Al_{12}Fe_3Si$ formed between stainless steel reinforcement and AlSi12 foam. (element distribution - left; TEM and diffraction pattern - right).

The reinforcements prevent liquid foam from collapsing on cooling and thus they have also an additional stabilizing effect. Moreover the reinforcements increase the thickness of surface skin (Figure 5), simplify joining of foamed parts (welding is possible) and enable limited shaping after the foaming process. The foamed components can be reinforced selectively and anisotropically according to expected load, thus saving the weight and the reinforcement costs. The shaped foamed parts can be strengthened via the same way. One of the attractions of this process is that these composites are prepared in a single operation that simplifies manufacturing and lowers cost.

Experimental

Preparation of specimens for testing

Alulight panels with variable porosity based on AlSi12- and AlMg1Si0.6-alloys were reinforced at one or both surfaces with expanded metal made of stainless steel. The mesh opening of expanded metal was 6 x 3 mm, with a weight 3.4 kgm^{-2}. The specimens for bending tests, performed in this investigation, were cut from the reinforced panel in two directions; with longitudinal and transverse orientation of expanded metal along the length of the specimen (see Fig. 5). It is obvious, that the load-bearing cross section is higher in longitudinal orientation and therefore better properties can be expected in this direction at the same overall weight of the specimen.

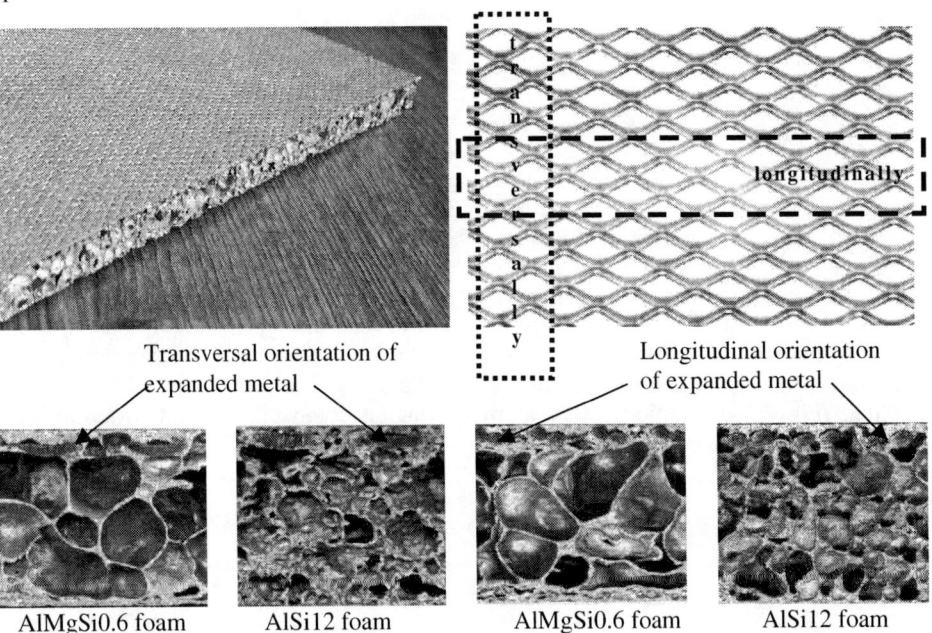

Figure 5: Typical cross sections of testing specimens (10 x 12 mm) cut from aluminum foam panel (top left) reinforced in upper surface with expanded metal sheet (top right).

Bending test

Four-point bending tests were performed on beams cut from reinforced panels. The size of specimens was 10 x 12 x 100 mm (thickness x width x length). The 4-point bending tests was used, as it is more suitable for materials with a heterogeneous structure due to the more convenient distribution of the bending moment during application of the bending loads. The reinforcements were located either in tensile- or compression-loaded foam surface. The beams made of plain foam were also tested for comparison. The deformation of selected specimens was recorded on CCD camera, to observe the initiation of the cracks in the porous structure. The bending stiffness B was calculated from the slope of the load-deflection curve recorded during four-point-bending test according to:

$$B = EJ = \frac{1}{48}\frac{Fc}{y_\varepsilon}(3l^2 - 4c^2)$$ (1)

where $E[MPa]$ refers to the elastic modulus of the material and $J\ [mm^4]$ is the cross sectional moment of inertia, $F[N]$ is the load causing the deflection $y_\varepsilon\ [mm]$, l and c refer to the span between the supports and to the distance between the bending force and the supports, respectively ($l=70$ and $c=20\ mm$ in this study). In order to obtain a linear part of the load-deflection curve the short unloading was applied at the strain of $\varepsilon \approx 0.1$. The deflection was measured in the center of the specimen by electronic extensometer; ram speed was kept constant at 2 mm/min. The bending stress σ was calculated according to:

$$\sigma = 6\ Fc\ /\ bh^2$$ (2)

where b refers to the width and h to the thickness of the test sample.

Properties of Reinforced Foams

Fig. 6 shows the bending behavior of AlSi12-foams with and without reinforcement. The porosity of the two foams were approximately the same (75-85%). The reinforcement was placed either in the top or in the bottom of the bending specimen. The expanded metal sheet increases the bending stiffness of the foam almost twice, although the weight due to the reinforcement increases only about 30%. The most efficient use of reinforcement is in bottom part of the specimens, i.e. in tensile loaded surface. In this case the sample does not break during the whole loading cycle. Foam shearing can be observed at higher deformation (Fig. 7a) in parts above the reinforcement. If the reinforcement is placed on top (compression loaded surface), the sample fractures with sudden failure (Fig. 7b) at significantly lower stress, which is comparable with the fracture stress of the foam without reinforcement.

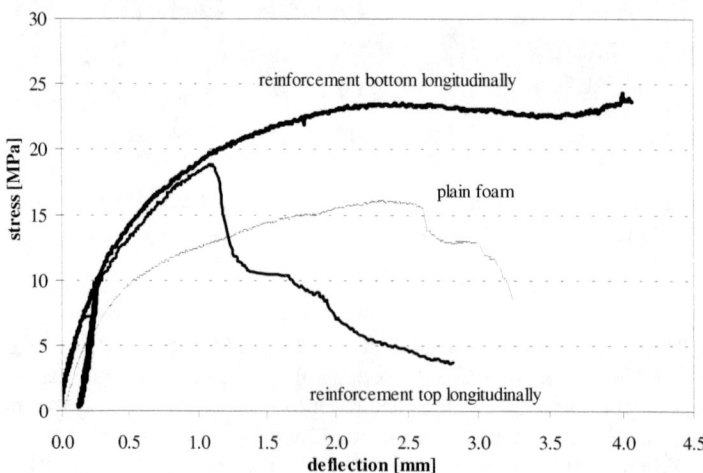

Figure 6: Effect of the reinforcement on bending behavior of foamed beams based on AlSi12 alloy at the same porosity of the foam - 81%.

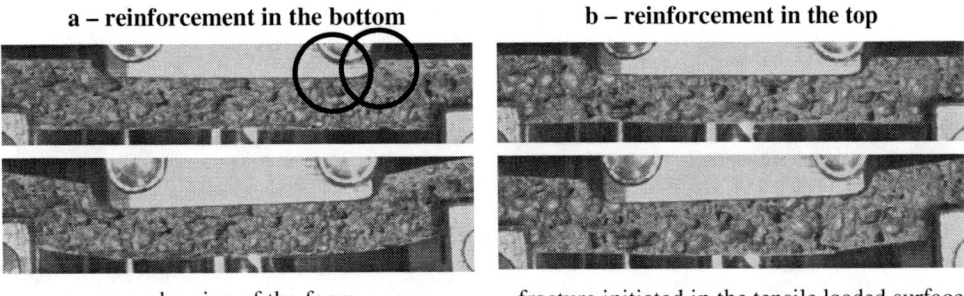

Figure 7: Failure mode of reinforced AlSi12 foams with different position of reinforcing expanded metal (porosity of the foam - 81%).

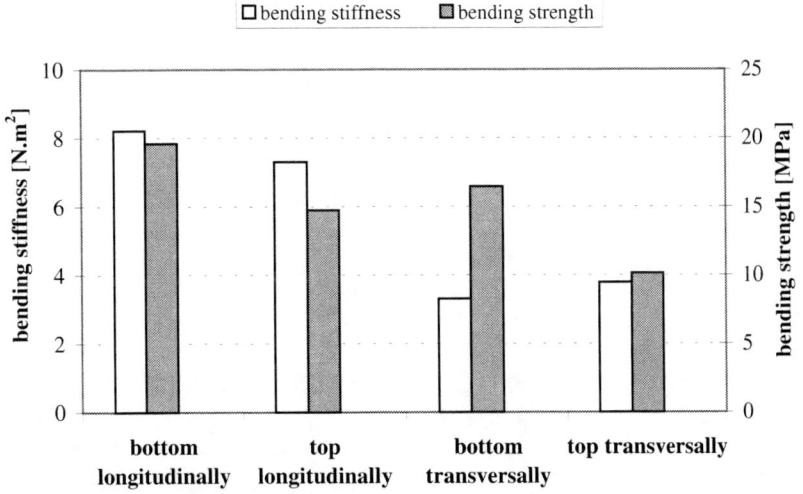

Figure 8: Effect of the orientation and position of reinforcing expanded metal on the properties of foamed beams based on AlSi12 alloy at the same porosity of the foam - 75%.

Fig. 8 shows the effect of the position and orientation of expanded metal on the performance of reinforced beams based on the brittle AlSi12 foams. Although the weight of all beams was almost the same, their properties were found considerably different. The bending strength of reinforced beams is always higher when the reinforcements are placed in the bottom of the specimen (in tensile loaded surface). The expanded metal is stronger in longitudinal direction because of larger load-bearing cross-section (see also Fig. 5). This leads to higher bending strength of reinforced samples in all cases.

Also the bending stiffness of the reinforced specimens depends on the orientation of the expanded sheet; is always higher in longitudinal direction. However, the bending stiffness, contrary to bending strength is not affected by the position of the reinforcement; it is almost the same for top- or bottom placements.

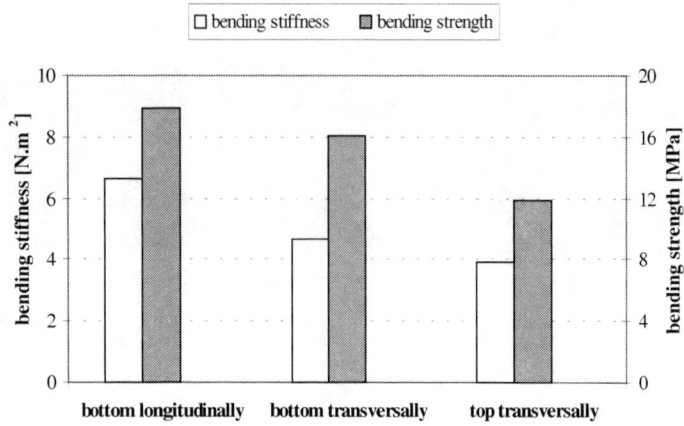

Figure 9: Effect of the orientation and position of reinforcing expanded metal on the properties of foamed beams based on AlMg1Si0.6 alloy at the same porosity of the foam - 76%.

Similar behavior was observed also for the reinforced beams based on more ductile AlMg1Si0.6 alloy (Fig. 9). However, in this case the plastic bending without visible fracture was observed also for specimens with reinforcing elements placed in compression-loaded surface. At similar porosity the bending strength of reinforced AlSi12-foam is slightly better than that of AlMg1Si0.6, however this difference is not significant (compare also Fig. 8 and 9).

The effect of the position of reinforcement on fracture toughness is clearly visible in Fig.10, which shows the final deformation of the foamed beams after impact test. While almost no impact energy is absorbed with plain (unreinforced) foam, the reinforced foams are able to convert impact energy into the deformation of foam. The remaining capability to absorb further energy can be estimated from the bending angle. It can be seen in Fig. 10, that the most efficient use of the reinforcement is again in the tensile loaded surface.

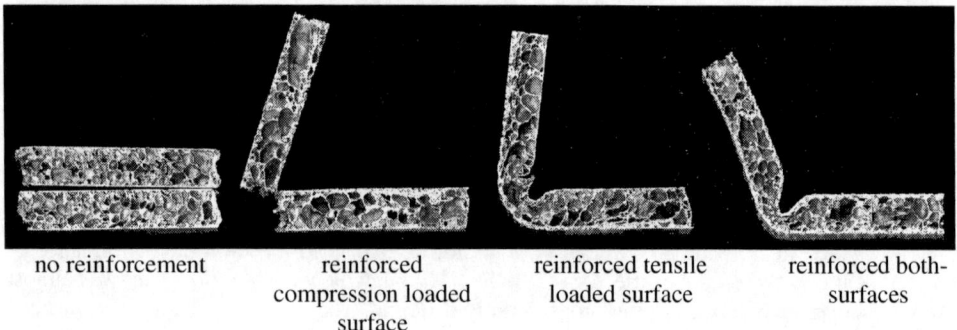

Figure 10: Deformation of beams based on AlMg1Si0.6 foam (15x15x120mm, porosity 85%) after impact test (max energy 15 J) as a function of the position of the reinforcements.

Fig. 11 illustrates the changes of bending properties for various uses of aluminum in beams of the same weight. As can be seen the beam made of plain foam possesses considerably higher stiffness than bulk Al-sheet. However, no benefits were observed for bending strength, and the resistance to fracture was even worse. Therefore, for practical use of aluminum foams in structural applications, it will be necessary to reinforce at least the tensile loaded surface of the foam. Similar effect can be obtained if the coversheet is applied [8]. However the use of the reinforcements is more efficient, because of their lower weight, which enables an increase in the thickness of the foam. The reinforcements have an additional stabilizing effect - they prevent liquid foam from collapsing thus allowing higher final porosity of the foam.

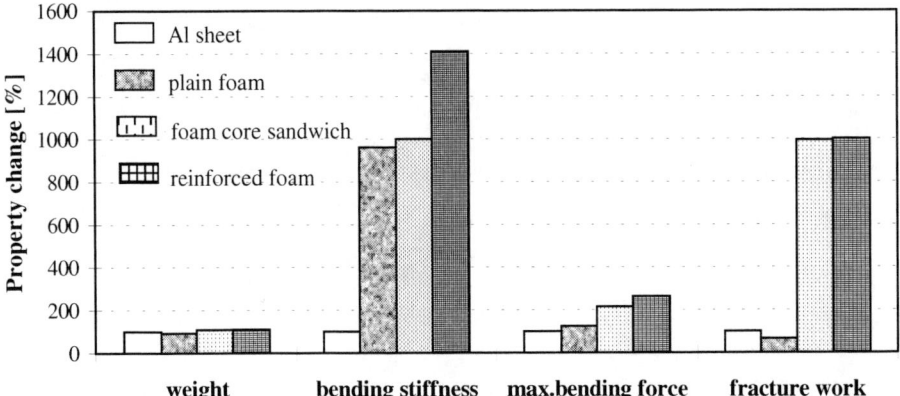

Figure 11: Properties of reinforced aluminum foam AlSi12 (thickness 15 mm, porosity 85%) in comparison with: Al-sheet (thickness 3 mm), AlSi12-foam (thickness 15 mm, porosity 80%) and AlSi12-foam sandwich (foam thickness 11 mm, porosity 80%, Al-coversheet 1 mm at one side).

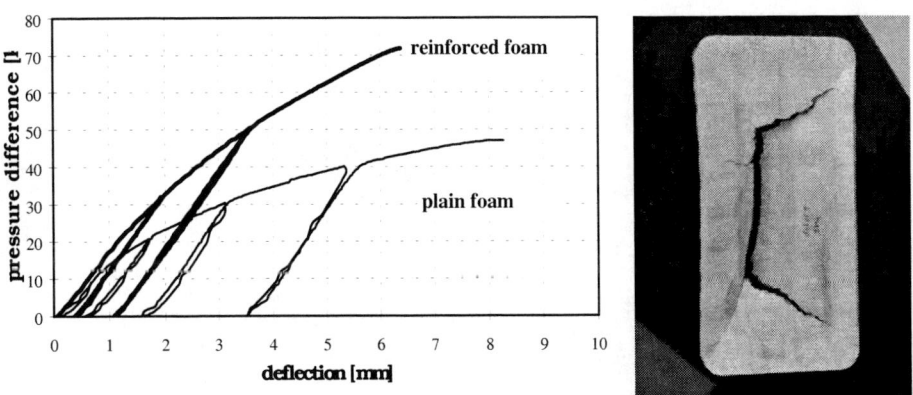

Figure 12: Bending of foamed panels under surface loading (right- broken plain foam panel).

The benefit of reinforcements was confirmed also in bending of foamed panels (565x285x15 mm) [9]. In this case the panels were loaded at the whole surface by applying a pressure difference between two chambers separated by the specimen. The reinforced AlSi12-foam panels withstand the pressure difference of 70 kPa while plain foams suddenly breaks at pressure of approx. 40- 45 kPa (Fig. 12). With increasing load the permanent deformation occurs due to yielding or fracture of weakest cell walls. The reinforcements allowed the cyclic loading of the panel with amplitude of 50 kPa. After the first loading the stiffness and permanent deformation of panel remained nearly unchanged during the whole experiments (up to 1500 cycles).

Conclusions

The stress to cause yielding or fracture of surface skin of aluminum foams can be increased by reinforcements incorporated into the foam surface, preferably on the tensile loaded side of the foam. In this way the significant improvement of the strength, plasticity, energy absorption capacity and damage tolerance of foamed part can be achieved with only slight weight increase (ca. 20-30%). The possibility to reinforce the foamed part selectively and anisotropically according to the expected load can enhance property-to-weight ratio. The reinforcements can be effectively used also for improving of bending stiffness when the thickness of the foam is limited.

The reinforcements increase the thickness of surface skin, simplify joining of foamed parts (make welding possible) and enable limited shaping after the foaming process.

Easy manufacturing allowing reasonable production costs can make the reinforced foams very attractive for the transport industry, especially for lightweight stiff body structures of future cars, busses, trains, ships, airplanes, etc.

Of course, the use of reinforcements is accompanied also with some difficulties, which should be addressed in a future. More complicated recycling, the residual internal stresses in the structure and lower resistance to electrolytic corrosion are some of these issues.

References

1. L.J.Gibson and M.F.Ashby, Cellular Solids, (Oxford: Pergamon Press, 1988).

2. A.Sosnik, US Patent 2434775, 1948.

3. J.Banhart, J.Baumeister, "Production methods for metallic foams", MRS Symposium Proceedings, Vol. 521, (ed. by D.S.Schwartz, D.S.Shih, A.G.Evans and H.N.G.Wadley, Materials Research Society, Pennsylvania: Warrendale, 1998), p. 121.

4. F.Simancik, K.Behulova, L.Bors, "Effect of ambient atmosphere on the foam expansion", Cellular Metals and Metal Foaming Technology, (ed. by J.Banhart, M.F.Ashby, N.A.Fleck, Bremen: Verlag MIT Publishing, 2001), p. 89-92.

5. B.C. Allen, US Patent 3087 807, 1963.

6. F.Simancik, W.Rajner and R.Laag, "Alulight - Aluminum foam for lightweight construction", SAE Technical Paper Series 2000-01-0337. (Warrendale: SAE, Inc., 2000), p. 31-38.

7. F.Simancik, H.Worz and E.Wolfsgruber, AT Patent 408317B, 2001.

8. F.Simančík, J.Kováčik, N.Mináriková, "Bending properties of foamed aluminum panels and sandwiches", Porous and Cellular Materials for Structural Applications, (MRS Symposium Proceedings, Vol. 521. Warrendale, MRS, 1998), p. 91-96.

9. F.Simancik, L.Lucan, J.Jerz, "Reinforced aluminium foams", Cellular Metals and Metal Foaming Technology, (ed. by J.Banhart, M.F.Ashby, N.A.Fleck, Bremen: Verlag MIT Publishing, 2001), p. 365-368.

CONSTRUCTED CELLULAR METALS

David J. Sypeck

University of Virginia
Department of Materials Science and Engineering
116 Engineer's Way
Charlottesville, VA 22904

Abstract

Superior properties with cellular solids can be achieved when good control over the distribution of material at the cell level is gained. It has been known for many years that cellular designs having the highest weight specific stiffness and strength are truss-like when the solid volume fraction is small. Imitating decades of work by structural engineers, researchers have now begun to investigate open cell periodic structures (miniature trusses) for load bearing applications amongst many other functionalities. These contain highly controlled distributions of material at the cell level and are similar in design and appearance to the large engineering type structures (e.g., skyscraper framing, bridges, transmission towers, etc.) that have already been made and tested for many years. Here, efficient miniature truss architectures are shown, methods for affordably making them from metals are presented and properties are assessed.

Introduction

Cellular solids are made up of interconnected networks of intersecting support members or plates that form the edges and faces of cells [1]. They can be made from a variety of materials, including metals, ceramics, polymers, composites and semiconductors, and can be designed to possess useful combinations of mechanical, chemical and thermophysical properties. These can be tailored by adjusting the relative density, cellular architecture or material type of the cellular solid. Such materials can be useful, for example, as building materials, heat exchangers, thermal insulators, filters, electromagnetic shields, battery components, to provide buoyancy, act as a host for the in-growth of biological tissue, provide impact and blast protection or support catalysts amongst many other applications [1-6]. There is growing interest in using materials of this type for applications where more than one function is required [7]. Designers of these multifunctional materials seek to tailor a main property of interest (e.g., stiffness and strength) in the most efficient way (e.g., least weight) through adjustment of the cellular architecture and base material properties where upon the intervening space can then be used for other functionalities (e.g., cooling fluid flow, power storage). Such designs are useful where weight savings are a concern (e.g., aerospace applications). Since virtually all man-made objects must be transported at one time or another, this equates to energy savings.

The porosity within a cellular solid makes it particularly attractive for a variety of low density, multifunctional applications. Several classes of these materials have been developed (normally manufactured by batch processes), including honeycombs, gasars, consolidated powder products, vapor deposited materials, hollow sphere structures, stochastic foams and lattice block or truss materials [1,2]. The least expensive are stochastic in nature, made by variants of foaming in the liquid, solid or semi-solid state. Stochastic metal foams can be made by directional foaming, which produces cellular architectures that are predominantly closed cell [2]. Closed cell stochastic metal foams are useful for sound attenuation and mechanical impact energy absorption. They can also be buoyant in fluids as they do not allow throughput. Fluids do however, flow through open cell stochastic foams because of the interconnected porosity. Metal foams of this type are made using reticulated polymer foams that serve as investment casting patterns or as templates for vapor or powder slurry deposition [2]. These materials are useful, for example, as lightweight heat exchangers or as electrodes in nickel metal hydride batteries. Examples of closed and open cell stochastic metal foams are shown in Figure 1.

With current foaming methods, it is difficult to gain good control over the final distribution of material at the cell level and properties (these can vary from place to place within a given cellular object) normally reside well below theoretical bounds [8]. Limits also exist to the types of materials that can be produced. Consider the Young's modulus and compressive yield strength for an open cell stochastic foam. These generally vary with relative density according to the following power law relations [1]:

$$E/E_s = (\rho/\rho_s)^2 \qquad (1)$$

$$\sigma_c/\sigma_{ys} = 0.3(\rho/\rho_s)^{3/2} \qquad (2)$$

where ρ is the density of the foam and ρ_s is that of its base material. E is the Young's modulus of the foam, E_s is the Young's modulus of its base material, σ_c is the compressive yield strength of the foam and σ_{ys} is the yield strength of its base material. The power law dependence upon relative density indicates a rapid property loss with decreased density owing to ligament bending (even for ideal microstructures that are free from porosity) [1,2].

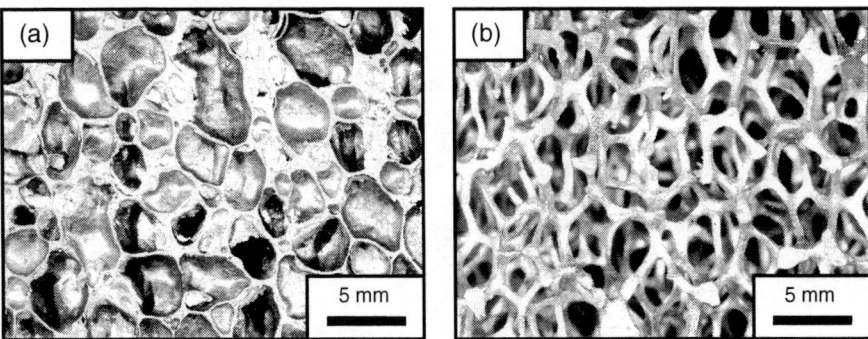

Figure 1: Stochastic aluminum foams. (a) Closed cell (Alporas®, Shinko Wire Co., Japan). (b) Open cell (Duocel®, ERG Inc., USA).

Superior properties with cellular solids can be achieved when good control over the distribution of material at the cell level is gained. It has been known for many years that cellular designs having the highest weight specific stiffness and strength are truss-like when the solid volume fraction is small [9,10]. Imitating decades of work by structural engineers, researchers have now begun to investigate open cell periodic structures (miniature trusses) for load bearing applications amongst many other functionalities. These contain highly controlled distributions of material at the cell level and are similar in design and appearance to the large engineering type structures (e.g., skyscraper framing, bridges, transmission towers, etc.) that have already been made and tested for many years. Miniature truss structures of this type have been made using injection molding, rapid prototyping, investment casting and electrodeposition techniques [11-15]. These are known as lattice block or truss materials. Individual cells can be on the order of a few millimeters or less. By manipulating the truss architecture, properties are modified. For example, the Young's and shear moduli along with the tensile, compressive and shear yield strengths for truss materials can vary with relative density in a substantially linear way when the architecture is designed for members to be in tension or compression only with no bending [16]. This linear dependence is especially important at low relative density where properties far exceed those of stochastic foams. But, current fabrication approaches are relatively expensive and subject to properties knockdown. Limits exist to the types of materials that can be produced and whether or not they respond to post processing (e.g., heat treatment). Hence, there remains a long-felt need in the art for a multifunctional cellular solid that is characterized by the low production costs of stochastic foam structures, yet possesses the improved properties and multifunctional characteristics found in many miniature periodic truss structures over a wider range of materials choices. Accordingly, cellular metals that provide unique combinations of properties and characteristics that may overcome many of these disadvantages are discussed below.

Metal Textile Laminates

A potentially inexpensive approach to fabricating miniature metal truss type structures is based upon a recognition that high structural quality porous metal objects have already been made for many years by taking wrought metal wires and weaving them into bundles [9]. Woven wire cloth, braided cables and knitted metal fabrics are good examples of this. The textile approaches used to create such articles are very well established and relatively inexpensive. Furthermore, a host of metal choices are available (virtually all metals can be drawn into wire and then woven, braided or knitted) and in a variety of filament arrangements (dutch weave, hexagonal mesh, 3-D weave, crimped mesh, triaxial weave, etc.) [17]. However, these metal textiles are unsuited for most multifunctional applications because the metal contacts are not normally bonded and the articles are thin (exceptions include 3-D woven articles). One plausible remedy for these shortcomings involves laminating [18]. Consider the lay-up and bonding of plain square woven metal cloth at points of wire and facesheet contact, Fig. 2.

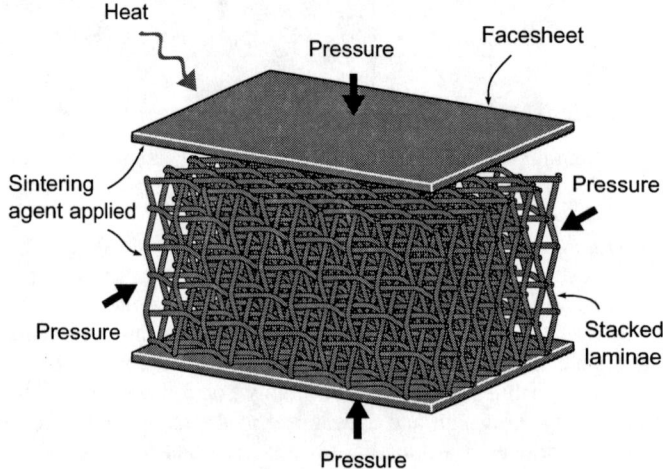

Figure 2: Laminate core and sandwich fabrication. See [18] for more details.

This metal topology was designed for efficient in-plane load support, good mechanical impact energy absorption and high convective heat transfer with low pumping requirements. Neglecting the bonding material and weave angle, this structure has $\rho/\rho_s = \pi d/4(w+d)$ where d is the wire diameter and w is the opening width (distance between wires). When axially compressed along a direction parallel to a set of wires, the stiffness and strength vary nearly as

$$E/E_s = 0.5(\rho/\rho_s) \qquad (3)$$

$$\sigma_c/\sigma_{ys} = 0.5(\rho/\rho_s) \qquad (4)$$

since about one-half of all wires are fully loaded while the other one-half are barely at all. This linear behavior outperforms closed and open cell stochastic foams at low relative density. Arranged tubular members (e.g. hypodermic tubing) improve upon the idea with the possibility of also utilizing the space within the tubes (e.g., cooling fluid flow, fuel storage, etc.).

Metal textile laminates have the needed open space for plastic collapse and absorb large amounts of mechanical energy when compressed. They can be designed for their ligaments to bend, plastically buckle, compress and stretch, absorbing energy at a nearly constant (plateau) stress followed by high strain densification. Furthermore, since their plateau stress is surmised to be higher than stochastic foams of similar density and base material, a higher specific energy absorptive capacity is anticipated. Titanium alloys may be a good choice for lightweight absorbers owing to their high weight specific yield strength and good corrosion resistance. But they are relatively expensive and difficult to process. I have made CP-Ti laminates using Ag-9Ga-9Pd as the transient liquid phase in protective heating (Ar gas with Ti getter). However, diffusion bonding may be the preferred route because of the limited joint ductility and corrosion resistance obtained with many current Ti brazing alloys. Monel is widely used for marine applications while stainless or low carbon steels with Ni-Cr bond coatings may be sufficient lower cost alternatives. The high strain compaction behavior of a metal textile laminate made from woven nichrome (Ni-24Fe-16Cr) mesh is shown in Fig. 3. This sample was bonded using a Ni-25Cr-10P transient liquid phase and had $\rho/\rho_s = 17\%$ prior to quasistatic crushing.

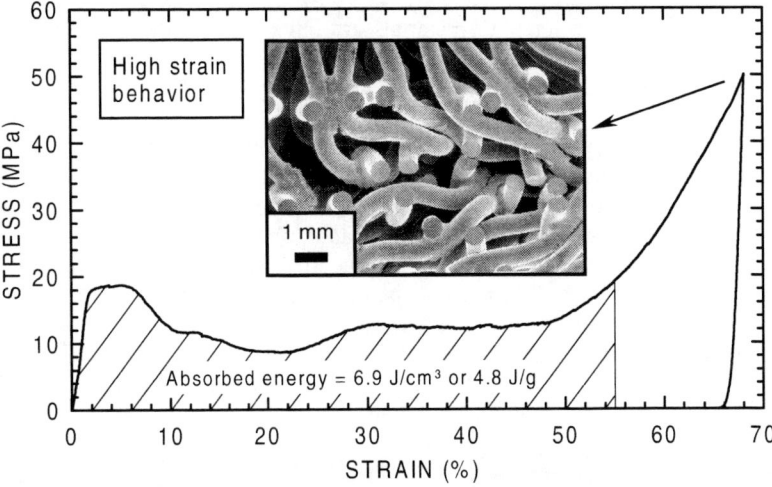

Figure 3: Quasistatic compaction behavior of a nichrome laminate. Note the wire plasticity and associated mechanical energy absorption. See [18] for more details.

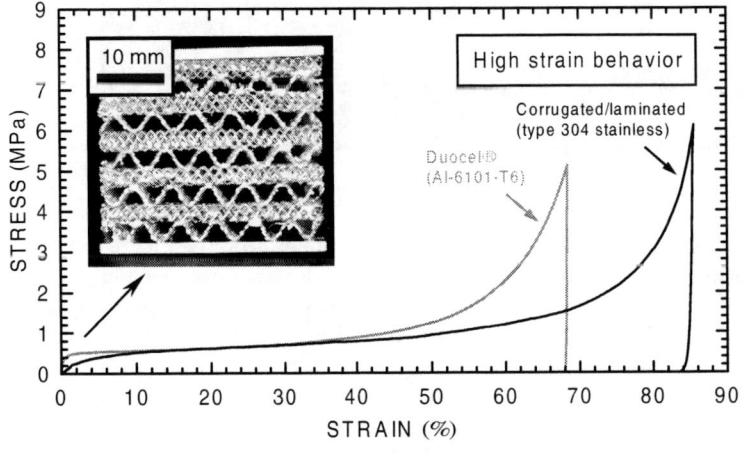

Figure 4: Quasistatic compaction behavior of a corrugated/laminated sample as compared to Duocel®. The photograph is a typical view prior to compaction.

Wires within conventional weaves can easily shift and move as the mesh relative density becomes small making it difficult to fabricate high quality laminates from them. One can however, shape woven wire mesh in the out-of-plane direction to produce some very interesting lightweight structures. For example, type 304 stainless steel insect netting (18 mesh/in with $\rho/\rho_s = 13\%$) was corrugated (resembles cardboard core), laminated and bonded using a Ni-25Cr-10P transient liquid phase, Fig. 4.

The 0-90° layup within the core facilitates a small contact area between the individual laminae and a high concentration of stress when loaded. These contacts are thought to yield almost immediately, thereby blunting the nose of the stress-strain curve and circumventing a potentially hazardous initial peak stress. The metal textile core had $\rho/\rho_s = 4\%$ and $\sigma_{ys} = 205$ MPa while the 10 pore/in Duocel® foam had $\rho/\rho_s = 6\%$ and $\sigma_{ys} = 194$ MPa. Although the density, base alloy yield strength and plateau stress for both materials are quite similar, the corrugated/laminated structure has a higher quasistatic energy absorptive capacity owing to its higher densification strain. This is likely related to enhanced cell architecture efficiency.

Cellular metals can also be used to facilitate heat exchange between a structure and a fluid [2,7]. Dense, highly conductive open cell architectures of large accessible surface area are preferred. The surface area density for the laminated metal textiles is about $\alpha_A = \pi/(w+d)$ [18]. A turbulent flow of tortuous path also aids convection but requires additional fluid pumping power. Materials of choice for heat exchangers include aluminum and copper for their high thermal conductivity whereas nichrome is often used for electrical resistive heating. For compact applications (e.g., power electronics heat sinks, air conditioners, etc.) a small overall size is desired. To create a heat exhanger prototype, plain square woven alloy 110 copper mesh (10 mesh/in with $\rho/\rho_s = 20\%$) were laminated and bonded using a Ni-25Cr-10P transient liquid phase, Fig. 5. The alloy 110 copper facesheets were added during a second heating.

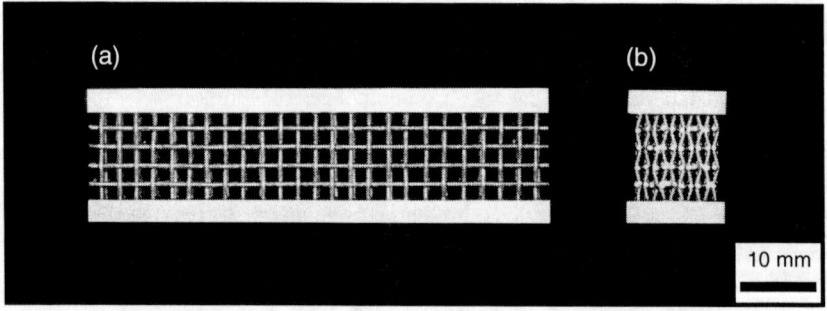

Figure 5: Copper heat exchanger prototype. (a) Front view. (b) Side view.

During bonding, the Ni-25Cr-10P melt readily flowed over the copper, to form a hard, oxidation and corrosion resistant coating. The high thermal conductivity of copper remains within. When fluids pass through a structure like that shown in Fig. 5 (a), the situation resembles a bank of aligned cylinders in cross flow. However, for flow along the other direction, Fig. 5 (b), the drag coefficient is expected to be higher.

Thin sandwich panels can also be made using metal textiles. One very simple and inexpensive approach follows many successful years of cardboard applications. When a single layer of plain square woven type 304 stainless steel insect netting is corrugated and bonded between thin facesheets, a lightweight multifunctional sandwich panel is made, Fig. 6. Besides providing for fluid throughput in all directions, cores of this type can be draped (prior to bonding) making them easy to form into complex curved shapes. Since wire mesh is available in long, wide rolls, continuous automated manufacturing processes could inexpensively manufacture these panels in high volume (as opposed to batch processes). Preliminary data on flat panels of this type indicates core relative shear strengths in the neighborhood of one-half those of hexagonal honeycomb. A triangular shaped corrugation might improve upon this.

Recently, metal textile laminate core sandwich panels have been made with relative shear strengths that are twice that of hexagonal honeycomb [19]. These panels have their plane square weave cores rotated 45° within the facesheets. The result is a diamond-like truss orientation that utilizes more of the wires upon panel bending (and core shearing). This differs from the core orientation shown in Fig. 5 (a) which acts as a mechanism when sheared (absent the minimal support of the node material) and is a poor choice for panels in bending. Without facesheet constraint however, the diamond oriented core acts as a mechanism when compressed. Further detailed studies are planned to assess advantages and disadvantages.

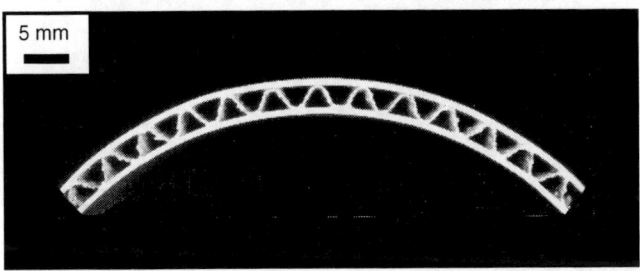

Figure 6: Curved multifunctional sandwich panel. The core had $\rho/\rho_s = 4\%$ and was bonded using a Ni-25Cr-10P transient liquid phase.

Tetrahedral Truss Core Structures

The favorable strength-to-weight geometry of cellular geodesic domes [20] was extended to rectangular prismatic forms by way of the octahedral-tetrahedral truss [21]. These stiff, strong designs are based upon a triangulated architecture wherein truss members are in tension/ compression only with no bending. Under this mode of deformation, the stiffness and strength have a linear dependence upon density making them the preferred cellular topology for lightweight load support applications. Recent studies have investigated miniaturized versions of similar cellular structures as the cores of all metal sandwich panels [12-14]. They have weight specific properties that are predicted to compete with honeycomb core designs [12] and were made with legs of circular cross-section by investment casting.

Consider a tetrahedral truss core made up of triad units with legs of length L and rectangular cross-section dimensions w and h. The triad height is about $L\sqrt{(2/3)}$ and the core relative density is $\rho_c/\rho_s = 3\sqrt{2}wh/L^2$. The elastic behavior of a pin-connected tetrahedral truss is isotropic with a relative shear modulus given by [13]

$$\frac{G_c}{E_s} = \frac{1}{9}\left(\frac{\rho_c}{\rho_s}\right) \tag{5}$$

where G_c is the shear modulus of the core and E_s is the Young's modulus of its base material. Provided members are thick enough to avert elastic buckling, failure by yielding occurs at relative shear strength [13]

$$\frac{1}{3\sqrt{2}}\left(\frac{\rho_c}{\rho_s}\right) \leq \frac{\tau_{cy}}{\sigma_{ys}} \leq \frac{\sqrt{6}}{9}\left(\frac{\rho_c}{\rho_s}\right) \tag{6}$$

where τ_{cy} is the core shear yield strength. Here, the minimum occurs when shearing is parallel to the projection of one set of members onto a base-plane (e.g., facesheet). The maximum occurs when shearing is oriented 30° to this projection.

Wrought metal tetrahedral truss core structures with characteristics similar to those described above have been made using inexpensive perforation, shaping and bonding approaches [22]. The core in Fig. 7 (a) was made by shaping hexagonal perforated type 304 stainless steel sheet. These cores readily bend at the nodes allowing for complex curved structures to be made. It was bonded between thin type 304 stainless steel facesheets using a Ni-25Cr-10P transient liquid phase, Fig. 7 (b). Ductile joints of large curvature radius were obtained.

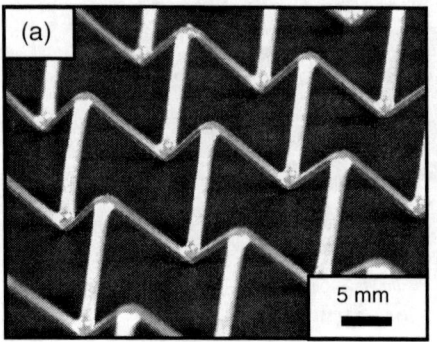
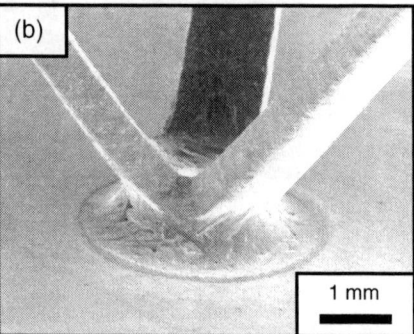

Figure 7: (a) Truss core after shaping. (b) Core/facesheet bond. See [22] for more details.

A midspan loading procedure [14] was used to determine the overall flexure response. An unload-reload scheme revealed a beam stiffness, $F/\delta \approx 2500$ N/mm, where F and δ are the midspan load and deflection. The core shear modulus was estimated to be $G_c \approx 1.01$ GPa, which corresponded to $G_c/E_s \approx 0.0053$ where $E_s = 193$ GPa. During beam failure, the set of truss members on the left side of the beam (with facesheet projection parallel to the shearing direction) first yielded in compression and then plastically buckled about their thinnest cross-sections, Fig. 8. A plastic hinge developed near the outer support on the left side. On the right side, members of this same set yielded in tension and work hardened. Neither plastic buckling or a hinge were observed there. Similar antisymmetric behavior has been seen in comparable cast systems [15]. The peak load was $F = 1470$ N and the core yield strength was estimated as $\tau_{cy} \approx 0.95$ MPa which corresponds to a relative core shear strength of about $\tau_{cy}/\sigma_{ys} \approx 0.0046$. After testing, all core/facesheet bonds appeared intact with no visually observed cracking. In Fig. 9, Hexcel Composites (Pleasanton, CA) Al-5056-H39 honeycomb shear properties (plate shear test) [24] are compared with those of the tetrahedral truss core and model predictions. The relative shear modulus and strength of the tetrahedral truss core compare favorably with those of similar relative density honeycomb. Note however, that predictions for the truss core shear modulus are based upon pin-connected members. In practice, the added rotational resistance afforded by the nodes and braze alloy fillets could lead to stiffness enhancement. Plate shear test confirmation would be helpful since core shear values obtained from flexure tests are often higher than those obtained from plate shear tests [23] and inversion of flexure data for the shear modulus is sensitive to measurement

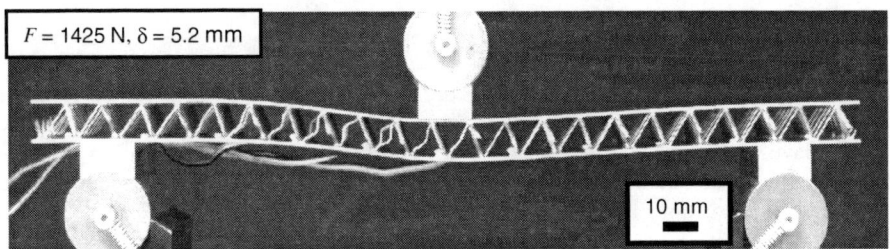

Figure 8: Visual observations. The aligned pores suggest bending to shape the cores.

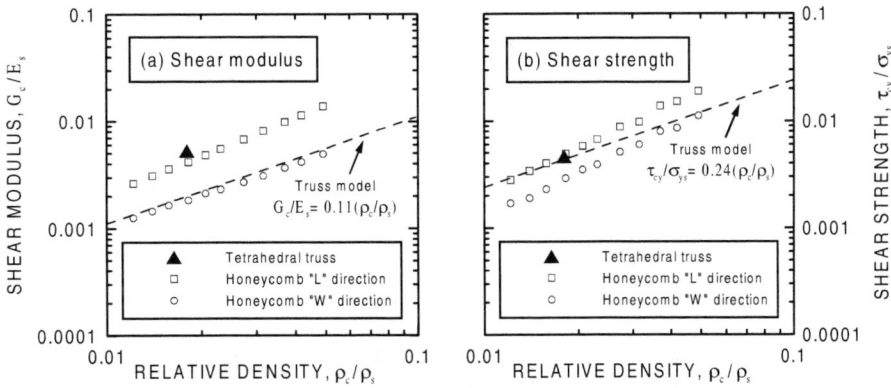

Figure 9: Relative shear properties comparison. (a) Modulus. (b) Strength.

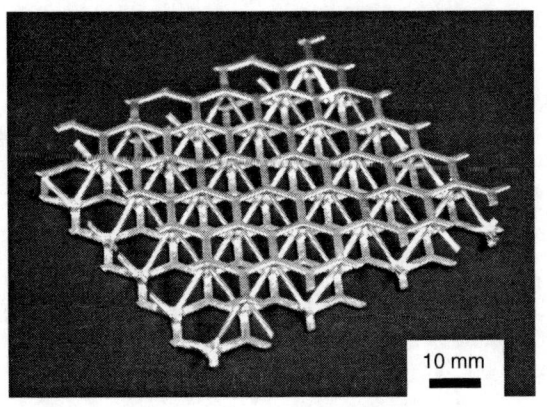

Figure 10: Tetrahedral truss core with hexagonal perforated facesheets

error. Nonetheless, the relative properties of the tetrahedral truss core are close to model predictions and show much promise when compared to commercial honeycomb products. Perforated facesheets can also be bonded to the cores creating a wide variety of inexpensive, miniature periodic truss structures. In Fig. 10, the same hexagonal perforated sheets that were used to make the above described cores have been transient liquid phase bonded as facesheets. Multi-laminate and/or hierarchical constructions are readily made in this way. If the facesheets have triangular perforations, miniature octahedral-tetrahedral trusses emerge.

Summary

A wide variety of inexpensive miniature truss architectures have been constructed from wrought metals. Superb control over the distribution of material at the cell level has been gained. The structures have very good base metal properties and can be fabricated from a variety of high temperature and heat treatable alloys. Future challenges are likely to involve inexpensive methods for automated manufacture. These materials and fabrication techniques are not necessarily limited to metals.

Acknowledgements

The author is grateful to H.N.G. Wadley for his support and encouragement. Also, A.G. Evans, M.F. Ashby, N.A. Fleck, J.A. Hutchinson, L.J. Gibson, B. Budiansky, D.T. Queheillalt, K.P. Dharmasena, D.M. Elzey, S.R. Chiras, D.R. Mumm, H. Bart-Smith, Y. Sugimura, S. Hyun, J. Grenestedt, S. Koehler, T.J. Lu and V.K. Deshpande for many productive interactions. Support for this work was provided by DARPA and ONR through research grants; N00014-96-1-1028 (program manager, S. Fishman) and N00014-01-1-0517 (program managers, L. Christodoulou and S. Fishman).

REFERENCES

1. L.J. Gibson and M.F. Ashby, Cellular Solids Structure and Properties - Second Edition (Cambridge University Press, Cambridge, 1997).
2. M.F. Ashby, A.G. Evans, N.A. Fleck, L.J. Gibson, J.W. Hutchinson and H.N.G. Wadley, Metal Foams: A Design Guide (Butterworth-Heinemann, Boston, MA, 2000).
3. G.J. Davies and S. Zhen, J. Mat. Sci., 18, (1983), 1899.
4. I. Jin, L.D. Kenny and H. Sang, U.S. Patent No. 4,973,358 (27 Nov 1990).
5. H.-D. Kunze, J. Baumeister, J. Banhart and M. Weber, pmi, 25 (4), (1993), 182.
6. R.B. Kaplan, U.S. Patent No. 5,282,861 (1 Feb 1994).
7. A.G. Evans, J.W. Hutchinson, N.A. Fleck, M.F. Ashby and H.N.G. Wadley, Prog. Mater. Sci., 46, (2001), 309.
8. H. Bart-Smith, A.-F. Bastawros, D.R. Mumm, A.G. Evans, D.J. Sypeck and H.N.G. Wadley, Acta mater., 46 (10), (1998), 3583.
9. A. Kelly and N.H. Macmillan, Strong Solids (Clarendon Press, Oxford 1986), 3rd ed.
10. R. Lakes, Nature 361, (1993), 511.
11. M.L. Renauld, A.F. Giamei, M.S. Thompson and J. Priluck, in Porous and Cellular Materials for Structural Applications (Mat. Res. Soc. Symp. Proc. 521, Pittsburg, PA, 1998), 109.
12. N. Wicks and J.W. Hutchinson, Int. J. Solids and Structures, 38, (2001), 5165.

13. V.S. Deshpande and N.A. Fleck, Int. J. Solids and Structures, 38, (2001), 6275.
14. S. Chiras, D.R. Mumm, A.G. Evans, N. Wicks, J.W. Hutchinson, K. Dharmasena, H.N.G. Wadley and S. Fichter, Int. J. Solids and Structures, In Press (2001).
15. S.T. Brittain, Y. Sugimura, O. J. A. Schueller, A.G. Evans and G.M. Whitesides, J. Microelectromechanical Systems, 10 (1), (2001), 113.
16. V.S. Deshpande, M.F. Ashby and N.A. Fleck, Acta mater. 49, (2001), 1035.
17. F.K. Ko, in Textile Structural Composites, ed. T.-W. Chou and F.K. Ko, (Elsevier, Amsterdam, 1989) 129.
18. D.J. Sypeck and H.N.G. Wadley, J. Mater. Res. 16 (3), (2001), 890.
19. D.R. Mumm, S. Chiras, A.G. Evans, J.W. Hutchinson, D.J. Sypeck and H.N.G. Wadley, Int. J. Solids and Structures, Submitted (2001).
20. Fuller, R.B., U.S. Patent #2,682,235 (29 June 1954).
21. Fuller, R.B., U.S. Patent #2,986,241 (30 May 1961).
22. D.J. Sypeck and H.N.G. Wadley, in Proceedings of the 2nd International Conference on Cellular Metals and Metal Foaming Technology (MetFoam 2001), ed. J. Banhart, M.F. Ashby and N.A. Fleck, (MIT-Verlag, Bremen, 2001), 381.
HexWeb™ Honeycomb Attributes and Properties, Publication No. TSB 120, Hexcel Composites, Pleasanton, CA (1999).

CELLULAR STRUCTURE BY DIRECT MATERIALS DEPOSITION

Eric Stiles and Jyoti Mazumder

Department of Mechanical Engineering
University of Michigan
Ann Arbor, MI 48109-2125

Abstract

This paper reports an investigation made into the ability of Direct Metal Deposition (DMD) to fabricate multi-material objects designed to meet specific engineering purposes (Designed Materials). The experiment tested the use of DMD to build a structure designed using Homogenization Design Method, a structural optimization technique. The structure of interest had an overall negative coefficient of thermal expansion as its design characteristic. Two structures were fabricated from chromium and nickel using DMD. Both showed initial contraction, followed by resumption of expansion after the temperature reached 75 °C in the x-direction, and over 200 °C in the y-direction.

Introduction

Direct Metal Deposition, or DMD, is a layer based additive manufacturing process that uses a high powered laser to melt powdered metals and make deposits, with the objective of making fully dense three dimensional objects. The laser beam is focused just above a metal substrate surface, where the deposition is to occur. A coaxial powder stream is focused into the melt pool formed at the substrate surface. This powder is melted upon entry into the melt pool. The substrate is attached to a CNC multi-axis system, and by moving it around, a two dimensional layer can be deposited. By building successive layers on top of one another, a three-dimensional object is formed. Any designed cellular structure can be fabricated layer by layer from a digital database.

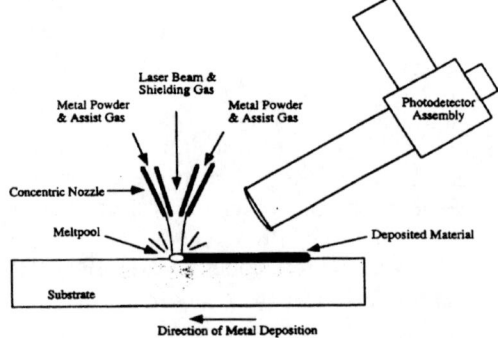

Figure 1: Direct Metal Deposition.

The goal of this research was to use the unique manufacturing abilities of DMD to make structures created with the Homogenization Design Method, a structural optimization technique. This technique allows the user to design a structure "in reverse"- instead of designing the geometry of the structure and then determining its overall mechanical properties, the properties are given as an input, and the geometry is created so that the structure will have the desired properties. The research done in this investigation on the DMD side was to push the limit of how small and intricate features could be made. Since the structures were made from multiple materials, how the materials would co-exist was an issue as well. The particular structure to be made was a five by five matrix of cells, whose desired property was a negative coefficient of thermal expansion. This was referred to as the Negative CTE structure, designed by Dr. Bing Chen and N. Kikuchi of the University of Michigan.

Background

Homogenization Design Method
Another area of current research is the field of Numerical Topology Optimization, which can be used to design composite material microstructures. Dr. Kikuchi at the University of Michigan has developed one such technique, the Homogenization Design Method. It can create optimal microstructures of composites based on a given set of desired properties and constraints[1,2,3]. It has been used to control Young's moduli, shear moduli, coefficient of thermal expansion, and Poisson's ratios[4,5,6,7]. Sigmund also demonstrated the fabrication of a micro-mechanical structure with a negative Poisson ratio using Plasma Enhanced Chemical Vapor Deposition[8].

Direct Metal Deposition

As a manufacturing process, Direct Metal Deposition has the distinct advantage of being able to form complex shapes in a single processing step that would take multiple steps for standard machining processes. It allows the part designer to go directly from the 3-D model to generation of a solid object. This reduction in the number of steps can cut down on lead times for parts, especially for molds and dies. This is especially evident on parts made from materials that are challenging for conventional manufacturing processes, such as titanium. The AeroMet Corporation claims a reduction in leadtime for complex structural titanium parts of 50-75%[9]. The quality of deposited parts can meet or exceed that of cast components. Components can be made with mechanical properties similar to plate material and almost fully dense in near net shape, even for oxide formers such as aluminum and copper[10]. Keicher claims an increase in tensile strength of 85% and an increase in ductility of 30% for 316 stainless steel. This is due to grain size refinement associated with rapid cooling[11]. Titanium components made by AeroMet pass ASTM standards for commercially pure titanium[9].

Both CO_2 and Nd:YAG lasers are commonly used for DMD processes, much the same as with laser cladding. Once distinct difference between the workstations of different institutions is the use of an inert gas chamber for processing. About half of the groups use an inert gas chamber, and the rest do the processing in open atmosphere. It is not necessary to use inert environments, but the choice of materials may dictate the use of a chamber (for example, with titanium). Careful control of process parameters is the key to the success of this process [10-14].

System Control
Research done at CLAIM has added to the process control capability of Direct Metal Deposition. Schifferer investigated the effects of increasing the number of height control cameras and varying the reset time of the laser input signal[14]. He found that by using three cameras at 120 degrees spacing, all effects due to the direction of movement of the CNC table were eliminated. He also found that the porosity of samples made using the DMD height control was minimized at a laser input signal length of 20 ms, or 50 Hz.

Hesse implemented the first DMD system at CLAIM to be controlled by PC software. This gave the ability to use information from the height control system to create a control of powder feed rate. The powder feed rate was then varied so that an optimal laser pulse duty cycle would be achieved. Hesse found that by maintaining a target laser duty cycle, the amount of power imparted on the melt pool could be kept at a constant level[15]. This led to more uniform layer thickness, elimination of cracking at the substrate, and reduction of porosity.

Experimental Procedure

Overview of DMD Workstation. All experiments were carried out using the CLAIM DMD workstation. The laser source is a Trumpf TLF 5000 5 kW carbon dioxide laser, which has an M^2 value of 8.7. The CNC workstation is an Allen-Bradley 8400 MP, with x-, y- and z-axes. The three axes are accurate to 0.0001 inches [0.0025 mm]. The beam is focused by a set of copper mirrors, with a focal length of 150 mm. The focal spot size is 600 μm. The normal efficiency of the mirror system is around 88% power transmitted.

All control of the CNC is carried out using CNC programs, which are generated from CAD files on a PC via a postprocessor. The CAD/CAM package used for DMD is SurfCAM, which can be used to create new geometry or read files from other packages. In the program, three-dimensional geometry can be sliced into layers for the DMD process. Each layer is then converted to CNC code, which consists of G- and M-commands. All G-codes correspond to different CNC functions, and all M-codes are used to control external devices (for example, to turn on/off the laser or process gasses.) These text files are transferred from the PC to the Allen-Bradley via the RS-232 port.

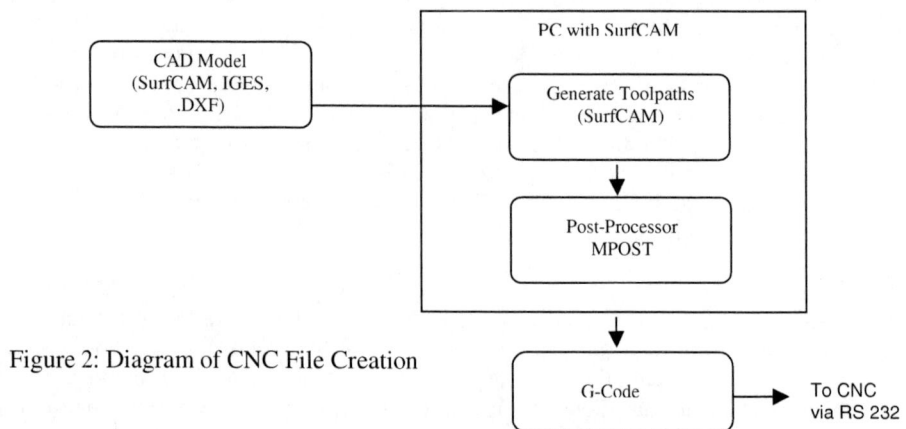

Figure 2: Diagram of CNC File Creation

The powder delivery system is partially controlled by the Allen-Bradley. Each powder hopper is run by a control box, which contains the driver card for each hopper's motor. These boxes are controlled by both the PC and the Allen-Bradley controller. The PC controls the speed of the motors via an analog signal, and the Allen-Bradley turns the motors on or off by controlling their power supply.

The DMD camera system is used to determine whether the deposition height of each layer is at the desired value. Each camera consists of a photodetector and a focusing lens, plus a CCD camera to allow the user to adjust the system. Two cameras are mounted on the z-axis frame, 90 degrees to one another, and do not move relative to the laser spot. They are pointed at the melt pool area, and depending on the height of the deposition, the photodetector may or may not detect light. The light passing through the lens tube is split at the lens cube, so the light is seen by both photodetector and CCD camera. The photodetectors are connected to the height control circuit, which changes the input signal to the laser depending on the status of the photodetector

Negative CTE Experiments. The Negative CTE structure experiment was done in conjunction with three other research groups from the University of Michigan. A student provided the optimized design and another student converted it from a series of elements to lines defining the regions of the two different materials. This model, in IGES format, could then be imported into SurfCAM and tool paths could be generated. The last group tested the structure for its coefficient of thermal expansion. Three challenges remained before the structure could be built- first, the two materials had to be chosen, second, a tool path needed to be created, and third, optimal process parameters needed to found by experiment.

Material Selection
The key to the functioning of the Negative CTE structure is the difference in the thermal expansion coefficient of the two constituents. The two materials act as a type of bi-metallic strip, when one material expands more than the other, the net result is bending. This in turn leads to a

net contraction, as shown in Figure 4. During the process of choosing two materials for the structure, their particular identity was of less importance than their CTE. The greater the difference in CTE, the more pronounced the contraction would be. The target ratio of the two materials CTE's was between two and three. Also taken into consideration was the miscibility between the materials, because some mixing would be necessary at the interface in order to ensure a solid bond. Table I lists materials that were considered as candidates, either because they were previously used for DMD or they were considered attractive for the project.

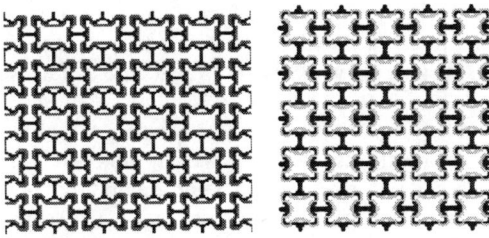

Figure 3: Negative CTE Structure;
Left, HDM model,
Right, IGES file after smoothing (Grey-Nickel, Dark Grey – Chromium)

From the table, it can be seen that many materials that are commonly used for DMD, such as copper, steel, and nickel, all have very similar CTE's. Aluminum has a much higher thermal expansion coefficient, but forms brittle intermetallics with many of the other materials. Invar has a very low CTE of course, but has problems with cracking during DMD processing. Chromium is the only material that remains. It is miscible in excess of 40 percent by weight at 1300 C with nickel, and has a CTE that is less than half of nickel. Therefore it was chosen for the low-expansion phase of the structure, and nickel was chosen for the high-expansion phase.

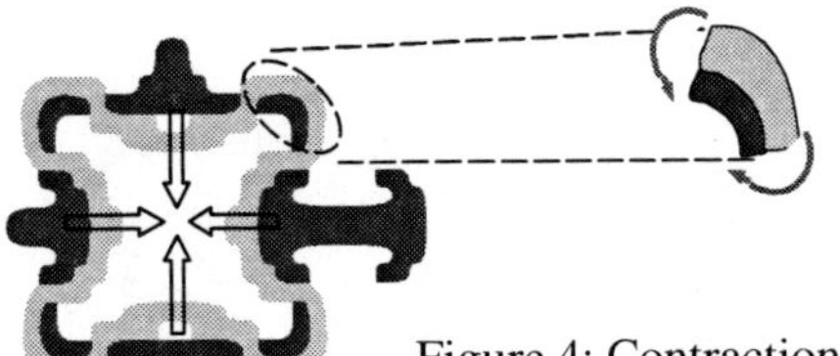

Figure 4: Contraction Mechanism of Negative CTE Structure

Table I: Coefficients of Thermal Expansion, DMD Materials

Material	Coefficient of Thermal Expansion (10^{-6} cm/cm *C)
Invar	1.54
Chromium	**6.0**
1020 Steel	12.0
Nickel	**13.0**
Copper	16.6
Aluminum	25.0

The specifications of the nickel and chromium powders used, and their suppliers are listed in Table II.

Table II: Negative CTE Powder Specifications

Material	Supplier	Purity	Particle Size
Nickel	Atlantic Equipment Engineers	99.9%	+325 / -100
Chromium	Atlantic Equipment Engineers	99.8%	Plasma Spray Grade; -100

Tool Path Design

The Negative CTE design can be scaled to any size, but the goal was to make the structure as small as possible. From the perspective of the DMD process, the smallest dimension is one DMD track wide. So, dimensions of the structure could be fixed by looking for small features and by assigning them a width of one deposition track, and then using this scale to measure the overall structure. The width of one DMD track can vary depending on the material and the process parameters, so a series of parametric studies were carried out to measure the height and width of a single-track deposit for nickel and chromium at varying process parameters. For both nickel and chromium, the track width was found to be 600 microns in the range of process parameters used. From this information, the overall structure's dimensions were determined to be 60 mm x 60 mm.

Several tool paths were attempted during the trials before arriving at the best result. Figure 5 shows two of the key dimensions of the unit cell tool paths labeled. These two dimensions are the spacing between the nickel and chromium tool paths around the perimeter of the cell itself, and the spacing between the chromium tool paths in the actual cell and the bridge tool paths. These values for the five different tool paths are also listed in Table III.

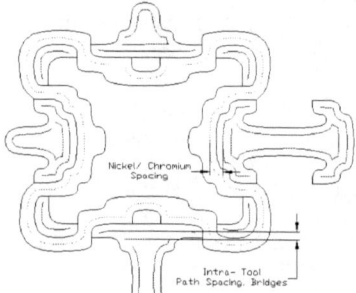

Figure 5: Two Primary Dimensions of Unit Cell Tool Paths

Table III: Descriptions of Five Tool Paths

Tool Path Number	Design	Chromium Intra-Path Spacing (in)	Chromium Tool Spacing (in)	Chromium-Nickel Spacing (in)
1	1	0.012		0.024
2	2	0.014		0.024
3	2	0.014		0.024
4	2	0.0174		0.025
5	2	0.0174		0.030

To put the values in perspective, for a 600 micron wide deposition track, spacing of 0.012 inches is equivalent to 50% overlap between the tracks, and 0.024 inches of spacing means there is no overlap.

Experimental Results and Discussion

Parametric Studies

The results from the parametric studies using statistically designed experiments revealed that significant process parameters for the two materials (Ni,Cr) were fairly consistent with each other. In terms of which effects were significant and which weren't, the two materials were fairly consistent with each other. Traverse Speed and Mass Flow Rate were significant for both width and height except for nickel, which was found to be unaffected by Traverse Speed for width. Both materials were also showed an interaction between Traverse Speed and Mass Flow Rate for height.

From these results, a rough idea of what levels the process parameters should be set at IV and V show the average values of height and width vs. the process parameters.

The desired layer height for the Unit Cell experiments was either 0.005 or 0.01 inches, which is equivalent to 1270 or 2540 microns. The desired width was approximately 600 microns. From the average values, it can be seen that these were within the realm of possibility. The desired layer height for the Unit Cell experiments was either 0.005 or 0.01 inches, which is equivalent to 1270 or 2540 microns. The desired width was approximately 600 microns. From the average values, it can be seen that these were within the realm of possibility.

Table IV: Averaged Values of Experimental Results, Nickel

Factors			Averages (microns)	
Traverse Speed (ipm)	Mass Flow Rate (g/min)	Nominal Power (W)	Width	Height
10	2	300	550	1450
10	2	450	556	1942.5
10	4	300	576.6	3523.3
10	4	450	640	3520
20	2	300	443.3	816.66
20	2	450	466.6	986.66
20	4	300	555	2105
20	4	450	606.6	2336.6

Table V: Averaged Values of Experimental Results, Chromium

Factors			Averages (microns)	
Traverse Speed (ipm)	Mass Flow Rate (g/min)	Nominal Power (W)	Width	Height
10	4	400	525	1795
10	4	550	586.6	1866.6
10	6	400	680	4143.3
10	6	550	810	4055
20	4	400	450	990
20	4	550	485	980
20	6	400	503.3	1703.3
20	6	550	545	1425

Combined Trials

When the two materials were deposited together, there were some combinational effects. This could have been caused by interference by another structure with the powder flow. The chromium portion of the unit cell had been deposited first, and then the nickel portion. Since the two are very close together, it was possible that the chromium deposit was getting in the way of the powder stream. So the nickel portion of the tool path was modified to give more spacing between the two materials. The new design was Tool Path #5 (see Figure 6).

No changes were made to any of the chromium tool path. The spacing between chromium and nickel tool paths was changed from 0.025 to 0.030 inches. This design was successful; sagging was not a problem on any of the combined material unit cell trials. The decision was made to not use the height control system during fabrication of the combined unit cell or the Negative CTE structure. The main problem with its use is the large difference between the intensity of light given off during processing of chromium and of nickel.

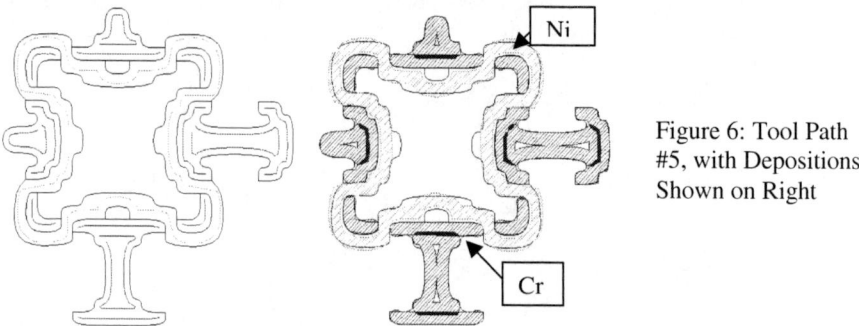

Figure 6: Tool Path #5, with Depositions Shown on Right

Chromium is optically bright, whereas nickel is less bright. When the height control system is calibrated, it is adjusted based on one material. To effectively work with two materials as different as nickel and chromium, a system has to be developed to change the attenuation in the camera system between materials. Recently a new system with the capability to change the filter is being designed and built for this purpose. The final test of the unit cell design was to prove that no fatal cracks existed. The combined materials unit cell was cut from the substrate with the diamond saw, and remained in one piece. So Design #2 with Tool Path #5 was a success.

The combined materials unit cell was cut and examined along a nickel-chromium interface using EDAX. The objective was to see if a significant mixed chromium/nickel third phase was forming at the interface area. This had not been accounted for in the original Negative CTE model, so it was important to see if it was present. SEM photographs were taken, as well as an EDAX line scan. The SEM photo is shown in Fig. 7, and the results of the line scan in Fig. 8. Some mixing was seen in the two line scans on the nickel side of the interface, between 5-10%. For the most part, the interface between the two materials is well distinguished.

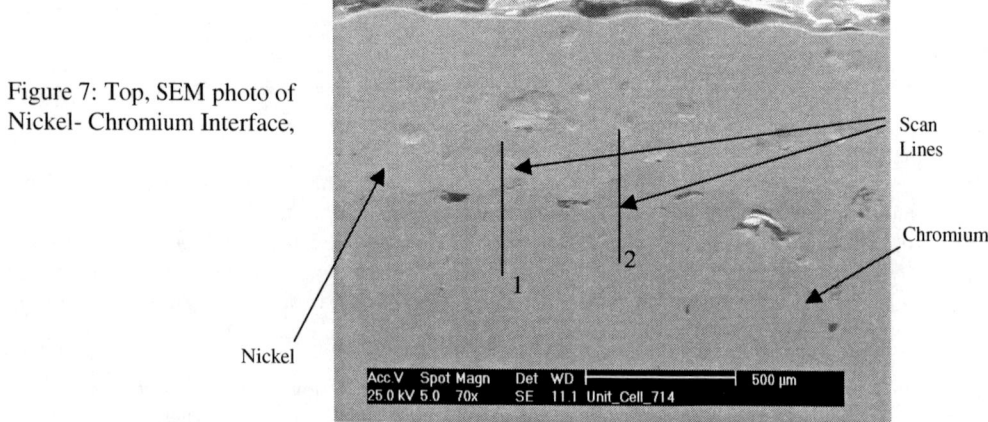

Figure 7: Top, SEM photo of Nickel- Chromium Interface,

Two Negative CTE structures were fabricated. The first was built 0.5 inches tall, and the second was 0.4 inches. After finishing the first one, with reduced power for the nickel layers, it was observed that the spacing between nickel and chromium was too much. Gaps could be seen in many locations. For the second structure tool Path #4 was used, which had 0.005 inches less of chromium-nickel tool path spacing. All of the process parameters were kept the same for the second structure. The reduction in tool path spacing did eliminate the gaps between the materials, but the problems of rounded corners on the nickel unit cells showed up again, and worse. Even the low power levels did not get rid of the problem. Again, faced with a problem that could stop the experiment, the process parameters would have to be modified. The traverse speed for the nickel layers was reduced to 8 ipm, and the power level to 300 W. This solved the problem (see Fig. 9).

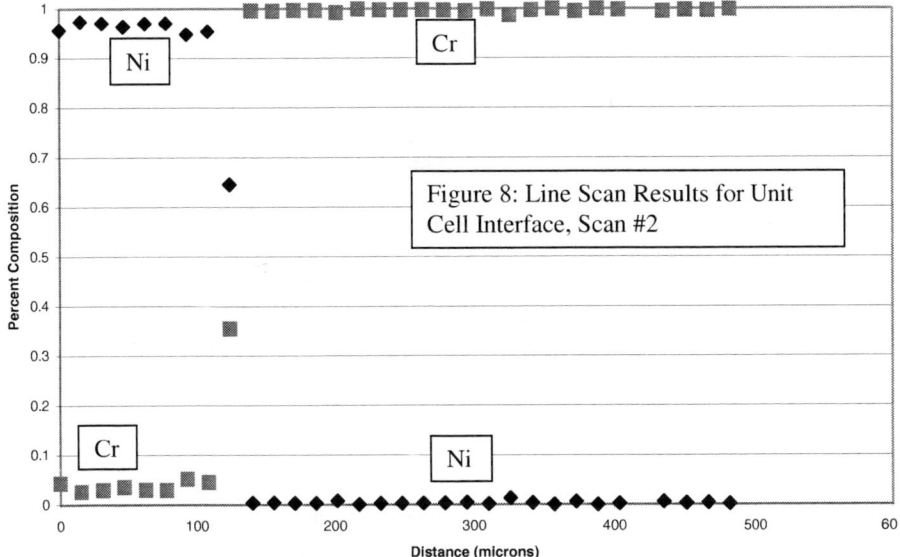

Figure 8: Line Scan Results for Unit Cell Interface, Scan #2

A ceramic abrasive cut-off wheel was used to cut through the steel substrate just below the surface of the specimen. Then any remaining substrate was ground off. The structures were tested for their expansion properties at Orton Refractory Testing and Research Center in Westerville, Ohio. At Orton Refractory, the specimens were placed in a precision controlled furnace, and

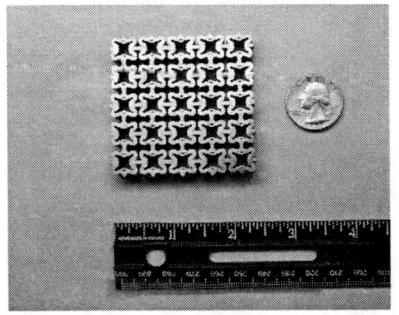

Figure 9: Negative CTE structure #1

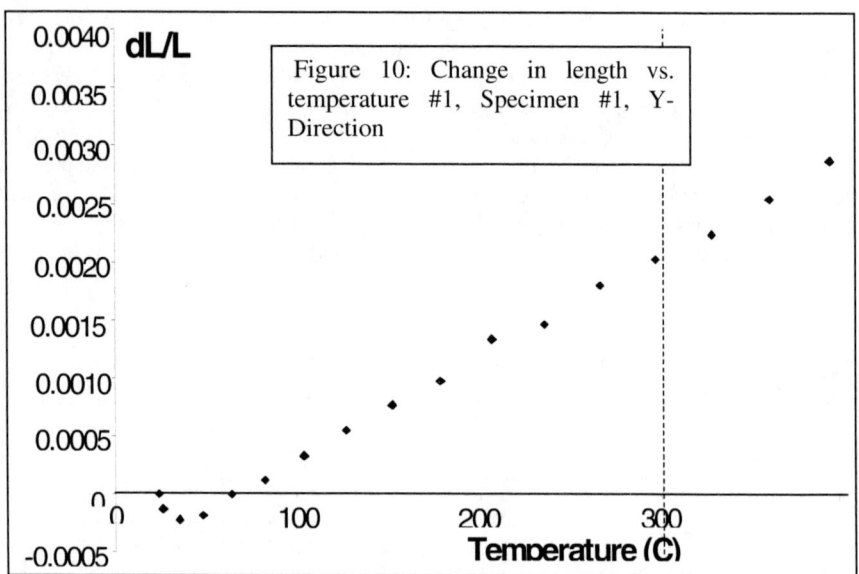

Figure 10: Change in length vs. temperature #1, Specimen #1, Y-Direction

measured during the heating cycle using a laser extensometer system. The furnace was heated at a rate of three degrees Celsius per minute, up to a maximum of 530C. Negative thermal expansion was seen in Fig. 10.

Conclusions

The experiments for fabrication of the negative Co-efficient of Thermal Expansion (CTE) structure from a topological design indicated that in the application of DMD to very small intricate structures many processing conditions need careful control. They are summarized below. For structures only one tool path wide, the tool path design is quite critical. It was found that when segments overlapped too much, parts of one segment (such as an end,) would not deposit properly due to a high spot. It was decided that when filling in odd shaped regions, having some void present was preferable because when overlapping occurred it was non-uniform, and high points would form. The materials involved also played a big part in the design issues. It was necessary for the DMD nozzle to adequate inert gas shielding for the prevention of oxidation of chromium. The general brittleness of chromium led to the decision to use Unit Cell Design #2, which had continuous nickel reinforcement rather than a continuous chromium phase. Nickel also has challenges due the ease with which thin walled nickel structures distorted. The power and layer thickness had to be kept low.

Two complete versions of the 5x5 matrix were produced. The first had spacing between the chromium and nickel tool paths of 0.030 inches (760 microns,) and the second had spacing of 0.025 inches (635 microns.) Both structures showed initial contraction in both directions. The net strain in Y-Direction on the samples reached positive territory somewhere between 200-270 C during each of the tests In the x-direction, the samples were in the negative strain region until 70 C. In the end, it appeared that the Negative CTE contraction mechanism worked for a period of time. This result was consistent between the two samples, suggesting that the unique imperfections in each structure from the manufacturing process did not cause widely varying results from one sample to the next. The departure from the intended designed performance can be explained by the fact that the interfaces of the fabricated sample has alloyed zone whereas the design considers no mixing between Ni and Cr. Currently effort is underway to include the interface alloying in the Homogenization design model.

Acknowledgement

This work was made possible by a grant from Defense Advance Research Project Agency administered through NIST [Grant No 70NANB8H0068]. Drs. Ram Sriram and Kevin Lyons are the project monitors. The authors also acknowledge the cooperation from Professors A. Ghosh, D. Dutta and N. Kikuchi in materials testing, CAD/CAM and Homogenization Design respectively.

References

1. R.V. Kohn, and G. Strang, Optimal Design and Relaxation of Variational Problems, Comm. Pure Appl. Math., 39, 1986, 1-25 (Part 1,) 139-182 (Part 2,) and 353-377 (Part 3).
2. M.P.Bendsoe, and N. Kikuchi, Generating Optimal Topologies in Structural Design Using a Homogenization Method, Comput. Methods Appl. Mech. Engrg., 71, 1988, 197-224.
3. K. Suzuki, and N. Kikuchi, A Homogenization Method for Shape and Topology Optimization, Comput. Methods Appl. Mech. Engrg., 93, 1991, 291-318.
4. G.W. Milton, Composite Materials with Poisson's Ratio Close to −1, Journal of Mechanics and Physics of Solids, 40, 1992, 1105-1137.
5. R. Lakes, Foam Structures with Negative Poisson's Ratio, Science, 23, 1987, 1038-1040.
6. O. Sigmund, Materials with Prescribed Constitutive Parameters: An Inverse Homo-genization Problem, International Journal of Solids and Structures, 31, 1994, 2313-2329.
7. J.S.O. Fonseca, and N. Kikuchi, Design of Microstructures of Periodic Composite Materials, Comput. Methods Appl. Mech. Engrg., submitted Feb. 1997.
8. U.D. Larsen, O. Sigmund and S. Bouwstra, Design and Fabrication of Compliant Micromechanisms and Structures with Negative Poisson's Ratio, Journal of Microelectromechanical Systems, Vol. 6, no. 2, June 1997.
9. D.H. Abbott, and F.G. Arcella, Laser Forming Titanium Components, Advanced Materials and Processes, May 1998 pp. 29-30.
10. J. Mazumder, J. Koch, K. Nagarthnam and J. Choi, Rapid Manufacturing by Laser Aided Direct Metal Deposition of Metals, Powder Metallurgy Conference (PM2TEC), vol. 15:107-118, July, 1996.
11. D.M. Keicher, and J.E. Smugeresky, The Laser Forming of Metallic Components Using Particulate Materials, Jourmal of Materials, May 1997.
12. W.M. Steen, M.A. McLean, and G.J. Shannon, Laser Direct Casting of High Nickel Alloy Components, Laser Research Group, University of Liverpool, Department of Mechanical Engineering, Liverpool U.K., private communication.
13. J. Lin, and W.M. Steen, Design Characteristics and Development of a Nozzle for Coaxial Laser Cladding, Journal of Laser Applications, Vol. 10 n2 April 1998.
14. J. Mazumder, Schifferer, A., and Choi, J., "Direct Materials Deposition: Designed Macro and Microstructure," Materials Research Innovations, Vol. 3, pp. 118-131, 1999.
15. T. Hesse, "Automation, Process Control and the Effects of Process Parameters on Laser Aided Direct Metal Deposition of H13 Tool Steel", M.S.Thesis, University of Michigan, Ann Arbor March 2000.

SECTION II

NOVEL AND LOW-COST PROCESSES AND PRINCIPLES

PROCESSING OF CONTROLLED POROSITY TITANIUM-BASED MATERIALS

D. Kupp, D. Claar and K. Flemmig
Fraunhofer USA - Delaware
Center for Manufacturing and Advanced Materials
501 Wyoming Road
Newark, DE 19716

U. Waag and H. Goehler
Fraunhofer IFAM, Dresden, Germany

Abstract

Alloys and intermetallic compounds based on titanium are used extensively in aerospace and biomedical applications, based on their high strength, light weight, and excellent corrosion resistance. To extend the range of applications, designers and engineers are now seeking titanium-based materials with controlled porosity. Ultra-lightweight, porous titanium alloys and intermetallics are being fabricated based on technologies being developed at the Fraunhofer Institute and Fraunhofer USA Center for Manufacturing and Advanced Materials, located in Dresden, Germany and Newark, DE, respectively. Powder metallurgy processing routes are being utilized to fabricate hollow metal spheres and open-celled foams in titanium, Ti-6Al-4V alloy, and TiAl and $TiAl_3$ intermetallics. The processing, microstructures, and selected properties of these controlled-porosity materials will be presented, along with a discussion of potential applications in the aerospace, biotechnology, and other industries.

Introduction

Cellular structures based on metals, ceramics or polymers are being used with increasing frequency as a means of reducing weight in aerospace, naval and land-based applications, for such novel components as energy absorbing media in personnel armor, aircraft structures, vehicular blast panels, vibration damping structures, and as biomedical implant materials designed to replicate natural bone tissue structure. By tailoring pore structure (size, shape and volume fraction) performance of these materials can be engineered to meet the requirements of each application. In aerospace applications, for example, lightweight components with a high specific stiffness (stiffness to weight ratio) are desirable. The industry workhorse -honeycomb structures - suffer from relatively high cost, adhesive bonding delamination, anisotropic properties and lack of heat resistance. To meet new and demanding requirements, a wide range of alternative cellular metal structures have been developed and tested extensively. Cellular components can be produced by various means, including metal casting, powder metallurgy, electro-deposition, and chemical and physical vapor deposition [1,2].

Hollow sphere structures

Due to the availability of structures in a wide range of compositions and highly controllable pore characteristics, the fabrication of hollow metal spheres via a P/M process is one increasingly important technology for fabricating cellular metal components. Hollow metal spheres are manufactured by coating expanded polystyrene (EPS) spheres with a metal powder suspension as the spheres move turbulently through a fluidized bed. This movement provides a uniform coating on the EPS spheres and allows almost instant drying of the slurry liquid. The coated spheres are then heated to pyrolyze the binder and EPS, followed by sintering of the metal powder particles to form a dense shell [3]. Other substrate materials can be used, but EPS has the advantage of being low cost and very low density with clean pyrolysis. A schematic of the process to produce hollow metal spheres is shown in Figure 1 [3].

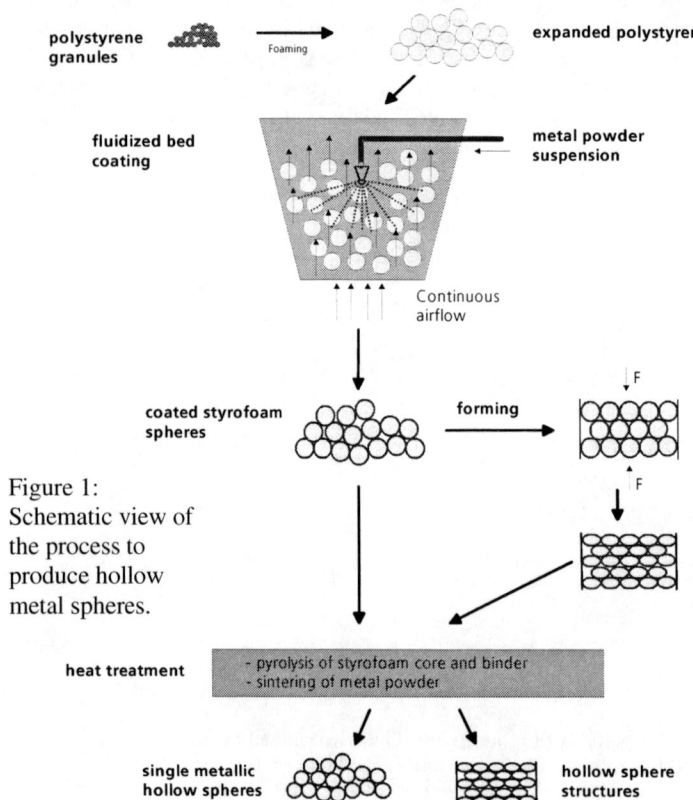

Figure 1: Schematic view of the process to produce hollow metal spheres.

The hollow sphere technology has been used to produce metallic structures in Ti, Ti-6Al-4V and TiAl as well as Fe, Ni, 316L stainless steel, Inconel 718 and other materials. Several examples of spheres and test components in Ti-based alloys and other compositions are shown in Figure 2.

Figure 2: Examples of hollow sphere structures in various compositions.

In recent work, highly porous titanium aluminide structures were successfully fabricated and evaluated for use as aerospace thermal protection components. Spheres produced from titanium aluminide pre-alloyed powders (Crucible Research, Pittsburgh, PA, USA) were sinter bonded into 50 mm x 50 mm x 25 mm thick plates. Sintered bulk porosities of >90 volume % were routinely reached using 2 mm diameter spheres with wall thicknesses ~60-80 µm. By increasing the sphere diameter to ~ 3mm, thereby increasing the diameter to wall thickness ratio, bulk densities >95 volume % were ultimately achieved. A close-up of one of these structures is shown in Figure 3.

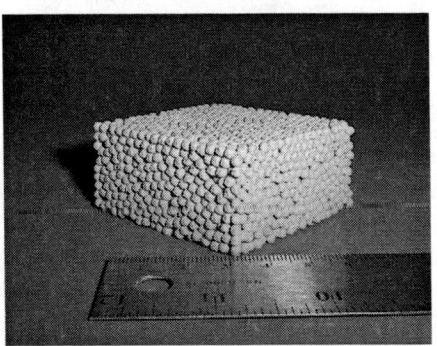

Figure 3: Sintered titanium aluminide hollow sphere structure, fabricated from 3 mm diameter spheres.

Reactive Sintering of Porous Titanium Aluminide

An alternative approach to fabricating porous titanium aluminide structures has also been investigated, relying on the well documented macrostructural swelling that takes place during pressureless reactive sintering of Ti + Al and other intermetallic powder mixtures. During production of dense titanium aluminide alloys, this swelling behavior prevents full and homogeneous densification during reactive sintering, but can be used to generate desirable porosity in sintered Ti+Al components. An increase in porosity is achieved by redistribution and penetration of molten Al into Ti-Ti contacts causing a displacement of Ti particle centers AND the formation of a highly porous $TiAl_3$ layer. This process is shown schematically in Figure 4 [4].

A direct method for the fabrication of porous Ti-Al components was evaluated, using reactive sintering of die-pressed compacts of elemental Ti and Al powder. This approach relies on the high-temperature reaction mechanisms previously described to develop a porous microstructure throughout the die-pressed component. By pre-blending relatively fine Ti and Al powders, then pressing to the maximum achievable green density, a homogeneous microstructure is produced which supported the development of porosity during reactive sintering. The Al powder (-75 μm) was provided by Eckart America L.P. (Louisville, KY, USA), and Ti powder (-45 μm) was provided by Pyrogenesis Inc. (Montreal, Quebec, Canada) as shown in Figures 5 and 6 respectively.

Figure 4: Schematic representation of phase formation and macroscopic swelling behavior during reactive sintering of Ti-Al.

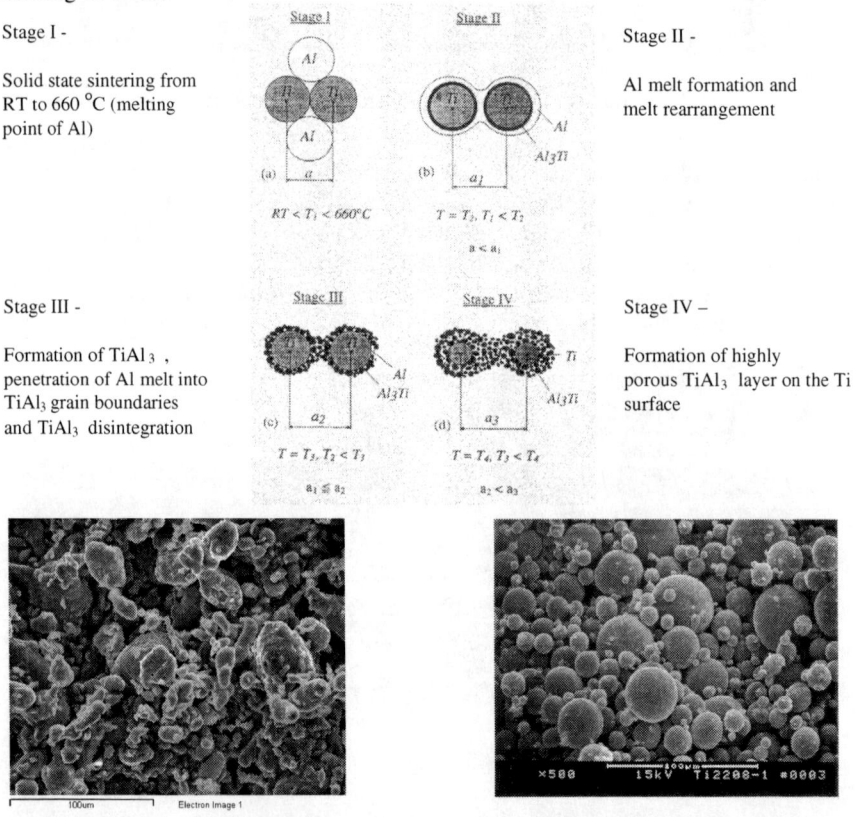

Stage I -

Solid state sintering from RT to 660 °C (melting point of Al)

Stage II -

Al melt formation and melt rearrangement

Stage III -

Formation of $TiAl_3$, penetration of Al melt into $TiAl_3$ grain boundaries and $TiAl_3$ disintegration

Stage IV –

Formation of highly porous $TiAl_3$ layer on the Ti surface

Figure 5: Al powder (-75μm) from Eckart America, L.P. used in the production of reactively sintered Ti-Al

Figure 6: Ti powder (-45μm) from Pyrogenesis Inc. used in the production of reactively sintered Ti-Al components

Using a 50 atomic % mixture of each material without binder or lubricants, Ti and Al powders were pressed at 138 MPa (20 kpsi) into ~50 mm square billets, approximately 12 mm thick, then heat treated at 1300°C in a flowing Ar atmosphere. Green densities after pressing averaged about 71% of theoretical density. Samples of the die compacted, reactively sintered titanium aluminide components are shown in Figure 7.

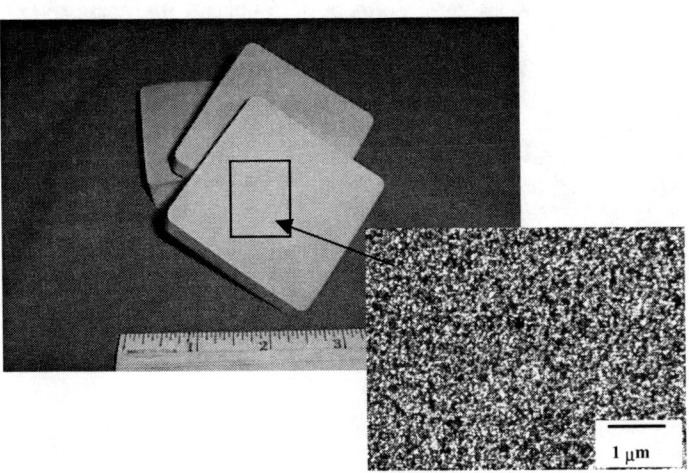

Figure 7: Reaction sintered Ti-Al plates with a close-up of the as-sintered surface (inset).

Figure 8 shows a typical microstructure of the die compacted Ti-Al components displaying the porosity resulting from low-pressure compaction and reactive sintering. Linear expansion in the pressing direction averaged about 15%, and about 12% perpendicular to the pressing direction. The bulk porosity increased from ~29 volume % in the "as-pressed" state to ~ 54 volume % after reactive sintering. X-ray diffraction performed on a representative section of the sintered components shows a mixture of TiAl, $TiAl_2$, $TiAl_3$ and Ti_3Al in addition to minor mixed oxide phases. Free Al was detected, but no pure Ti was observed via XRD. Further analysis of the samples was not carried out, nor were any other attempts made to optimize the structure or composition of the component.

Fabrication of Ti-based Components using Polymer Scaffolds

For orthopedic implants, in addition to the requirement for biocompatibility, load-bearing capabilities, fatigue strength and corrosion resistance, the need for durable implant-bone fixation is of paramount importance. A common approach for enhancing tissue-to-implant fixation is the use of surface-textured implants. Numerous fabrication approaches have been developed for producing porous surfaces on orthopedic implants, including sintering of metal or ceramic powders or beads onto the outer surface of the implant device, plasma spraying, sintering of added chopped metal fibers, pressing/sintering of long metal fibers or metal felts, among others.

The functional requirements of such porous layers are to provide internal porosity for bone in-growth and fixation to the implants, and to demonstrate sufficient shear strength and mechanical integrity for long-term stability. Pilliar [5] has reviewed various techniques used for surface modification to create macroscopic surface features. Examples for machined macroscopic anchorage elements are millimeter-sized threads, vents, fins and surface grooves or serrations. Microscopic surface features are typical of sub-micron or micron dimensions, and are formed by: (i) removal of material from an implant surface by grit blasting, or chemical or ion etching, and (ii) the addition of material onto the implant surface through plasma spraying or sintering of particles onto the surface. The thickness of porous surface coatings on solid metallic substrates ranges from 10-30 μm for plasma spraying and up to mm or greater for sintered metal powder layers. Although porous sintered layers of metal powders are among the most common surface modifications used in orthopedic implants, they are far from being ideal. Aseptic loosening, especially 10 years after implantation, is a principal concern, particularly as younger patients with active lifestyles and increased life expectancy are undergoing joint replacement procedures. Thus, the metal foam technology detailed here may represent a more desirable surface modification to promote improved bone in-growth compared to existing techniques.

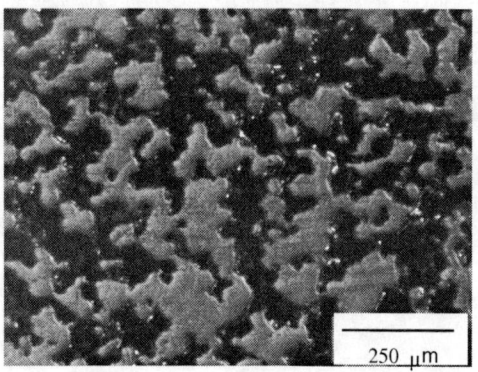

Figure 8: Polished cross-section of reactively sintered Ti-Al specimen (porosity ~ 54 volume %).

Powder metallurgical processing techniques have been used for nearly three decades for development of surgical implants. The sintering of metal powders onto a metallic core implant material, produced by machining or casting a solid metal piece, creates a porous surface layer with uniform distribution of interconnected pores. Implants produced in this manner have surface areas five or more times greater than as-machined or as-cast metal implants [6]. As a result, higher rates of metal ion release due to corrosion of the implant by biochemical fluids are expected. Thus, the consideration of implant biocompatibility during material selection and implant preparation is crucial. The optimal pore size of the sintered surface layer for human use has not been well established, although it is generally believed that pore openings in the range of 100 to 500 μm are suitable [7]. A varied surface pore size can be achieved by using more than one particle layer. The resulting multi-layered particulate surface zones provide extensive interconnected porous networks.

Pilliar et al. [8] reported the importance of eliminating local contaminants in the surface zone in order to avoid tissue inflammatory responses, and to establish initial implant stability. Sintered porous surface structures have demonstrated more favorable conditions for mineralization and bone formulation than other surface designs that do not have three-dimensional pore interconnectivity. The work of Simmons et al. [9] on implants placed in rabbits demonstrated that the pullout force of sintered porous-surfaced implants was 5 to 10 times greater that that for plasma sprayed implants, which had no interconnected pores. Furthermore, the rate of implant fixation was also much enhanced in the case of porous-surfaced implants. There is also *in-vitro* evidence suggesting that sintered porous coatings on metallic surfaces promote osteoblast (i.e. bone forming cells) adhesion [10,11] as well as maturation and matrix mineralization [12,13]. This suggests that the long-term stability of the interface between bone and the porous surface may depend on the influence of the macroscopic/microscopic surface characteristics on cellular function and activities. Finally, bone in-growth into the sintered porous-surfaced region is believed to provide effective force transfer between implant and bone, and hence, implant stability. Such mechanical coupling can be beneficial or undesirable, depending on the implant stress relative to the normal physiological stress level and the occurrence of implant "stress shielding" [14]. This observation points to the importance of proper design of the porous implants for optimum load transfer and long-term implant fixation.

A number of metallic systems have been processed with sintered porous surfaces for orthopedic applications. The most commonly accepted materials are Co-Cr-Mo, Ti-6Al-4V and commercially pure (i.e. CP grade) Ti. The sintering temperatures are 1300°C for Co-based alloys and 1250°C for Ti-based alloys. Non-oxidizing environments (Ar atmosphere for Co-Cr-Mo alloys and high vacuum for Ti alloys) are needed. Co-Cr-Mo alloys have a long history of applications and excellent wear resistance. The formation of continuous carbide-containing zones during sintering, however, can make implants susceptible to particle debonding and, hence, wear due to third-body particles, as well as crack propagation in the solid core material. The fatigue strengths of Co-Cr-Mo cast or forged porous-coated implants range from 400 to 650 MPa [15].

Both vacuum sintering and pressure sintering have been used for fabricating Ti and Ti alloy implants. Because of the reactive nature of these materials, the furnace atmospheres need to be sufficiently non-oxidizing, and hydrogen embrittlement needs to be prevented. High vacuum sintering ($<10^{-3}$ Torr) is required for Ti-6Al-4V powders and fibers. Appropriate selection of sintering time and temperature is essential for ensuring the formation of sufficiently large sintered neck regions between adjoining particles, while maintaining the desired open-pored structure. It has been reported that sintered neck diameters of 0.4 times the particle diameter give suitable combination of strength and porosity [5].

Ti-6Al-4V is used extensively for orthopedic devices because of its excellent *in vivo* corrosion resistance due to the passive surface oxide layer formed in oxidizing environments, and high fatigue strength. Titanium is also reported to promote more efficient tissue and bone in-growth than Co-based alloys. Sintering Ti alloys at 1250°C may modify the fine-grained, equiaxed structure to a lamellar structure, resulting in a drop in fatigue strength of about 15%. However, a much more severe drop in fatigue strength is associated with stress concentrations that result at the particle-substrate sintered neck region [16]. Reductions in fatigue strength occur for textured Ti alloys surface produced by grit blasting, plasma spray coating, and ion-etching, due to the similar development of stress concentrators. Thus, it has been recommended that proper design of Ti-6Al-4V implants is required to ensure that stresses are maintained below the fatigue endurance limit of 200 MPa and that porous surface regions be avoided at implant surfaces subjected to high tensile stress.

In summary, there are a number technological barriers and challenges in the fabrication of metallic implants with sintered surfaces, including the following:
- The selection of powders of uniform size is essential for achieving a continuous open-pore surface structure. Smaller pore sizes can be accomplished with finer particles, but at higher costs and at greater processing risk due to the enhanced reactivity of the finer powder.
- The diameter of sintered neck region between adjoining particles may control the strength of the surface-pored layer. Fracture at the neck region gives rise to loose metal particles and third-body wear.
- Typical sintered metal powder surfaces cannot be fabricated with high porosity levels (>60 volume %), anticipated to be required to achieve fixation affecting shear strength and integrity.

Reticulated Open Cell Metal Foams as Porous Coatings.

In view of some of the limitations inherent to metallic implants with sintered porous surfaces, it is desirable to explore the technological advantages of metal foam materials for porous-surfaced implants. Metal foam materials can be categorized as one form of *cellular materials*. A cellular solid is one made up of an interconnected network of solid struts or plates which form the edges and faces of cells. *Honeycombs* are typical 2-D cellular materials while *foams* are commonly referred to as 3-D cellular solids. For foams, distinctions are made between *open* and *closed* cell materials, which are inherently different.

Almost all solids can be foamed. Techniques now exist for making three-dimensional cellular solids out of polymers, metals, ceramics and glasses. Different techniques are used for foaming different types of solids. Polymers can be foamed by introducing gas bubbles into the liquid monomer or hot polymer. Metallic foams can be made by either liquid or solid state processing. Glass foams are made by methods paralleling those for polymers. Carbon foams are made by graphitizing polymeric foams in a carefully controlled environment. Ceramic foams are made by infiltrating a polymer foam with a slip, or can be made by chemical vapor deposition onto a substrate of reticulated carbon foam. Microcellular silica foams, with cell sizes less than 100 nm and densities as low as 4 kg/m^3, have been made by the sol-gel polymerization of alkoxy silanes [17].

Porous cellular metals (i.e. metal foams) can be produced by various processes, including metal casting, powder metallurgy, electro-deposition, and chemical or physical vapor deposition [18-20]. Metal foams can be fabricated with either open-cell or closed-cell microstructures. The selection of appropriate fabrication process is based on the desired characteristics of the cellular metal foam, such as volume fraction of porosity, average pore size and pore size distribution, component geometry, and functional requirements.

Cellular materials are found in everyday uses for their lightweight, high specific stiffness, and other properties. Their applications range from lightweight construction and packaging, to thermal insulation, vibration damping, and chemical filtration. Metallic cellular materials, namely metal foams, in particular are becoming a new class of engineering materials. These properties should make metal foam an attractive material for orthopedic implants. The medical use of these foams has not been fully explored.

<ins>Fabrication of metal foams using P/M processing routes</ins>
The microstructure and cell morphology of the metal foams can be modeled after the reticulated open cell polymer foams that are commercially available in a broad range of pore cell sizes. The total porosity of the foam structures can be varied from 50 to as high as 95 volume%. The pore sizes can be varied over the range 200 to 500 μm consistent with bone in-growth requirements.

To prepare open celled Ti implants by powder infiltration of polymer scaffolds, sections of reticulated polymer foam are cut, replicating the net shape of the desired porous component. A solvent/binder solution is prepared using about 2-5 weight% PVA or PEG in alcohol. Aqueous solutions can also be used if compatible with the selected metal powder, but the use of volatile solvents reduces the drying time between powder coating operations. The polymer scaffold is immersed in the solvent polymer solution to fully wet the structure, then removed. At the point during the drying process when the majority of the liquid has evaporated yet remains "tacky", the polymer scaffold is coated by applying metal powder in a fine flowing stream or by covering the substrate with a powder bed. The coated scaffold is then removed from the powder bed and light agitated to free any powder not bonded to the polymer scaffold ligaments. Upon complete drying, the process is repeated until a sufficiently thick layer of metal powder is developed. Once the desired pore structure has been fabricated, the coated scaffold is thermal debindered and sintered to final density using standard P/M processing conditions. Figure 9 shows the technique of coating reticulated polymer foams with powdered metal/ binder systems being developed at Fraunhofer USA. Figures 10 and 11 show the porous microstructure of the sintered Ti materials fabricated using this process.

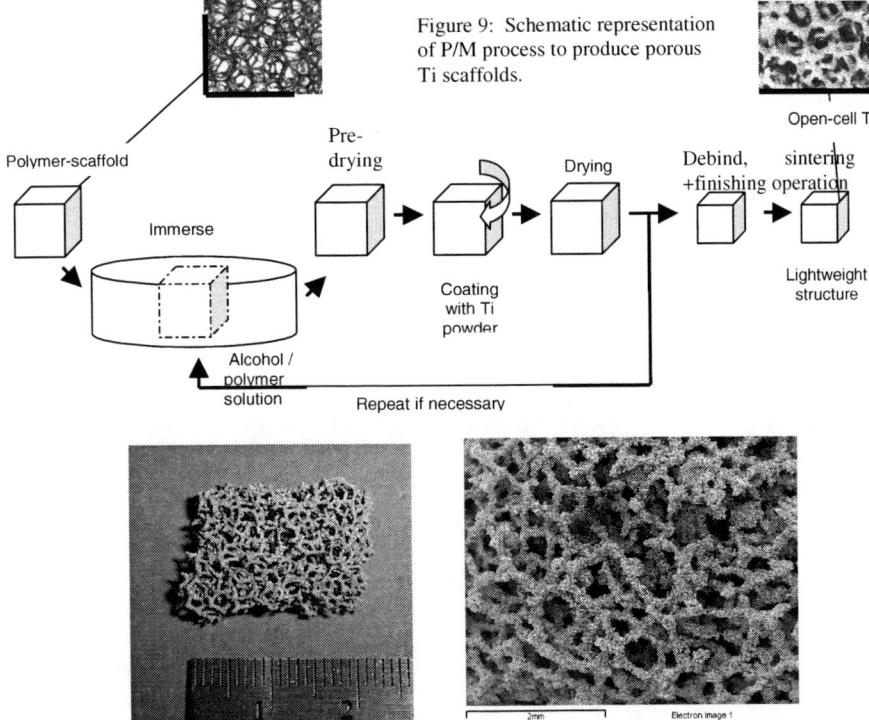

Figure 9: Schematic representation of P/M process to produce porous Ti scaffolds.

Figure 10 and 11. Optical (left) and SEM (right) images of sintered Ti scaffolds.

Potential Advantages of Reticulated Open Cell Metal Foams

The objective of current research at Fraunhofer is to develop a porous metal foam layer having an improved reticulated microstructure that more closely simulates the structure of natural bone, thus providing greater strength and long-term reliability, as well as the "material bed" with adequate interconnected porosity necessary for tissue in-growth. More specifically, the metal foam surface layer/core metal combination will eliminate or minimize most of the technical problems encountered in implants with sintered porous-surfaced layers, as stated earlier. This structural similarity should enhance the functional compatibility between the synthetic cast open cell foam implant surface and natural bone.

Conclusions

Several novel techniques for producing porous Ti-alloy materials for a wide range of applications have been demonstrated including 1) fabrication of hollow metal spheres and structures, 2) reactively sintering Ti+Al powder compacts, and 3) infiltration coating of reticulated polymer foam scaffolds. All three processes utilize common P/M processes for powder handling, debinding and sintering.

These processes also facilitate the tailoring of bulk density (i.e. total porosity) and pore structure so that specific performance levels for such varied applications as aerospace and orthopaedic implants might be attained. Furthermore, the use of P/M approaches for forming and downstream process also permit the fabrication of components from a wide range of alloys depending on the application and performance requirements.

Acknowledgment

The authors would like to thank the NASA Langley Research Center, especially Dr. Stephen Hales, for their financial and technical support in the preparation of the titanium aluminide hollow spheres and reactively sintered titanium aluminide.

References

1. G.L. Davies and S. Zhen, "Review of Metallic Foams: Their Production, Properties and Applications", *Journal of Materials Science,* vol. 18, 1983, p. 1899.
2. J. Banhart and J. Baumeister, "Production Methods for Metallic Foams", *Proceedings of 1998 MRS Metal Foam Symposium,* vol. 521, 1998, pp. 121-132.
3. G. Stephani, U. Waag, P. Löthman, O. Andersen, L. Schneider, F. Bretschneider, H. Schneidereit, "New Lightweight Structures Based on Low-cost Metallic Hollow Spheres", Proceedings of 2000 International Conference on Powder Metallurgy and Particulate Materials, New York, June 2000.
4. A. Böhm and B. Kiebeck, "Investigation of Swelling Behavior of Ti-Al Elemental Powder Mixtures During Reaction Sintering", Z. Metallkd. 89 (1998) 2, pp. 90-95..
5. R.M. Pilliar, "P/M Processing of Surgical Implants: Sintered Porous Surfaces for Tissue-to-Implant Fixation", Int. J. of Powder Metallurgy, (1998), 34: 33-45.
6. R.M. Pilliar, H.U. Cameron and I. Macnab, "Porous Surface Layered Prosthetic Devices", Biomed Eng J. 10(4), 1975, 126-131.
7. G. Heimke, "Osseointegrated Implants", Vol I, Basics: Materials and Joint Replacement. CRC Press, Boca Raton, 1990.
8. R.M. Pilliar, D. Deporter, and P. Waston, "Tissue-Implant Interface in Materials in Clinical Applications", ed. Vincenzini, P., Techna Srl, Faenza, Italy, 1995, p 569.
9. C.A. Simmons, S.Valiquette, S. Carter and R.M. Pilliar, Porous Coated Implants Provide Better Initial Tissue Integration and Stability Compared with Plasma Sprayed Implants", Trans Orthop. Res. Soc., New Orleans, 1998.
10. R.K. Sinha and R.S. Tuan, "Surface Composition of Orthopaedic Implant Metals Regulates Cell Attachment, Spreading And Cytoskeletal Organization of Primary Human Osteoblasts In Vitro", Clin. Orthop. Rel. Res. 305, 1994, pp 258-272.

11. R.K. Sinha, L.A. Brigandi and R.S. Tuan, "Regulation of Human Osteoblast Integrin Expression by Orthopaedic Implant Materials", Bone 18, 1996, pp. 451-457.
12. B. Groessner-Schreiber and R.S. Tuan, "Enhanced Extracellular Matrix Production and Mineralization by Calvarial Osteoblasts Cultured on Titanium Surfaces in Vitro", J. Cell. Sci. 101, 1992, pp. 209-217.
13. B. Groessner-Schreiber and R.S. Tuan, "Bone Cell Response to Hydroxyapatite-coated Surfaces In-vitro", Sem Arthrop. 2, 1991, pp. 260-267.
14. J.D. Bobyn, H.U. Cameron, D. Abdulla, R.M. Pilliar and G.C. Weatherly, "Biologic Fixation and Bone Remodeling with an Unconstrained Canine Total Knee Prosthesis", Clin. Orthop. Rel. Res., 166, 1982, pp. 301-312.
15. K.K. Wang, L.J. Gustavson and J.H. Dubleton, "A New Description Strengthened Vitallium Alloy Developed for Medical Implants", Trans. Soc. for Biomaterials Annual Meeting, Lake Buena Vista, FL, 1989, p. 154.
16. D. Wolfarth and P. Ducheyne, "Effect of Change in Interfacial Geometry on the Fatigue Strength of Porous Coated Ti-6Al-4V", J. Biomed. Mater. Res. 28, 1994, pp. 417-425.
17. L.J. Gibson and M.F. Ashby, "Cellular Solids, Structural and Properties", 2nd edition, Cambridge University Press, Cambridge, UK, 1997.
18. G.J. Davies and S. Zhen, "Metallic Foams: Their Production, Properties and Applications", Journal of Materials Science, 18, 1983, pp. 1899-1911.
19. J. Banhart and J. Baumeister, "Production Methods for Metallic Foams", Proceedings of 1998 MRS Metal Foam Symposium, 1998, pp. 121-132.
20. O. Anderson, U. Waag, L. Schneider, G. Stephani, B. Kieback, "Novel Metallic Hollow Sphere Structures: Processing And Properties", Proceedings of Metal Foam, 1999, pp.183-188.

NOVEL LIGHTWEIGHT OPEN CELL METAL FOAM PROCESS AND RESULTING PROPERTIES

A. K. Ghosh and X. Tong
Department of Materials Science and Engineering
The University of Michigan, Ann Arbor, MI 48109-2136

Abstract

A new low cost process has been developed for fabricating both stochastic and periodic open cell metal foam structures by direct deposition of solidifying liquid wires, droplets and spray, and simultaneously welding them in-situ. The process is capable of producing a wide range of pore spaces from microns to millimeters and amenable to computer-aided manufacturing technique. Open cell structures with closed outer surfaces and solid face sheets have been produced. The process is also capable of producing graded porosity, and various internal geometries when deposition is carried out over removable geometric inserts. Samples with a variety of internal geometries were produced from several aluminum alloys to create multiscale porosity levels, and tested under compression for elastic modulus and strength. The results show that plateau strength and modulus can be influenced by the choice of internal geometry, which are not uniquely predicted by ligament bending-based models. Simple models demonstrate that the orientation of internal columns and walls are directly connected with the measured deformation behavior.

Introduction

Porous or cellular metals are emerging as a new class of engineering materials that offer unusual combination of physical and mechanical properties. They can be utilized in a variety of applications, e.g. for their load carrying ability in structural sandwich cores, air and fluid permeability, unusual acoustic properties, impact energy absorption capacity, improved thermal conductivity and good insulating properties. Their applications include shock and impact absorbers, auto bumpers, cushions, core structures for high strength aircraft wing panels, matrix for embedding sensors and actuators, sound and vibration dampers, heat sinks and exchangers, porous electrodes, catalyst supporters, etc. Currently, there is interest in using porous metals and alloys (e.g., Al, Mg, steel and Ti) for automotive, railway, aerospace, and chemical applications where weight reduction, chemical pollutant minimization, and improvement in comfort and safety are needed. The field of applications of porous metals and alloys is growing steadily [1-3].

The properties of these metals depend on the characteristics of the pore structure, including the volume fraction of pores, pore size and shape, distribution and morphology (i.e. open/closed cell), and the properties of the matrix. Many processes have been developed to produce porous metals [1-6]. Currently a great deal of interest exists in the fabrication of open cell foams for their multifunctional properties, such as fluid flow, heat transfer catalysis, etc. Open cell metallic foams are usually made by infiltrating an open cell polymer foam such as polyurethane with ceramic slurry, and then burning off the polymer by heating, and then casting liquid metal into the resulting form and then removing the ceramic core material [4]. Closed cell foams are currently made by several liquid metal based processes, that require addition of second phase particles and foaming agents [5-7]. Contamination so developed often renders the foam metal brittle. Contamination and high cost can also be a problem with the polymer precursor routes for metallic foams. It is thus of interest to develop foam fabrication methods which do not introduce embrittling constituents into the alloy and those that can reduce fabrication cost.

In this report, a novel liquid metal-based process has been described for economically producing open cell porous metals without using additives. This method, in principle, can be applied to all metals without the possibility of contamination. Additionally, it was intended for the process to be flexible enough to create pore sizes in the micron to millimeter range, with pore volume in the 20-85% range; and make process compatible with subsequent processing methods, e.g., shape rolling, forging etc. and spray deposition of a cover sheet.

Previous comparisons of the tensile and compressive strengths of metallic foams have yielded different results for open and closed cell metal foams. Andrews et al.[8] derived strength and elastic modulus of open and closed cell foams based on a mechanism of simple bending of ligaments. They predicted that open and closed cell foams follow distinctly different but well-defined functional dependence of mechanical properties as a function of density. In the new open-cell foam fabrication technology described here, which permitted the formation of a variety of porous structures with different orientation of the solid ligaments, evaluation of mechanical properties and comparison with existing models is needed.

Process Design And Experimental Procedure

A. Processing

The new method utilizes a process of liquid extrusion through fine orifice to solidify and create wires or fibers of those metals whose melting temperature is above the temperature outside of the extrusion die. Liquid stream ejected from the orifice undergoes solidification to form wire (or fiber). This is opposite of the ink jet break-up process as it relies on maintaining integrity of the liquid stream. Generally the heated reservoir containing liquid metal is pressurized by inert gas to eject liquid through the orifice, propelled by extrusion pressure and gravity. The orifice diameter is controlled by a small size hypodermic needle of metal or ceramic (75-500 μm, typically), and a gas control valve is used to turn liquid metal flow on or off. Fig. 1 shows a schematic diagram of the liquid extrusion process and an inert gas chamber for minimizing oxidation of ejected liquid. Gas cooling is applied to solidify liquid stream when orifice diameter is larger. When forced cooling is not employed, solidification of liquid stream can be delayed until the head of stream reaches a position well below the orifice (~250 mm or so), depending on the orifice size, extrusion pressure and liquid and orifice temperatures.

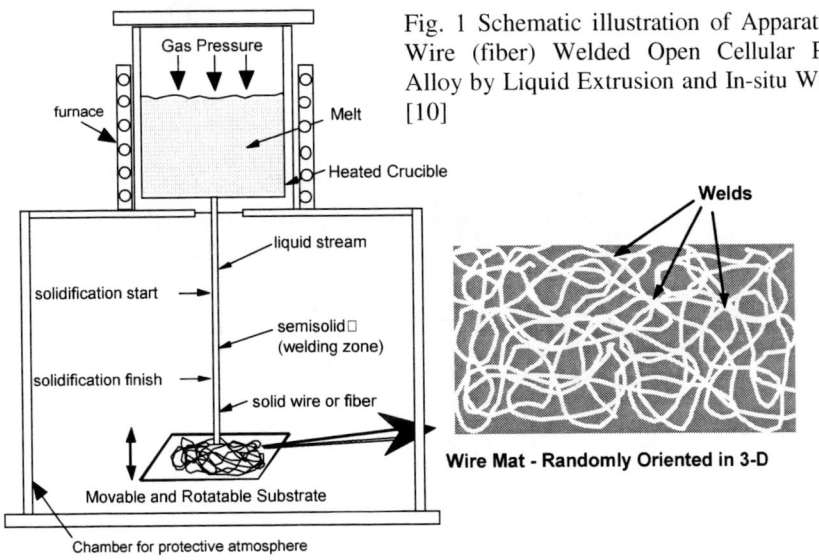

Fig. 1 Schematic illustration of Apparatus for Wire (fiber) Welded Open Cellular Porous Alloy by Liquid Extrusion and In-situ Welding [10]

The stream's outer skin freezes rapidly as it exits the container, and thus the upper portion of the stream is a solid-liquid mixture, and some distance below in the lower part the stream is replaced by a solid wire. Extruded wire is collected on a substrate, which is moved up and down between the solid and the liquid zones of the stream to permit welding of the solid wire segments instantly. This solid-liquid zone is referred to as the wire "welding zone" as deposition within this zone allows the internal liquid to flow outward and mix with the just-solidified material to form an integral weld [10]. This is possible due to partial remelting of the surface of previously deposited wire. Movement of the substrate (or location where deposit is building) is carried out on an x-y-z table with simultaneous but varying amount of movements in three orthogonal directions to create a supportable structure in three dimensions. This process has been utilized to develop unique concepts for building wire-welded open cell mats, fiber-core(or foil-core) solids and graded porosity solids. Gradation of porosity is achieved by varying the amount and time of deposition coverage in certain areas. These solids can be produced by either building directly from a deposited solid substrate or unattached to the substrate as a free material. Furthermore, the process can be used to fill hollow objects by directly injecting metal fiber (or wire) mat into it. With properly controlled movement of depositing substrate, freeform fabricated shapes of complex geometry can be fabricated with porosity content varying from near-zero to about 95%. For higher pore content, the use of removable solid inserts of soluble substances is made to produce hollow cavities surrounded by wire-welded structures. Bimodal pore size distribution can be achieved in this manner.

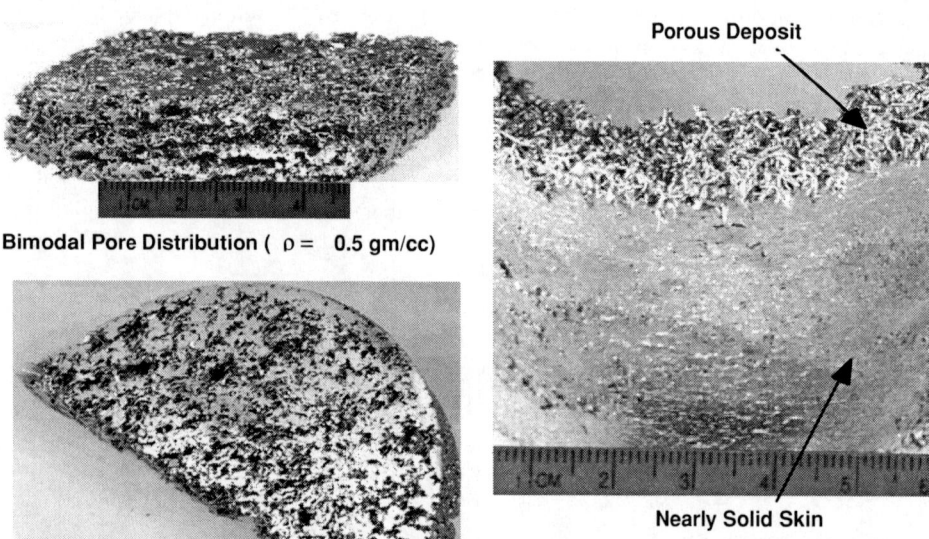

Fig. 2 Open Cell Micro- and Macrostructures Produced by U of M Deposition Process (Material: Al Alloys)

Thus no foaming agents or other chemical additives (or contamination) is necessary to create foam, and the integrity of alloy composition is maintained by operating in an inert gas environment. By combining the layered deposition process with stepwise deformation process, additional complexity of internal passages and external shapes are imparted. Furthermore, cooling tubes, actuators, sensors can be embedded if desired. At periodic intervals, solid stringers can be deposited in-situ for rigidity, and external surfaces can be closed up as skin in the last deposition step [10]. This type of structure is preferred over rod truss-connected geodesic structures which are coarse-framed and expensive to make by investment casting. Because deposition and buildup are performed in a single process, material and energy are conserved and multiple operations at multiple stations are avoided. This leads to an inexpensive fabrication process with high rate of manufacturing possible, and since the structural component is the desired end product, all costs and energy associated with an intermediate product shape, e.g. billet, plate, sheet etc. are eliminated.

Thus many energy intensive operations currently used in manufacturing operations, e.g. thermal soaking, hot forging, hot and cold rolling, coiling, solution treatment, etc. are avoided, and for part fabrication the separate steps of machining, welding, rivetting etc. are eliminated [10].

B. Materials

The starting materials for this study included several Al alloys, A-356, 6061Al, and 7075Al alloys. These were received in the form of 12-kg ingots which were cut into small pieces (about 60 mm thick, long, and wide) and then melted in a graphite crucible using an electrical resistance heating furnace before placing it in the heated container for extrusion to form the cellular structure. The chemical composition of the A356 and 7075 Al alloys used in the present work is given in Table I. The liquid temperature was maintained at around 720-740 C, typically a superheat of 150C with a lower orifice temperature. Gas pressure for extrusion was around 150-400 kPa. The traverse rate of substrate was varied in the range of 20-180 mm/s for x, y and z motion.

Table I. Chemical Composition of experimental alloys: A-356, 6061Al and 7075 Al alloys, in wt. percent

Material	Al	Si	Zn	Mg	Cu	Fe
A356	90.72	7.70	-	0.37	0.04	0.25
6061	97.93	0.56	0.08	0.77	0.25	0.59
7075	89.40	0.10	5.80	2.04	1.31	0.29

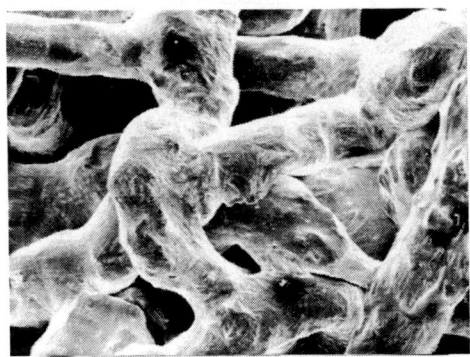

Fig. 3 Welded wire segments within stochastic part of the deposited open cellular material of A-356 Al

Multiscale Cellular Structures

Open cell wire-weld network fabrication process [10] has been used to produce multiscale periodic structures, with overall pore volumes of 75- 90%. These materials contained hollow structural features, e.g. embedded hollow spherical cavities, hollow triangular prisms, etc. These were fabricated by depositing the semisolid wire network covering over specific periodic locations and surrounding removable geometric inserts. The combined periodic and network structure permitted the evaluation of mechanical properties of such materials with several different geometric features and in which the orientation and size of the solid and hollow features were varied. The most unique feature of the process is that deposition allows creation of face sheets integral to the cellular core structure thereby creating a bond stronger than adhesively bonded face sheets and provides a lower fabrication cost by eliminating a separate manufacturing step. Droplet deposition was carried to a micron-size range by atomizing liquid metal into fine spray and depositing such spray to fabricate uniform, pore-free sheets as a base substrate or cover sheet. Multiscale cellular specimens were fabricated by liquid metal process in configurations with hollow spheres, hollow triangular prisms, column-supported panels with face sheets, etc. The alloys used in the above process were 6061 Al, 7075 Al and A-356 cast aluminum alloy. Figs. 2-5 show photographs of a few of the wire-welded microporous aluminum samples fabricated in this study. Fig. 4 shows a schematic diagram and an actual sample of spherical inserts (later removed) created a regular arrangement of hollow cavities within the wire-welded network, combining features of both stochastic and periodic foam structures.

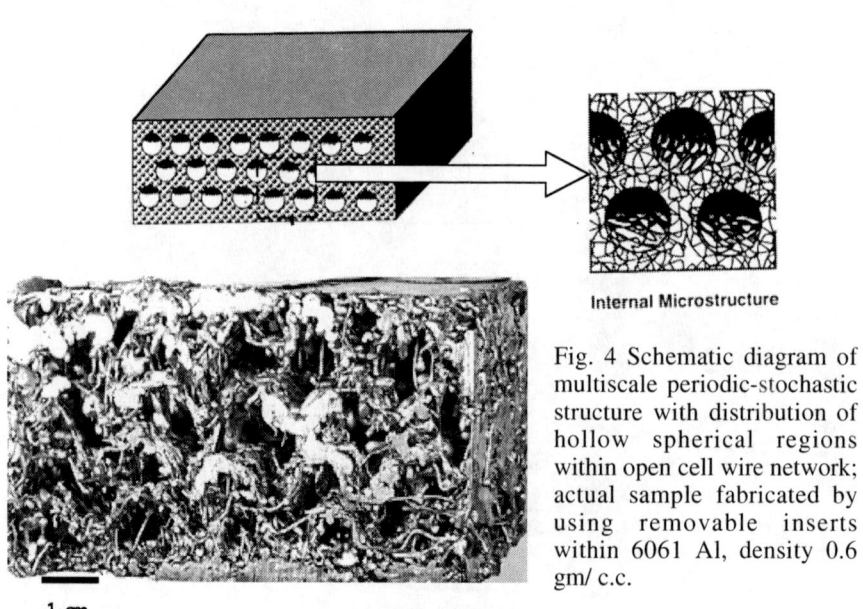

Fig. 4 Schematic diagram of multiscale periodic-stochastic structure with distribution of hollow spherical regions within open cell wire network; actual sample fabricated by using removable inserts within 6061 Al, density 0.6 gm/ c.c.

Mechanical Test Results

Mechanical tests were carried out on compression test specimens machined from fabricated and heat treated alloys. Machining was done by wire electro-discharge machining process to minimize damage to the specimen ligaments. These specimens revealed many interesting trends in the behavior of cellular and porous metals. Fig.5 shows microstructure and test orientation for column-supported sandwich in directions both perpendicular and parallel to the face sheets. Fig. 6 shows compression stress-strain curves. It is seen that yield strength perpendicular to the face sheets is rather low, but parallel to the face sheets the yield strength value reaches 10-15 MPa for an overall pore density of 75% (material density, including face sheets ~ 0.7 gm/cc). At this strength level these open cellular aluminum materials compete well as structural materials. The elastic modulus of these materials are in the range of 10-15% of aluminum's, and thus they are superior to the lighter and bulkier (high specific volume) structural metal foams. These properties are comparable to that of hardwood, except that these materials are also excellent sound dampeners, heat exchangers, impact energy absorbers and not flammable. The stress-strain

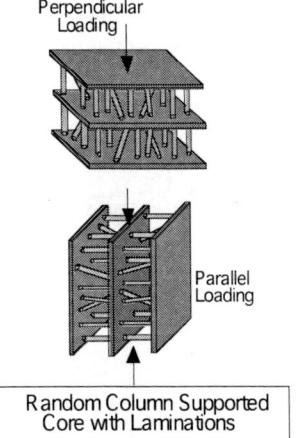

1 cm

Fig. 5 Photograph of column supported face sheet structure with internal open passages, produced by droplet and spray deposition from 6061 Al, overall density = 0.7 gm/c.c. Compression tests perpendicular and parallel to face sheet shown

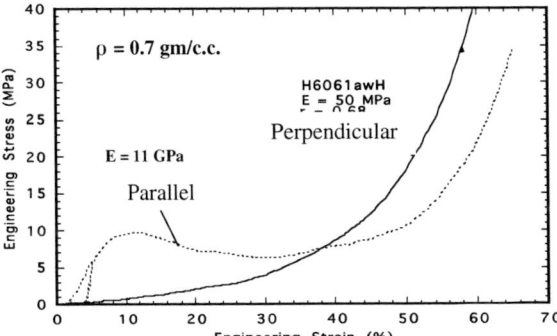

Fig. 6 Stress-strain curves in compression for the structure in Fig. 5, in two orientations shown above. Note that thin columns under compression or bending yield more easily, but solid support sheets under compression do not yield easily, indicating ways to take advantage of the structural alignment in design.

curves (Fig. 6) indicated 50-60% deformation at constant or falling stress before stress rises due to compaction, similar to other metallic foams, indicating a good potential for their use as an energy absorber. However the orientation of the samples affect this behavior strongly. For example, the yield strength of the parallel face sheet configuration is higher than that for most metallic foams of similar density because the vertical orientation of the face sheets resist buckling when compared with the bent elements in the core of the stochastic closed or open cell porous metal. However, because the face sheets carry a large fraction of the compressive load, when buckling eventually occurs, it causes a loss of stability leading to a falling load. Considering the data perpendicular to the face sheets which produces a higher hardening rate, it seems reasonable to precompress the panels for enhancing their yield strength as needed for application. Precompressed column supported material is more isotropic. Column support provides high stiffness and high peak strength parallel to face sheet, followed by strain softening. Another design lesson learned from this test is that the increase in the vertical, stringer-like segments in the wire network or other weak wall porous structures can have a significant effect on compressive strength.

Fig. 7 shows similar stress-strain curves for hollow spherical cavity material described in Fig. 4, for 7075 Al in the heat-treated (T4) condition. The solid line is for a material having larger hollow cavities, which has lower density and lower strength. Smaller cavity size (dotted curve) produces has a higher density and higher strength. Lower density material also shows lower elastic modulus compared to the one with higher density.

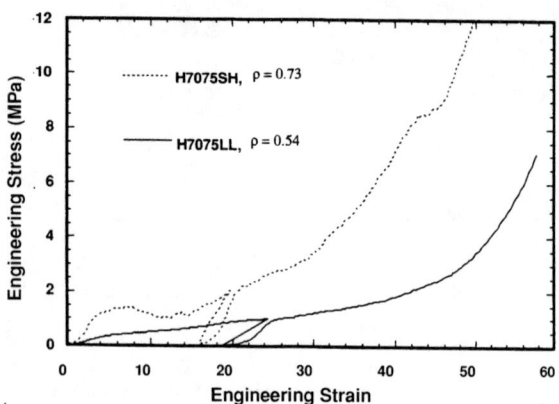

Fig. 7 Stress-strain curves in compression for 7075 Al wire-weld porous structures with hollow periodic spherical regions as showm in Fig. 4. Solid line is for a material with a density of 0.54 gm/c.c. (20% of solid density), Dotted line is for a material with a density of 0.73 gm/c.c. (26% of solid density)

Fig. 8 shows normalized yield strength vs. normalized density of several materials from the literature compared with the results obtained in this study, and also theoretical estimates based on simplified models. Table 2 lists the data from the literature that was used to make this plot. The outlined elliptical bounds for different material groups are indicated to show how properties vary widely as a function of internal architecture of the materials, differing from those in simple models. We observe that data from the literature contain significant scatter. LB indicates lattice block Al alloy, and ERG or Fraunhofer indicate materials from those sources as reported in the literature. The shaded portion of the theoretical trends indicate the range possible for closed cell foams with and without wall imperfections. The open cell predictions are considerably smaller. In the present tests it was possible to test the ligament wires independently to determine the matrix material's strength and elastic modulus. Strain to failure of the wire ligaments were found to be around 5%, which is considerably lower than that of the wrought material of the same type. This is believed to be a result of the cast rather than wrought microstructure. The strength and stiffness of spherical cavity samples of 6061 and 7075 alloys were found to be less than the theoretical predictions. This result is possibly due to uneven wall thickness when inserts are used. Also, the data for 6061 Al is higher than that for 7075 Al, as opposed to that of the parent alloys. It is believed that this effect is due to embrittlement of the wires during processing, which has been verified by tensile tests performed on the constituent wires.

However, in the column-supported material a great degree of anisotropy was seen. The 6061 Al with perpendicular-to-the-face loading showed lower normalized modulus but its strength was spread in the range between open and closed cell foams. In the parallel to the face direction, very high normalized strength and stiffness were seen, well above the theoretical level for even closed cell foam, based on the simple bending model. In addition, Ti and Ni base alloy tubes were used to fabricate a tubular connected structure, as a linear solid like honeycomb core. Such a structure has high strength and stiffness (Fig. 8) and follows a rule-of-mixtures law. The transverse properties of these linear materials are close to that of the theoretical open-cell model. The high ductility of the wrought tubular solid also permits the attainment of a high strength without wall fracture. Thus, it is clear that by using vertical, stringer-like members to bear compressive load, and weaker-wall foam to fill between them, ideal property combinations can be achieved.

For both elastic modulus and strength, the spread of test data for the multiscale materials is substantially larger than that for the stochastic cellular materials, however the spread is directly connected with the orientation of the cell microstructure. Microstructures with predominance of walls parallel to the loading direction, as provided by the face sheets in column supported structures exhibit much higher stiffness and strength than those in which walls are perpendicular to the loading direction. Another effect was found is the role of forging deformation on the strength properties of A-356 Al alloy. From cast condition, elevated temperature forging increased the density of this material and its strength, but its entire strength curve follows along the upper bound of (perfect) closed-cell theoretical data (left plot in Fig. 8). It is not close to the open-cell data.

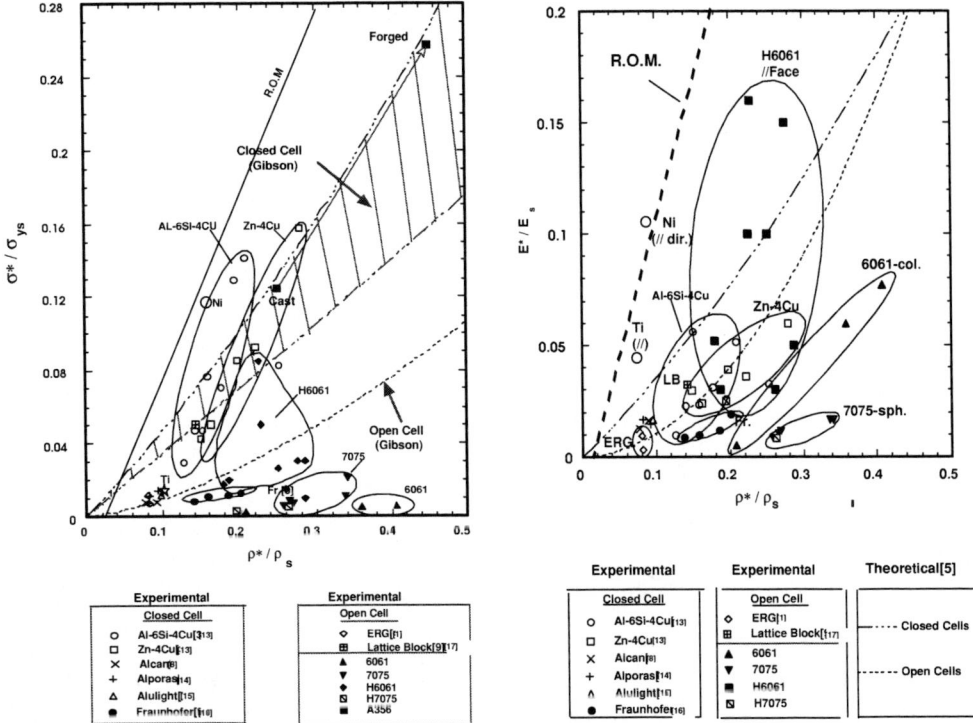

Fig. 8 Yield strength and elastic modulus of cellular metals, normalized with respect to those of 100% solid material, plotted as a function of normalized density. Data from this work, those in the literature and theoretical estimates based on simplified models are included. The outlined elliptical bounds for different material groups show how properties vary widely as a function of internal architecture of the materials, which differ from those in the simple models. For literature data, the numbers in the square bracket are listed in the reference list.

Table 2. The density and compressive properties of several porous metals from the literature

Material	Structure	ρ^*(g/cm³)	ρ_s(g/cm³)	σ_{Pl}(MPa)	σ_{ys} (MPa)	E* (GPa)	E_s (GPa)	Ref. #
Al-6wt%Si-4wt%Cu(Fr.) (Closed cell foam with skin)	Load $\parallel$ Skin	0.39 0.44 0.54 0.58	2.79 (Ref.[11])	8.00 13.0 22.0 24.0	170 (Ref.[11])	1.682 1.740 1.914 3.828	74 (Ref.[11])	[13]
	Load $\perp$ Skin	0.35 0.42 0.49 0.70	2.79 (Ref.[11])	5.00 8.00 12.0 14.0	170 (Ref.[11])	0.754 4.176 2.320 2.436	74 (Ref.[11])	[13]
Zn-4wt%Cu	Closed Cells	1.04 1.14 1.38	7.0 (Ref.[12])	6.00 7.00 12.0	140 (Ref.[12])	2.90 2.32 2.90	97 (Ref.[12])	
		1.55 1.95		13.0 22.0		3.48 5.80		
ERG[1]	Open-Cell	0.217	2.69	2.17	193	0.634	69	[1]
ERG[2]	Open-Cell	0.220	2.69	1.39	193	0.236	69	[1]
Alcan[1]	Closed-Cell	0.249	2.76	2.30	310	1.520	93	[8]
Alcan[2]	Closed-Cell	0.212	2.76	2.34	310	1.180	93	[8]
Alporas[1]	Closed-Cell	0.250	2.71	1.84	130	1.000	69	[14]
Alporas[2]	Closed-Cell	0.220	2.71	1.46	130	1.140	69	[14]
Alulight	Closed-Cell	0.274	2.86	3.01	250	1.170	69	[15]
Fraunhofer[1]	Closed-Cell	0.376	2.72	1.88	215	0.61	69	[16]
Fraunhofer[2]	Closed-Cell	0.429	2.72	2.30	215	0.69	69	[16]
Fraunhofer[3]	Closed-Cell	0.502	2.72	2.50	215	0.82	69	[16]
Fraunhofer[4]	Closed-Cell	0.544	2.72	2.82	215	1.33	69	[16]
Lattice Block	Closed-Cell	0.389	2.74	11.27	224	2.24	69	[17]

ERG(6061/T6): Al-0.6Mg-0.5Si; Alcan: Al-15..2vol%SiC; Alporas: Al-16.3vol%Al/Ti/Ca; Alulight: Al-26.0vol%Al/Si/Fe; Fraunhofer: Al-9.6v0l%Al/Si/Fe.

ρ^*=density of porous metal, ρ_s = density of 100% solid, σ_{Pl}=Plateau stress, σ_{ys} = Yield stress, E* =Elastic modulus of porous metal, E_s-Elastic Modulus of

Summary

A low cost process has been developed for fabricating open cell metal foam structures by direct deposition of solidifying liquid wires, droplets and spray, and simultaneously welding them in-situ. The process is capable of producing a wide range of pore spaces from microns to millimeters and amenable to computer-aided manufacturing technique to produce multiscale structures which can have both stochastic and periodic features. Open cell structures with closed outer surfaces and solid face sheets have been produced. The process is also capable of producing graded porosity, and various internal geometries when deposition is carried out over removable geometric inserts. Mechanical tests have revealed interesting information about these materials as well as design considerations for cellular metals.

Microstructures examined included wire-weld walls with embedded hollow spheres and triangular hollow prisms. Column supported structures with open pore or closed face sheets have also been examined. Column supports have low modulus and strength in the direction of column axes, but high modulus and strength are seen when loaded parallel to their face sheets. For a given density, larger spheres have higher strength and modulus than smaller hollow spheres.

The range of elastic modulus and strength achievable in multiscale cellular materials can be substantially larger than that predicted for the stochastic cellular materials. This spread is directly connected with the orientation of the cell microstructure and density, and can be controlled by the process. Microstructures with predominance of walls parallel to the loading direction, as provided by the face sheets in column supported structures exhibit much higher stiffness and strength than those in which walls are perpendicular to the loading direction.

A few design guidelines gained from this study are:
1. Solid plates or pipe-like structures parallel to the primary loading direction increase specific strength and stiffness
2. Supporting angular columns and trusses enhance properties, particularly when oriented at 30^0 to the loading axis
3. For loading perpendicular to pipe axis, the shape of the pipe (or prism) has smaller effect on specific strength and modulus.

The transverse properties of these linear materials are close to that of the theoretical open-cell model. The high ductility of the wrought tubular solid also permits the attainment of a high strength without wall fracture. Thus, it is clear that by using vertical, stringer-like members to bear compressive load, and weaker-wall foam to fill between them, ideal property combinations can be achieved.

Finally, forging deformation has important effect on the strength properties of deposited materials. Elevated temperature forging increased the density of this material and its strength, and its entire strength curve follows along the upper bound of (perfect) closed-cell theoretical data. This kind of open cell material does not follow the existing theoretical models.

Acknowledgments

This research was supported by the U.S. Office of Naval Research, under grant: DOD-G-N00014-97-1-0510, Program Manager: Dr. S. G. Fishman

References

1. G. J. Davies, S. Zhen: *J. Mat. Sci.*, 1983, vol.18, pp.1899-911.
2. V. Shapovalov: *MRS Bull.*, 1994, vol. 19(4), pp. 24-28.
3. A. E. Simone and L. J. Gibson: *Acta Mater.*, 1996, vol. 44, pp. 1437-47.
4. J. Banhart, J. Baumeister, Production methods for metallic foams, Porous and Cellular Materials for Structural Applications, MRS Symposia Proceedings, D. S. Schwartz, D. S. Shih, H. N. G. Wadley, A. G. Evans, (Eds.), vol. 521, MRS, Pittsburgh, PA, (1998), pp.121-132.
5. S. Sridhar and K. C. Russell: *Journal of Materials Synthesis and Processing,* 1995, vol.3, pp.215-222.
6. J. Baumeister: *German Patent* DE 40 18 360 C1, 1990.
7. J. W. Patten: US Patent 4099 961, 1978.
8. E. Andrews, W. Sanders, L. J. Gibson, Mater. Sci. & Eng. A270(1999)113-124.
9. R. B. Pond: US Patent 2976 590, 1961.
10. A.K. Ghosh, A Method for Producing Microporous Objects with Fiber or Foil Core and Microporous Cellular Objects, U.S.Patent Applied, #09/577,176, 2001
11. ASM Specialty Handbook: Aluminum and Aluminum Alloys, edited by J. R. Davis, ASM International, 1996.
12. Metals Handbook Ninth Edition, Volume 2, Properties and Selection: Nonferrous Alloys and Pure Metals, American Society for Metals, Metals Park, Ohio 44073, 1979.
13. J. Banhart, J. Baumeister, J. Mater. Sci., 33(1998)1431-1440.
14. A. E. Simone, L.J. Gibson, Acta Mater. 46 (1998) 3109-3123.
15. Y. Sugimura, J. Meyer, M. Y. He, H. Bart-Smith, J. Grenestedt, A. G. Evans, Acta Mater. 45 (1997) 5245-5259.
16. M. Weber, J. Baumeister, J. Banhart, H. D. Kunze, Proceedings of the Powder Metallurgy World Congress PM94, Paris, 1994, pp.585-588.
17. J. C. Wallach and L. J. Gibson, Mechanical Behaviour of a Three-Dimensional Truss Material, unpublished report, 2001.

PROCESSING OF IN-718 LATTICE BLOCK CASTINGS

Mohan G. Hebsur
Ohio Aerospace Institute
National Aeronautics and Space Administration
Glenn Research Center
Cleveland, Ohio 44135

Abstract

Recently a low cost casting method known as lattice block casting has been developed by JAM Corp., Wilmington, Massachusetts for engineering materials such as Aluminum and Stainless steels that has shown to provide very high stiffness and strength with only a fraction of the density of the parent alloy. NASA Glenn Research Center has initiated a research to investigate lattice block castings of high temperature Ni-base superalloys such as Inconel-718 (IN-718) for lightweight nozzle applications. Although difficulties were encountered throughout the manufacturing process, a successful investment casting procedure was eventually developed. Wax formulation and pattern assembly, shell mold processing, and counter gravity casting techniques were all developed.

Totally ten IN-718 lattice block castings (each measuring 15-cm wide by 30-cm long by 1.2-cm thick) have been successfully produced by Hitchiner Gas Turbine Division, Milford, New Hampshire using their patented counter gravity casting techniques. Details of the processing and resulting microstructures are discussed in this paper. Post casting processing and evaluation of system specific mechanical property of these specimens are in progress.

1.0 Introduction

Cellular metals are attractive class of materials with extremely low density and outstanding combination of mechanical, thermal and acoustic properties. They offer great potential for lightweight materials in energy absorption and thermal management structures.

Lattice block materials (LBM) are a family of cellular structural materials that consist of a highly ordered structure of internal triangles acting as a space frame or interlocked truss system throughout that bestows high strength and stiffness characters. The concept behind LBM can be applied to wide variety of parent materials such as engineering metals and ceramics. The proprietary manufacturing methods developed by JAM Corp. can be applied in most existing production facilities. Choosing appropriate element proportion (length to width ratio) and angles allows the LBM matrix density specified. Additionally, the matrix can be configured with layers of different geometries and densities. Overall shape be created by machining from solid LBM or by specifying custom production methods. JAM Corp. has demonstrated successfully LBM of various engineering materials such as Al base, Fe-base, Be-base and Cu-base alloys by a low cost investment casting technique. As with all forms of LBM, the variability of the internal lattice proportions allows for control of density, internal angles and overall dimensions. Cast LBM can be furnished in variety of shapes and curves and made with integral face sheets (Ref. 1).

Alloy 718 was developed by the international Nickel Company in the 1950's and is currently used in cast, wrought and powder forms. It is the most used by both GE and Pratt & Whitney aircraft engines for many critical applications such as disks, cases, shafts, blades, stators, seals, supports tubes and fasteners. This alloy has both business and technical characteristics that have enabled continued growth of its application base: (1) free license to manufacture has allowed a large material supplier base to evolve (2) Favorable precipitation kinetic that provides substantial processing advantage, particularly welding and castability (3) relative manufacturing ease has permitted multiple manufacturing process and products to develop over the years. Investment casting is extensively used successfully for manufacturing aircraft components cost effectively (Ref. 2).

Despite the enormous success of IN-718 several significant challenges must be addressed if this alloy is to retain its predominance well into 21st century. Chief among these is the need to improve the temperature capability and time dependent fracture behavior. Numerous attempts have been made over the past 20 years to improve temperature capability of IN-718 for existing applications through chemistry modification involving Co and Ta additions but none has been entirely successful (Ref. 3). In addition to the temperature capability, advanced applications such as NASA Glenn's Ultra-Efficient Engine Technology (UEET) for the next generation of the transport system require very light weight alloys to achieve high fuel efficiency and range. The concept implement is dictated by the tradeoff between performance and manufacturing cost. For attainment of minimum weight of a component four important factors, such as material selection, utilization of shape, topology optimization and multi functionality are essential.(Ref. 4). Short comings in material density can be circumvented by designing shaped components that optimize load capacity. The topologies can be used to achieve load capacity at yet lighter weights, as exemplified by truss structures and honeycomb panels. The truss topology (such as lattice block) has the further benefit that its open spaces can be used to impart functionalities in addition to load bearing, such as cooling. Therefore NASA Glenn has initiated a research to investigate lattice block castings of high temperature Ni-base superalloy such as IN-718. A lightweight exhaust nozzle was chosen as the first component with which to demonstrate the technology. The objective of this investigation is to develop a knowledge base of these LBM castings of Ni-base superalloys.

2.0 Experimental

NASA Glenn requested JAM Corp. to supply total of 10 LBM castings, 5 with and 5 without skin (Face Sheet) of the size 15-cm wide by 30-cm long by 1.2-cm thick. The patterns of this size and shape were prepared by JAM Corp. using their injection molding process.

2.1 Pattern Making

There have been a number of challenges in getting the production of Investment cast LBM off to a solid start. The bulk of earlier difficulties resulted from the injection and assembly of the wax patterns. To address these problems we took a number of steps, ultimately leading to a full in-house pattern making capability at JAM Corp. In the final analysis it was easier for us to figure out how to produce the patterns than it was to try and teach the foundries to do it for us. The primary advantage of making the patterns at JAM Corp. was the ability to experiment with dozens of different waxes, setting, and other variables such as gating and mold venting. This is just one of many crucial details that we never could have gotten right if we left it to the foundry wax rooms to figure out.

In total six sprues were made with two of each to be cast using the above-mentioned methods. Each sprue was containing one open lattice and one single face (Skin) lattice. The shells were produced at Hitchiner Gas Turbine Division (Milford, New Hampshire) using their automatic shell building technique. Several layers of ceramic slurry followed by a layer of sand known as stucco were added to built the shell and dried in about 8 hours.

2.2 Hitchiner Counter gravity investment casting process

Hitchiner Counter gravity casting processes (Ref. 5) were used to cast LBM of IN-718. These processes are particularly suitable for thin castings. Based on casting equipment and environment, used Hitchiner has at least three improved variations of the basic counter gravity casting process (here after called Type I, Type II and Type III). As these processes are still under patent review, we cannot describe here the details of these processes. All the three variations were tried to cast lattice blocks of IN-718. So far we have made 10 LBM castings of IN-718 using Hitchiner's counter gravity investment casting process.

The basic Hitchiner counter gravity technique originally developed in mid 1970's uses vacuum to provide the driving force to fill the castings. Entraining air into the molten alloy during stirring can adversely affect the filling of the mold. Ladle pouring of the metal common to sand, permanent mold and conventional investment casting can easily introduce turbulent flow as the metal plunges from ladle into the down sprue of the gating system. A hot metal is placed in a vacuum chamber with a fill pipe extending out the bottom. The fill pipe entrance is then lowered beneath the certain surface of the molten metal and a vacuum is applied to the casting chamber. Atmospheric pressure on the melt forces the metal to rise up the fill pipe and enter the mold. By tailoring the vacuum profile to the casting geometry and alloy, the result is metal drawn from the clean undisturbed zone under the melt surface filling the cavity in quiescent manner. A hot mold near the solid temperature for the material helps achieve the cell sizes.

3.0 Results and Discussions

3.1 Pattern Making

The design of the castings and supply of wax patterns was provided by the JAM Corp. Figure 1 shows the photograph of wax pattern that was the end result of about 4 months of continuous effort. A suitable glue that bonds with the slippery pink wax compound was identified and a prototype glue system was built to automate assembly. The first prototype system was an air driven glue nozzle, compressed air was used to feed glue out through a tip that delivered a measured drop for each node. The next generation will deliver an entire row of glue spots at a time, possible even a complete sheet in one quick operation. It is also found that the gate areas can be "stitched" together with a soldering iron. The iron creates holes through the multiple wax layers, which then fill with molten wax and harden as one. If hand operation is not sufficient, this can be improved with a multiple head system. The techniques developed will be generally useful for all LBM

Figure 1: Photograph of an assembled wax pattern for investment casting of IN-718.

investment casting patterns. Since the pattern making appears to be the technology bottleneck, and the production bottleneck as well, this is where we have focused our efforts. Not only did the molds filled easily with the new formula, also the parts came out of the molds without breaking and were easy to assemble the manufacture of 12-lattice mesh design structural castings. A twelve layer ceramic shell was then prepared at Hitchiner, Inc., and used in all the casting trials. This is particularly suitable for the counter gravity processes. However, it appears that a reduced layer shell is needed for Type I counter gravity process. The reduced shell hence increased permeability should improve mold fill out in this counter gravity process. There was penetration and flashing of metal into the shell in all casting processes. A review of the shell building process and technique is needed to reduce this occurrence.

3.2 Hitchiner Counter gravity casting process

The main advantages of this process over the conventional casting process are the following: (1) Cleaner metal with less slag and non metallic inclusions (2) Increased control of fluid flow minimizing lower metal and mold turbulence (3) controlled grain structures with improved mold temperatures with thin sections (4) increased efficiency. Several intricate and complex shaped gas turbine components such as combustor liner of Ni-base super alloys have been manufactured cost effectively by Hitchiner counter gravity process.

3.3 Casting Results

Each casting measured 15-cm wide by 30-cm long by 1.2-cm thick. The average ligament diameter was 1.58 mm with a 0.045 mm standard deviation and maximum/minimum values were 1.62 and 1.523 mm, respectively. For the five castings with a face sheet on one side, the average face sheet thickness was 0.074 mm with 0.0025 mm standard deviation and maximum/minimum values were 0.077 and 0.066 mm, respectively.

Figure 2 shows the photograph of IN-718 casting produced by Type I counter gravity process. Figure 2 indicates clearly that the metal did not fill the mold completely because this process used lower molten temperature and lower rate of rise. In order for the Type I process to be suitable for this casting the casting fill out has to be improved upon. So the Type I process requires the fastest rate of rise and high temperature in order to achieve the fill out.Type II Counter Gravity Process: Fig. 3. The Type II process also did not produce defect free casting. There was excess metal filling the lattices. This Type II process used very high molten metal temperature and fast rate of rise.

The Type III counter gravity process did not seem to have problems in filling the lattice castings Figure 4. The remaining six castings were produced by Type III process. For dimensional control the ring gate around the lattice was pulling the casting during solidification and distorted the casting. When allowed to flow back completely, there was no distortion to the casting. In future development this process needs to be continued or the ring gate needs to be broken into sections so it will not pull the casting.

3.4 Lattice Casting Defects

Out of 10 castings, only 6 castings produced by Type III process had minor defects. In future Type III process will be used to cast IN-718 and other Ni-base superalloys such as MAR- M 247. The remaining 4 castings had numerous defects such as (1) partially filled ligament due to non metallic

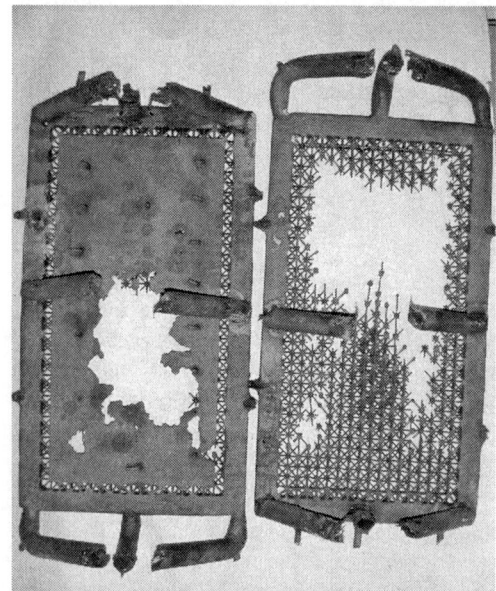

Figure 2: Photograph of IN-718 LBM casting produced by Type I counter gravity process.

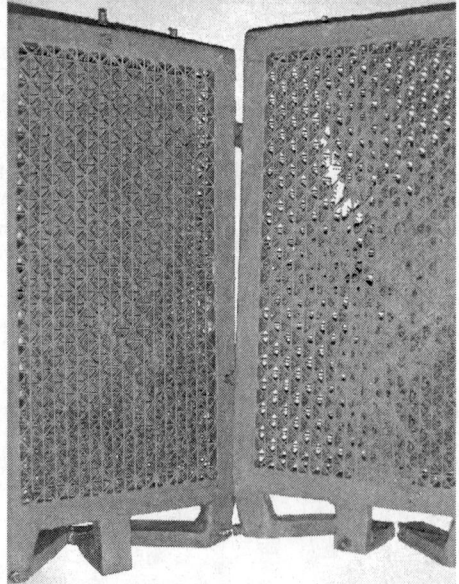

Figure 3: Photograph of IN-718 LBM casting produced by Type II counter gravity process.

inclusions (Fig. 5). The industrial practice for producing IN-718 ingot is to use triple melting consisting of vacuum induction melting followed by electro slag refining (ESR) and vacuum arc remelting (VAR). The size of nonmetallic inclusions is drastically reduced during electro slag refining. The number of non metallic inclusions and gaseous inclusions is reduced during vacuum arc remelting. In the present investigation the IN-718 was produced by vacuum induction melting only. The large sized inclusions present in the molten metal might have entered the thin section of

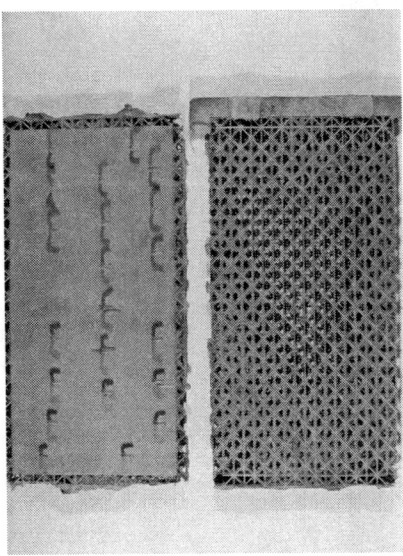

Figure 4: Photograph of IN-718 LBM casting produced by Type III Hitchiner process.

Figure 5: Photograph of partial filling of ligaments due to non metallic inclusions.

Figure 6: Photograph of excess metal.

Figure 7: Photograph of non fill.

LBM casting and thus created a defect. (2) Excess metal (Figure 6); (3) Non fill (Figure 7); (4) Shell fill (Figure 8); and (5) Metal entering through gate (Figure 9) and metal entering through flash (Figure 10). These defects can be minimized or totally avoided by choosing a proper counter gravity process and processing parameters for alloy IN-718.

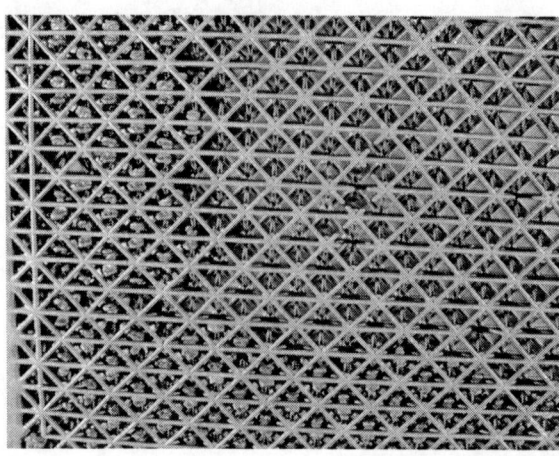

Figure 8: Photograph of shell fill.

Figure 9: Photograph of metal entering through gate.

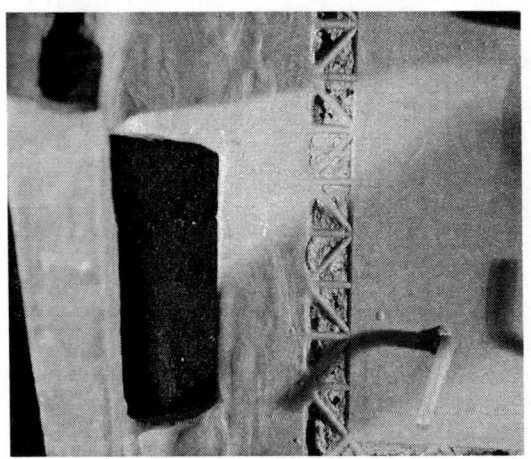

Figure 10: Photograph of metal entering through flash.

3.5 Chemistry and Microstructure of as-cast IN-718 LBM

The nominal composition of alloy IN-718 is given in Table I.

Table I Nominal Composition (in wt.%) of IN-718

Ni	Fe	Cr	Mo	Nb	Ti	Al	C	B
balance	18	18	3	5	1	0.5	0.04	0.004

The phases reportedly found in IN-718 are Nb, TiC, TiN, delta, laves, γ, γ', αCr, sigma and M_6C. The major strengthening phases in IN-718 are the γ' and γ phases which produce coherency strains in the γ matrix. The γ' phase is the main strengthening phase and has a DO_{22} (BCT) crystal structure. Because the γ'/γ phases grow with higher temperature and long time exposure at low temperatures the transition from γ to γ' to delta phase occurs slowly at lower temperatures the delta phase found in IN-718 is in coherent with the γ and has an orthorhombic crystal structure. During solidification the γ matrix rejects selectively large atoms of elements such as Nb, Mo, Ti to the interdendritic regions. These elements form the NbC, TiN and the Nb rich laves phase in the high Nb areas as the metal solidifies from about 1300°C to about 1050°C. Basically the cast IN-718 is heavily segregated and consists of two compositions: the dendrites which are high in Fe, Cr and Ni while interdendritic areas are rich in Nb, Ti and Mo. The dendritic Nb content is as low as 2 percent while the inter dendritic regions are rich in Nb (>12 percent) (Ref. 6). The homogenization cycle is the key to producing a uniform material. Typical as-cast microstructures of IN-718 LBM are shown in Figures 11(a) and (b). The SEM image in Fig. 11a shows a light and dark pattern whose size depends on the cooling rate. The light areas are high in Nb and the dark areas are low in Nb. Figure 11(b) shows the same sample with variety of phases present in the interdendritic areas. The Laves phase islands are dark due to their response to the sample preparation while delta plates, γ' and γ phases are light.

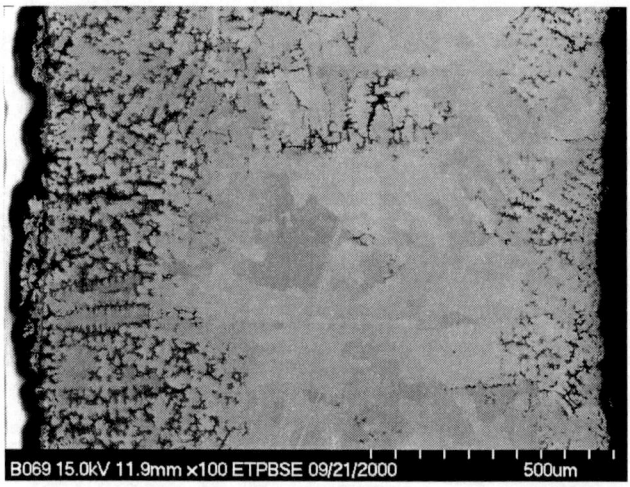

Figure 11(a): SEM micrograph of LBM cast IN-718.

4: Summary and Conclusions

Lattice block castings of IN-718 (15-cm long by 30-cm wide by 1.2-cm thick) has been successfully accomplished. Initially there were numerous problems in making the pattern of our design. This was over come by using a new wax having higher strength and lower thermal expansion. Hitchiner counter gravity processes were used to cast the lattice blocks of IN-718.

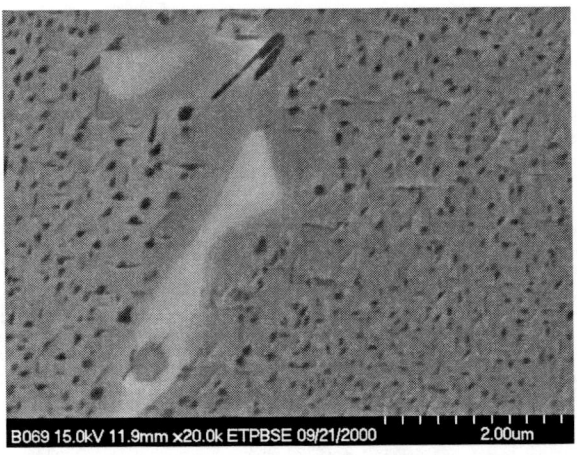

Figure 11(b): Same as (a), but at higher magnification.

Type I and Type II counter gravity processes did not yield a good quality casting. However, the Type III counter gravity process yielded a very good quality with few defects. So in future Type III counter gravity process will be used forecasting of lattice blocks of IN-718 as well as MAR-M 247. The main defects in lattice block castings are partial filling of ligaments due to non metallic inclusions, excess metal, non filling of the lattice, shell filling of the lattice and metal flashing at the gate. These defects can be minimized by carefully choosing the type of the process and the processing parameters. In fact, more castings of IN-718 with and without race sheets have now been produced successfully using Type III counter gravity processing. After hot isostatic pressing (HIP'ing) followed by solution and aging treatment these castings will be evaluated for system specific mechanical properties.

Acknowledgments

This work was supported by the Ultra-Efficient Engine Technology (UEET) program at NASA Glenn, Cleveland, Ohio. The author would like to express his sincere thanks to Jon Priluck and Scott Breckett of JAM Corp. and Blair King of Hitchiner Corp., for many technical contributions.

References

1. "Ultra Lightweight Metallic Structure," DARPA/ONR Report, Contract No. N00014–96–0400AM, Prepared by UTRC, East Hartford, Connecticut, 1999.

2. R.E. Schafrck, D.D. Ward, and J.R. Groh, "Superalloys 718, 625, 706 and Various derivates," ed. E.A. Loria, TMS Warrendale, Pennsylvania, June 2001, pp.1–12.

3. D.Paulonis and J.J. Schrra, "Superalloys 718, 625, 706 and Various Derivatives," ed. E.A. Loria, TMS Warrendale, Pennsylvania, June 2001, pp. 13–24.

4. A.G. Evans, "Lightweight Materials and Structures," MRS Bulletin/Vol. 36, 2001, pp. 790–797.

5. G.D. Chandley, "Apparatus and Process for Countergravity Casting of Metal with Air Exclusion," U.S. patent no. 5, 04, 2561, August 1991.

6. J.F. Radavich, "Superalloy 718 Metallurgy and Applications," ed. E.A. Loria, TMS, Warrendale, 1989 pp. 229–240.

ADVANCES IN THE MELT ROUTE PRODUCTION OF CLOSED CELL ALUMINIUM FOAMS USING GAS-GENERATING AGENTS

V. Gergely, D.C. Curran, and T.W. Clyne

Department of Materials Science & Metallurgy
University of Cambridge
Pembroke Street, Cambridge, CB2 3QZ, UK

Abstract

A brief survey is given of the current state of the art for hydride-based processing of closed cell foams, including the production of precursor material in which gas release has been delayed during dispersion in the melt by the introduction of surface coatings on the powder particles (the FORMGRIP process). A study is then presented of the factors involved in the search for gas-generating agents offering superior performance. These include kinetic and thermodynamic characteristics of potential decomposition reactions, the ease of dispersion of the powder in the melt, the nature and likely effect of decomposition products on melt flow and the availability and cost of the powder concerned. It is shown that there is at least one very promising candidate material, which can offer advantages compared to hydride powders in virtually all aspects of their performance. It is confirmed that foams can be produced having appreciably finer (<1 mm cell size) and more uniform cell structures than currently-available melt route foams, a lower ceramic content in the cell wall microstructure and dramatically reduced raw material costs. It is also shown that stress-strain curves during compressive loading of such material are much smoother and more progressive than those of previously-available melt route material, which is consistent with the finer and more homogeneous structure and may also be at least partly attributed to greater ductility exhibited by the cell walls.

Processing and Properties of Lightweight Cellular Metals and Structures
Edited by Amit Ghosh, Tom Sanders and Dennis Claar
TMS, 2002

Introduction

There is considerable current interest in closed cell aluminium foams for structural and energy absorption applications [1]. There is a strong incentive to develop economically attractive processing routes for such foams, creating material having a fine, uniform cell structure (preferably without coarse, embrittling constituents [2]). Routes based on handling of a melt are cheaper and more tolerant of impurities than those based on a powder compact. Techniques in which a gas-generating powder is dispersed throughout the melt produce finer and more uniform structures than those based on bubbling a gas into the liquid. A recently-developed procedure[3], in which a protective coating is created on the surface of the powder (TiH_2) particles, eliminates the problem of premature gas release during dispersion and allows a foamable precursor material to be created. This is subsequently baked in order to produce the foam. However, while this procedure works well, there is a need for a gas generating agent with superior dispersion, reaction and cell wall stabilisation characteristics. There is also a strong commercial incentive to find an agent which is cheaper and easier to handle than TiH_2 and similar hydrides. In the current paper, a first report is presented on a highly promising candidate gas-generating agent.

Experimental procedure

Foam processing

Material specification. DURALCAN F3S.10S type metal matrix composite, supplied by A/S Tempønik, Denmark, was used for foam preparation. The metal matrix, the chemical composition of which is listed in Table I, contains 10 vol. % of SiC particulate, with a mean particle size of about 13 µm. Two types of gas-generating agents were used for foaming of the molten composite. The titanium hydride (TiH_2) powder was supplied by Goodfellow Cambridge Limited, UK. The maximum particle size is 150 µm, the purity is 99.0+% and the density is 3.9 gcm^{-3}. The calcium carbonate ($CaCO_3$) has a nominal particle size of 5 µm and a purity of 99.5 %, and was supplied by Alfa Aesar Johnson Mathey GmbH.

Precursor preparation. The TiH_2 foaming agent was pretreated using a two-step thermal oxidation treatment (24 h@400 °C+1 h@500 °C) in air to generate a titanium-dioxide (rutile) diffusion barrier layer on the particulate surface [3]. The calcium carbonate powder was used in as-received form. The foaming agents were mixed with ~150 µm diameter Al-12Si powder (mass ratios Al-12Si : TiH_2 = 1:4 and Al-12Si:$CaCO_3$=1:2) before their incorporation into the melt. Foamable precursor materials were produced using the following procedure. The F3S.10S composite (~1 kg) was induction melted and heated to about 650 °C. The power was switched off and the melt was stirred for about 10 seconds, while its temperature was in the range 640 °C - 630 °C. After 10 seconds of stirring, the foaming agent / Al-12Si powder mixture was introduced into the melt and the melt was stirred (1200 rpm) for a further 40-90 s. Low porosity precursor material was produced by allowing the melt to solidify.

Table I Chemical Composition of the DURALCAN F3S.10S Composite Matrix.

Si	Fe	Cu	Mg	Ti	All Other	Al
8.5-9.5	0.2 max.	0.2 max.	0.45-0.65	0.2 max.	0.1 Tot.	REM.

Foam baking. Cylindrical specimens (d=32 mm and h=10-20 mm) machined from the precursors were placed in a stainless steel mould coated on with a suspension containing boron nitride. Foam baking was carried out in a conventional air furnace. The furnace was preheated to a temperature higher than the required foam baking temperature. The mould containing a precursor material was placed in the furnace and the set-up was heated above the liquidus temperature (~605 °C) of the composite matrix. This treatment led to decomposition of the foaming agent. The evolved gas (H_2 in the case of TiH_2 powder and CO_2 in the case of $CaCO_3$ particulate) foamed the melt. Baking temperatures were varied between 620 °C-700 °C and the foaming time between 2 min - 15 min, depending on the precursor type and the targeted foam porosity level. The foam baking profiles were recorded using a thermocouple located directly in the precursor material.

Material characterization

The particle size distribution of both types of foaming agents was measured using a Malvern MasterSizer E machine. Thermogravimetric analysis (TGA) was used to study the effect of the foaming agent type and its pretreatment onset temperature and kinetics of gas evolution in an argon atmosphere. The experiments were carried out STA-780 Series Thermal Analyser (Stanton Redcroft), using a standard platinum crucible and with Al_2O_3 powder as a reference material. The heating rate employed was 20 •C min^{-1} and the amounts of analyzed particulate were in the range of 16-25 mg.

The foamed specimens for mechanical testing and metallographical preparation were cut using an electro-discharge machine, in order to minimise cell wall damage. Open cells on a sample surface of specimens selected for microscopy and image analysis were filled with a cold polyester casting resin (KLEER-SET TYPE FF, MetPrep, Ltd., Coventry, UK) containing a blue dye. This procedure ensures that good contrast is achieved between the cell walls / edges and pores and also eliminates cell wall smearing during grinding and polishing. Conventional SiC papers were used for sample grinding (up to grit #1000). Polishing was carried out using 6 µm and 1 µm diamond paste. Metallographically prepared sections of foams were scanned at a resolution level of 1200 dpi for image analysis. Cell size and porosity were measured using a KS 400 Imaging System, (Carl Zeiss Vision GmbH). The mean cell size is represented by the diameter (d) of a circle having an equivalent area to a corresponding pore-section. A JSM 5800 LV scanning microscope equipped with X-ray detectors (EDS and WDS) was used for microstructural and chemical analysis.

Compression tests

Test specimens (cubes of side 23 mm x 9 mm) were inserted between two stainless steel platens, lubricated with anti-friction spray. Uniaxial compression tests were carried out on a servo-hydraulic testing machine equipped with a 10 kN load cell. The tests were conducted under displacement control, with a cross-head speed of 0.5-2.0 mm min^{-1}. The aim of these experiments was to examine the effect of specimen/cell size ratio and cell wall microstructure on the compression loading response of the materials.

Results and discussion

Foaming agent characteristics

Secondary electron micrographs of as-received foaming agents are shown in Figure 1(a) and (b). The measured average particle size of the hydride (Figure 1(a)) was ~40 µm and that of the carbonate (Figure 1(b)) was around 25 µm. It can be seen from the micrographs that carbonate particles tend to be slightly agglomerated. However, close inspection of the particles suggests

Figure 1: Secondary-electron micrographs showing the size and morphology of (a) titanium hydride and (b) calcium carbonate foaming agents.

that the clusters may have arisen by co-precipitation and growth of several rhombohedral crystallites rather than by physical agglomeration.

It was established previously [3] by the present authors that use of the titanium hydride foaming agent requires a thermal oxidation treatment for successful and flexible production of foamable precursors. This treatment generates a titanium dioxide diffusion barrier on the foaming agent surface and slightly changes the Ti/H ratio in the hydride [4]. As a result of this, thermally activated hydrogen release is delayed, which facilitates preparation of a low porosity foamable precursor by incorporation of the hydride into a melt, followed by solidifiaction. The effect of hydride pre-oxidation on the onset temperature of hydrogen evolution in argon is shown in Figure 2. The effect of the powder treatment on the porosity of the solidified castings was demonstrated in the original papers [3, 5].

The onset temperature of carbon dioxide evolution from the carbonate in argon is relatively high in comparison with hydrides (see Figure 2). Relatively slow kinetics of gas-release from the carbonate has proved to be also the case during its incorporation into a composite melt, allowing the casting to be solidified before initiation of melt foaming. The precursor void content is low (~15-25 %), with some evidence that the pores which do form are created adjacent to $CaCO_3$ agglomerates. The carbonate is significantly cheaper than the hydride and does not require any special pretreatment procedure.

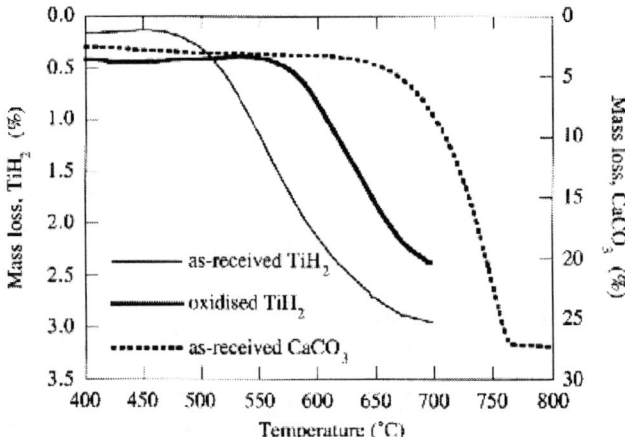

Figure 2: Thermogravimetric curves for heating of as-received calcium carbonate and of as-received and pre-oxidized (2 h at 400 °C+1 h at 500 °C) titanium hydride powder in argon, showing the effect of pre-treatment and powder type on the onset temperature of gas evolution. The heating rate was 20 °C·min^{-1}.

Reaction products generated during gas-release stabilize the cell-walls and facilitate production of very fine-celled foams (see below).

Mechanisms controlling the foam structure evolution

It is already well established [3, 6-8] that solid particles in liquid metal can play a key role in stabilizing metallic foams. Their presence in cell walls and Plateau borders (see Figure 3) raises

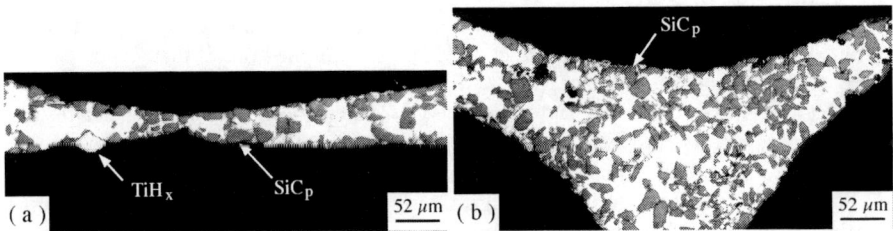

Figure 3: Optical micrographs of sections through (a) the cell-wall and (b) Plateau-border of hydride-based foams, showing the distribution of SiC_p (~13 µm).

the melt viscosity, slowing down cell coalescence and melt drainage in a foam column. It is, however, worth noting that enhancement of the melt viscosity by solid particles can be counterproductive if particles with unsuitable size and/or shape are used. For example, it has been shown [9] that an increase in the size of irregular, sharp-edged particles, from about 13 µm to ~63 µm, significantly accelerates cell coalescence and drainage processes. This effect has been attributed to particle-induced cell wall rupture during capillarity-driven wall thinning and/or stretching in foams undergoing expansion.

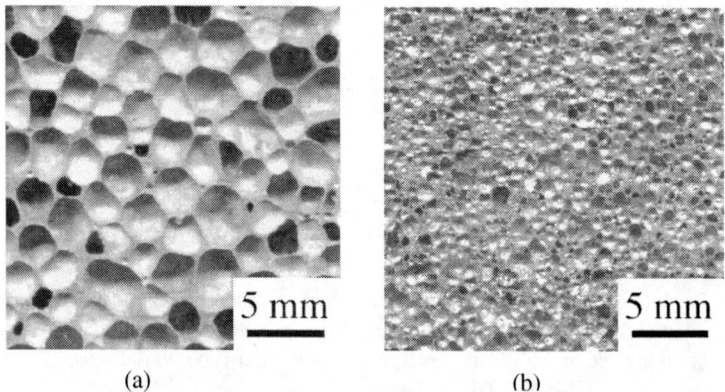

(a)   (b)

Figure 4: Optical micrographs of Al-9Si-0.5Mg/10 vol.% SiC_p foams created using (a) pre-treated TiH_2 and (b) as-received $CaCO_3$ gas-generating agents. The porosity of the foams is ~80 %. The cell size is much finer (~0.7 mm) in the foam produced using the calcium carbonate, compared with the hydride foam (~2.9 mm).

Several other important particle-related foam stabilizing mechanisms can operate in liquid foams, if the size, shape and volume fraction of particles and the wetting between the melt and particles (contact angle, θ) are optimized. It has been shown in the literature [10-12] on water-based foams that even large particles can have a stabilizing effect on the foam structure, provided their shape is spherical and the contact angle between the sphere and the liquid is $\leq 90°$.

On the other hand, even poorly-wetted particles ($\theta > 90°$) can have a stabilising effect on the cell face material. This was reported by Kumagi, et al. [13]. They argued that if many such particles are arranged at the gas-liquid interface, the radii of curvature of the gas-cell wall interface (between the particles) would be close to those of the gas-Plateau border interface. As a result of this, the capillarity-induced pressure difference (= γ/r, where γ is the melt surface tension and r is the radius of curvature of the Plateau border surface) between the cell walls and Plateau border decreases. This pressure difference is a driving-force for melt redistribution from cell walls to Plateau borders, which often leads to cell wall rupture and hence cell coarsening. The presence of particles at the liquid-gas interface also increases the surface viscosity, which further decreases the rate of liquid flow in cell walls and Plateau border channels (see [14]).

In the present work, a new concept for stabilizing of liquid metal foams is introduced. It is based on an in-situ generation of products as a result of reaction between the melt and the foaming gas released from the carbonate. The mechanisms responsible for foaming with $CaCO_3$ have not been studied before. Possible reactions include [15]

$$CaCO_{3(s)} \longleftrightarrow CaO_{(s)} + CO_{2(g)} \quad\quad \Delta G° (kJ \cdot mol^{-1}) = 554 - 0.655 \cdot T \quad\quad (1)$$

$$6\,Al_{(l)} + 3\,CO_{(g)} \iff Al_2O_{3\,(s)} + Al_4C_{3\,(s)} \quad \Delta G° \,(kJ \cdot mol^{-1}) = -1378.7 + 0.624 \cdot T \quad (2)$$

$$8\,Al_{(l)} + 3\,CO_{2\,(g)} \iff 2\,Al_2O_{3\,(s)} + Al_4C_{3\,(s)} \quad \Delta G° \,(kJ \cdot mol^{-1}) = -2715.7 + 0.683 \cdot T \quad (3)$$

Fine particles (e.g., Al_2O_3, CaO, Al_4C_3) generated in situ are expected to enhance the melt viscosity. The alumina formed by reaction with liquid aluminium is likely to decorate the cell wall surfaces, perhaps as a continuous thin film, and this is expected to be particularly effective in stabilizing the walls against rupture and cell coalescence. Typical foam cell structures created using pre-treated hydride and as-received calcium carbonate are illustrated in Figure 4. It can be seen that the cell size is much finer (~0.7 mm) in the foam produced using the calcium carbonate, compared with the hydride foam (~2.9 mm). The porosity is 80 % in both cases.

<u>The effect of macro and microstructure on compression response</u>

There several papers in the literature [16-18] addressing the effect of the ratio of specimen size to the cell size, D/d, on the compression loading response of metal foams. The most comprehensive study of the size effect was probably that carried out by Gibson and co-workers [18]. They showed that with D/d values lower than 5 the compressive strength (plastic collapse strength) starts to decrease rapidly. The measured plastic collapse strength was more or less constant for $D/d \geq 6$. The reduced strength at lower values of D/d was attributed to the reduced constraint of cell walls at the free surface as well as from the increasing area fraction of stress-free cell walls. These measurements were carried out on foams with relatively ductile cell walls (Alporas; Shinko Wire, Amagasaki, Japan).

Stress-strain curves obtained during uniaxial compression of foams produced using carbonate and hydride gas-releasing agents are shown in Figure 5. For the large-celled (hydride) foams, the plots (Figure 5(a)) exhibit the familiar plateau region, with considerable fluctuations in stress level. It should be noted that the specimen length in the loading direction only extended over about 5 cell diameters. The repeated load drops are thought to be associated with collapse of rows of cells, which thus generate substantial increments of strain and hence stress relaxation. For the finer (carbonate) foams, on the other hand, the stress-strain plots are much smoother (Figure 5(b)). This is particularly true for the specimen with the length/cell diameter ratio of about 19. With smaller specimen dimensions, where the ratio falls to 11, it can be seen that some fluctuations are starting to appear.

Conclusions

It is shown that a certain type of calcium carbonate powder offers substantial advantages compared to hydride powders, in terms of both control over the foam structure and the product cost. It is also confirmed that foams can be produced with finer (<1 mm cell size) and more uniform structures than currently-available Al-based closed-cell foams. Thermodynamic and kinetic features of the reactions which occur with the carbonate powder are favourable. It is also clear that cell wall rupture and cell coarsening are inhibited, as a result of enhancement of the melt viscosity and formation of oxide films. Stress-strain curves for compressive loading of the carbonate-based foams are significantly smoother and more progressive than those of previously-available melt route material, which is consistent with the finer and more homogeneous

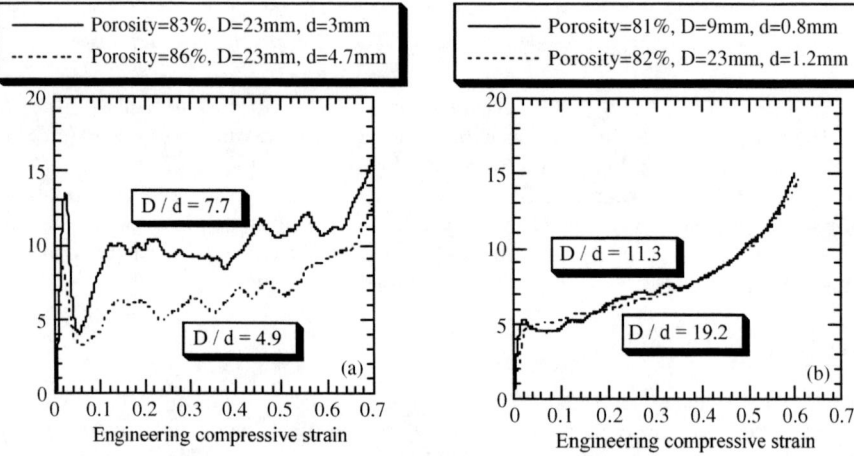

Figure 5: Stress-strain plots of foams produced using (a) hydride and (b) carbonate foaming agents, showing the effect of foam porosity and speciment length/cell diameter ratio (*D/d*) on compresion loading response.

structure.

Acknowledgements

The authors are grateful for support from the EPSRC, QinetiQ Ltd. (Farnborough, UK), and Cymat Corporation (Canada).

References

1. J.E. Seibels, "Derivation of Materials Energy Absorption Requirements from Crash Situations" (Paper presented at the International Conference on Metal Foams and Porous Metal Structures, Bremen, 14-16 June 1999), 13.
2. A.E. Markaki and T.W. Clyne, "The Effect of Cell Wall Microstructure on the Deformation and Fracture of Aluminium-Based Foams", Acta Mater., 49 (9) (2001), 1677-1686.
3. V. Gergely and T.W. Clyne, "The FORMGRIP Process: Foaming of Reinforced Metals by Gas Release In Precursors", Adv. Engng. Mater., 2 (4) (2000), 175-178.
4. V. Gergely, "Melt Route Processing for Production of Metallic Foams", (Ph.D. thesis, University of Cambridge, 2000), 64-77.
5. V. Gergely and T.W. Clyne, "The Effect of Oxide Layers on Gas-generating Hydride Particles during Production of Aluminium Foams", Porous and Cellular Materials for Structural Applications, vol. 521, ed. D.S. Schwartz, D.S. Shih, A.G. Evans, and H.N.G. Wadley (MRS, 1998), 139-144.

6. V. Gergely, H.P. Degischer and T.W. Clyne, "Recycling of MMCs and Production of Metallic Foams", Comprehensive Composite Materials, vol. 3 : Metal Matrix Composites, ed. T.W. Clyne (Elsevier, 2000), 797-820.
7. S.W. Ip, Y. Wang and J.M. Toguri, "Aluminum Foam Stabilization by Solid Particles", Canadian Metall. Quarterly, 38 (1999), 81-92.
8. O. Prakash, H. Sang and J.D. Embury, "Structure and Properties of Al-SiC Foam", Mater. Sci. Eng., A199 (1995), 195-203.
9. V. Gergely, R.L. Jones and T.W. Clyne, "The Effect of Capillarity-Driven Melt Flow and Size of Particles in Cell Faces on Metal Foam Structure Evolution", Trans. JWRI (Special Issue), 30 (2001), 371-376.
10. G.C. Frye and J.C. Berg, "Antifoam Action by Solid Particles", J. Colloid Interface Sci., 127 (1989), 222-238.
11. R. Aveyard, B.P. Binks, P.D.I. Fletcher and C.E. Rutherford, "Contact Angles in Relation to the Effects of Solids on Film and Foam Stability", J. Dispersion Sci. Technology, 15 (1994), 251-271.
12. A. Dippenaar, "The Destabilization of Froth by Solids: I. The Mechanism of Film Rupture", Int. J. Mineral Processing, 9 (1982), 1-14.
13. H. Kumagai, Y. Torikata, H. Yoshimara, M. Kato and T. Yano, "Estimation of the Stability of Foam Containing Hydrophobic Particles by Parameters in the Capillary Model", Agric. Biol. Chem., 55 (1991), 1823-1829.
14. D. Desai and R. Kumar, "Flow Through a Plateau Border of Cellular Foam", Chem. Engng. Sci., 37 (1982), 1361-1370.
15. I. Barin, O. Knacke and O. Kubaschewski, Thermodynamic properties of inorganic substances. Supplement 1977 ed. 1977, Berlin, Springer-Verlag.
16. Y. Sugimura, J. Meyer, M.Y. He, H. Bart-Smith, J. Grenestedt and A.G. Evans, "On the Mechanical Performance of Closed Cell Al Alloy Foams", Acta Mater., 45 (12) (1997), 5245-5259.
17. K.Y.G. McCullough, N.A. Fleck and M.F. Ashby, "Uniaxial Stress-Strain Behaviour of Aluminium Alloy Foams", Acta mater., 47 (8) (1999), 2323-2330.
18. E.W. Andrews, G. Gioux, P. Onck and L.J. Gibson, "Size Effects in Ductile Cellular Solids Part II: Experimental Results", Int. J. Mech. Sci., 48 (2001), 701-713.

RHEOLOGICAL CHANGES OF CELLULAR STRUCTURE IN FOAMING METAL

Zhenlun Song and Steven R. Nutt

University of Southern California,
Center for Composites, VHE-602,
Los Angeles, CA 90089-0241, U.S.A.

Abstract

In this work, rheological changes of molten aluminum during foaming were measured as a function of position within the mold cavity and correlated with changes in cell geometry. In this process, the aluminum charge is placed in a crucible and heated to the melting point. Mechanical agitation is introduced, and when a suitable viscosity is achieved, a granular foaming agent is injected. Foam expansion ensues rapidly, and control of the melt viscosity is critical to achieve stable foaming and to avoid slumping. Cell shapes are considered in three locations in the mold: (1) near the side-wall of the crucible, where cells experience shear deformation in the expansion direction, (2) near the center of the foaming melt, where cells are relatively large and experience compression deformation, and (3) between the center and side-wall, where cells are fine, equiaxed and well-developed. A qualitative model of the foaming process is proposed explain the observed phenomena, and matched to the experimental data.

1. Introduction

Aluminum foam made by melt processing generally has a closed cell structure. The unique structure of this material leads to unusual combinations of mechanical, thermal, electromagnetic and acoustic properties that provide opportunities for diverse structural and functional implementation [1,2]. These properties derive from the peculiar details of the cellular structure, which depend on the history of the foaming process [3].

Melt viscosity is a key process parameter in the production of metal foams by melt processing. Failure to control viscosity can result in foam collapse or inadequate expansion. In general, foam is fabricated by injecting a foaming agent into a thickened aluminum melt, causing foam expansion. The expanding foam is a structured fluid in which gas bubbles are separated by thin membranes. It is an unstable system. Normally, the liquid phase in stable foams (such as soap film foams) contain a surface-active agent that preferentially accumulates at the gas-liquid interface and imparts varying degrees of stability to the films. However, in molten aluminum, this is not the case. Consequently, thickening agents are typically added to avoid slumping. Foam expansion occurs rapidly, and control of the melt viscosity is critical to achieve foam stability. In this work, we investigate the rheological changes of an expanding aluminum melt and show how the insight gained can be used to control the structure of aluminum foam.

2. Experimental Methods

<u>The Foaming Process</u>

Aluminum foam synthesis involves the steps listed below, also shown in Figure 1.

1. Melting: A pure aluminum charge is placed in a crucible, heated in a resistance furnace to 973K, and melted.
2. Thickening: Pure calcium powder is added to the melt and stirred at a constant speed, thereby increasing the melt viscosity.
3. Mixing: When the viscosity of the melt reaches a critical value, the foaming agent powder is injected into the melt. At the same time, the mixture is stirred at higher speed to disperse and homogenize the powder in the melt.
4. Holding: The mixture temperature in the furnace is maintained to allow decomposition of the foaming agent. During the holding period, substantial changes take place in the structure of the melt. Bubbles nucleate and grow until a cellular structure forms. The structure of the final product is governed to a large extent by the rheological changes of the melt during this period.
5. Cooling: The foam is removed from the furnace and cooled to solidify the melt. During solidification, the foam shrinks in a manner that also depends on melt viscosity.

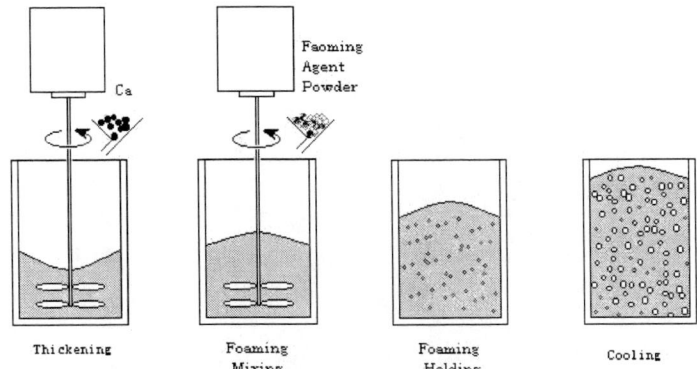

Figure 1. Foaming process.

3. Results

Measurement of foam expansion and shrinkage

A special non-contact device [4] was designed and employed to determine the real-time displacement of the melt surface during the foaming process. Stirring of the melt during the mixing period was employed for times of 20-180 seconds to determine the effects on expansion kinetics. These effects are shown in Figure 2 below. Note that longer stirring times generally result in slower expansion and smaller maximum volumes. Gradual shrinkage occurs after the peak height is reached in all cases.

The structure of typical foam produced by this process is shown in Figure 3. The relative density of the foam is ~0.1, and the pore sizes are typically 2-4 mm. Note that the pore sizes are approximately uniform, although some regional variations are apparent even at this relatively coarse magnification, and some evidence of pore elongation is apparent in selected regions.

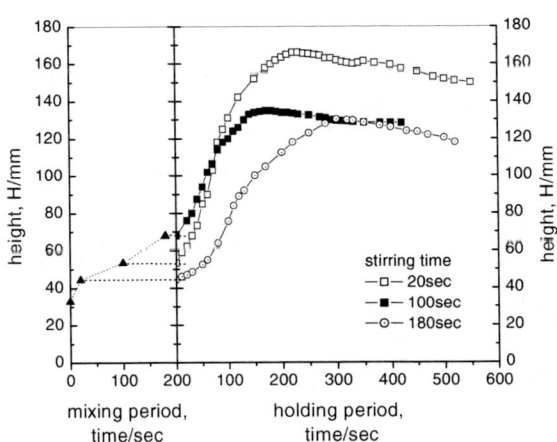

Figure 2

The relation between the height of the melt and holding time; (TiH_2: 1.4%wt, 300 mesh, initial height 33mm)

Figure 3. Longitudinal section of aluminum foam sample

4. Discussion

Models of Foam Rheology

The structure of metallic foam derives from the melt rheology, although the structural details can be somewhat difficult to quantify. Nevertheless, some elementary features can be readily identified. By observing multiple soap bubbles and soap films supported on wire frames, Plateau (1873) [5] recognized that when three films (walls) met, each with uniform total curvature, they formed dihedral angles of 120° (Figure 4). These wall junctions, called Plateau borders, determine the edges of polyhedral gas bubbles. When four Plateau borders meet, they join at angles of ~109.47°. Planar walls cannot satisfy the latter constraint because a planar polygon cannot have all angles equal to this value. This necessitates curved walls with complex curvatures and shapes. Kelvin (1887) showed that space could be partitioned into identical cells of equal volume and minimal surface area. Surprisingly, Matzke (1946) did not find such an identical cell during meticulous observations of 600 bubbles [6].

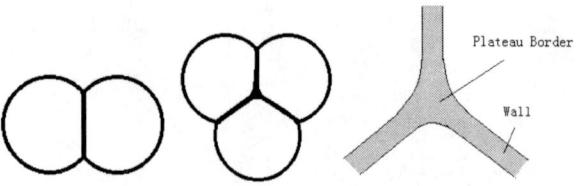

Figure 4 Schematic sketch of a Plateau border

Before describing models of metal foam rheology, it is useful to review theories that describe *dilute* gas-bubble suspensions, analogous to the beginning of the foaming process. As cells grow during metal foam fabrication, the dispersion passes through various volume-fraction regimes. Talyor (1932) [2] showed that when shear rates are small enough for drop deformation to be neglected, the effective viscosity μ_e of an emulsion of neutrally buoyant drops is given by

$$\mu_e = \mu_m \left[1 + f \cdot \frac{5r+2}{2(r+1)} \right]$$

where μ_m is the matrix viscosity, r is the viscosity ratio μ_d/μ_m (μ_d refers to the viscosity of the dispersed phase), and f is the volume fraction of the dispersed phase. The effective viscosity for a dilute suspension of spherical bubbles, for which $r \to 0$, is $\mu_e = \mu_m(1+f)$.

Experimental difficulties associated with the systematic measurement of foam flow are both obvious and subtle. A salient physical characteristic of foaming metal is the tendency to minimize free energy by reducing surface area. Thus, bubbles tend to increase in size with time for two reasons: to reduce surface area, and because of the progressive decomposition of foaming agent.

Certain characteristic features of foaming melt rheology relate to constitutive behavior and to the resulting foam structure. The foam possesses a yield stress, below which the deformation rate is zero and, therefore, the viscosity is infinite. When the shear stress exceeds the yield stress, the shear-rate-dependent viscosity μ_r can be represented by

$$\mu_r = \frac{\tau}{\dot{\gamma}} + \mu$$

where μ is a constitutive function that depends upon shear rate, τ is shear stress, $\dot{\gamma}$ is shear rate concerning with expanding of the foaming melt.

Another curious characteristic of foaming melt is "wall slip", in which cells can move relative to the adjacent crucible wall. This slip is merely a convenient macro-scale description of the wall boundary condition. The slip of cells near the wall depends on a thin fluid layer that does itself slip but spreads on the wall and lubricates the expanding foam.

Analysis

The longitudinal section of aluminum foam shown in Figure 2 reveals three kinds of cells: (1) equiaxed cells, which have smaller volumes and thicker walls, (2) vertical distortion cells near the side-wall of the crucible, which have small volumes, and are slightly elongated in the foaming direction, and (3) horizontal distortion cells at the center, with thinner walls and larger volumes.

In the foaming process, the melt expands rapidly once the granular foaming agent is injected. The top surface rises above the gas-liquid mixture as the foam structure is created within the melt. Movement within the foam occurs and can be regarded as flow. Foam flow involves two distinct regions in the melt - a solid-like plug region and a shear flow region, as shown in Figure 5. In the solid-like plug region, the shear stress does not exceed the yield stress and "foam viscosity" is infinite. In the shear flow region, the shearing stress exceeds the yield stress and foam viscosity is finite. In the wall region, the macroscopic foam velocity approaches a finite slip velocity.

On the microscopic scale, foam flow is accompanied by a change of structure. In the shear flow region, the foam can flow like a viscous fluid by inter-cell slip and cell slippage at the walls.

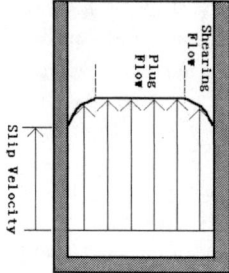

Figure 5 Schematic sketch of foam flow

Slip between the cells: The foam possesses a yield stress. When the shear stress exceeds the yield stress, cells deform and slip, as shown in Figure 6. The nonlinear elastic moduli are isotropic below the yield point. The viscous flow in the walls and Plateau-borders is time-dependent. These factors affect the lifetime of a bubble.

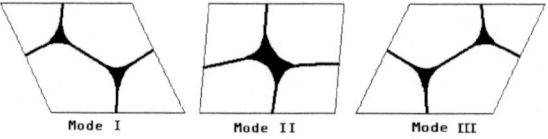

Figure 6 Schematic of slip between the cells

Macro-scale cell slippage at the walls is related to foam structure and viscous flow in the vicinity of the wall, as shown in Figure 7. The wall region can be regarded as Plateau-borders if the melt wets the wall.

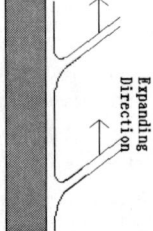

Figure 7 Schematic sketch of cell slippage at the wall

In the expanding direction, there is a speed gradient from the side-wall to the center of the mold. Near the side-wall, the speed is controlled by slip velocity and near the center, the speed is controlled by the plug flow. In the case of unbalanced speed, cell deformation takes place. From the profile of the foam, the shear flow region is about 15-20 mm thick. Thus, in the formula

$$\mu_r = \frac{\tau}{\dot{\gamma}} + \mu,$$

the shear rate ($\dot{\gamma}$) that can be detected by a device similar to the one used in Figure 2, and the constitutive function μ that can be estimated using $\mu_e = \mu_m(1+f)$. The remaining task it to determine the shear stress τ. On a macro-scale, a deformed cell will reveal some information about the shear stress acting on this cell. Normally, a cell is assumed to be equiaxed. However, under a shear stress, the cell will deform to an oval shape. The axial ratio (r_x) of this deformed cell can be regarded as a function of the shear stress, $\tau \propto f(r_x)$.

Unlike normal foam flow, the cell sizes in a foaming melt grow in during the foaming process, and the cell size is thus time-dependent. In the mold center, the temperature of the melt is higher than near the mold walls. Consequently, bubbles in the center grow more easily than those near the side. When two adjoining bubbles reach a critical size, the thin membrane between them will be annihilated. Thus, viscosity plays an important role in this period by changing the life of the membrane.

5. Conclusions

Controlling melt viscosity is a key process parameter in the production of metal foams by melt processing. During the foaming process, control of melt viscosity is essential in order to achieve stable foaming and to avoid slumping. Three types of cell shapes in the foam are identified in three regions: (1) near the side-wall of the crucible, where cells experience shear deformation in the expanding direction, (2) near the center of the foaming melt, where cells are large and experience compression deformation, and (3) between the center and side-wall of the foaming melt, where cells are fine and equiaxed.

Acknowledgement

Support from Honda R&D Americas through a Honda Initiation Grant is gratefully acknowledged.

References

[1] G.J.Davies, Shu Zhen; Metallic foams: their production, properties and applications (review); J. Mater. Sci., 18 (1983) 1899.
[2] M. F. Ashby, A.G. Evans, N.A. Fleck, L.J. Gibson, J.W. Hutchison, and H.N.G. Wadley, *Metal Foams: A Design Guide*, (Butterworth-Heinemann, Boston, 2000).
[3] J. Banhart, D. Bellman, and H. Clemens; Investigation of Metal Foam Formation by Microscopy and Ultra Small-Angle Neutron Scattering; Acta Mater. 49 (2001) 3409.
[4] Z.L. Song, J.S.Zhu, L.Q.Ma, and D.P. He; Evolution of foamed aluminum structure in foaming process; Mater. Sci. & Eng. A, 298(1-2)(2001)137
[5] J.A.F.Plateau; Statique Experimentale et Theorique des Liquides Soumisaux Seules Force Moleculaires, (1873) Ghent.
[6] E.B. Matzke; The three dimensional shape of bubbles in foam— an analysis of the role of surface in three dimensional cell shape determination, American J. Botany, 33(1946)

ROLE OF WETTING BEHAVIOR IN THE STABILIZATION OF LIQUID METAL FOAMS BY CERAMIC PARTICLES: AN EXPERIMENTAL SIMULATION STUDY

T. Gao and Y.Q. Sun
Department of Materials Science and Engineering
University of Illinois, Urbana, IL 61801

Abstract

Stabilization of liquid metal foams by ceramic particles is studied by experimental simulations, employing a liquid-particle system for which the wetting angle is continuously adjustable. The objective is to determine the role of the wetting behavior in optimum foam stabilization. Water-ethanol solution is used to mimic the liquid metal; inert polymer particles of Teflon, polyethylene and PVC are used to mimic ceramic particles. Varying the wetting angle, by varying the liquid concentration, is shown to drastically change the foam stability; a wetting angular range of 75 to 85° is identified for optimum foam stabilization. The foam stability is shown to be unrelated with liquid viscosity which remains unchanged with the wetting angle. Foams formed in the optimum wetting condition exhibits a slow decay, a final stable volume that persists for a long time, and a fine cell structure in the micrometer range.

Introduction

A lot of interests have been drawn towards the metal foams as evidenced by recent research activities, most of which are focused on the mechanical and thermal properties (for reviews, see [1-3]). In comparison, the underlying foaming physics, which is crucial for the production of high performance metal foams, is understudied and not well understood [4]. An understanding of foam stabilization mechanism is needed to develop the ability to control the foam structure such as cell size and porosity that are critical to properties [1, 2, 5]. An ability to control metal foam properties by processing is crucial to the engineering applications of metal foams, for example as lightweight sandwich panels, thermal and sound insulation materials, heat exchangers, filters, catalyst carriers and bone scaffolds for tissue growth [1, 5-8].

The aim of this work is to determine the roles of ceramic particles in the stabilization of liquid meal foams. Many issues are involved in the successful production of metal foams via the liquid state processing route, but the prerequisite is the stabilization of the liquid metal foam. For the liquid foam to solidify into a solid foam, it must obviously be stable or at least meta-stable. A liquid foam is however unstable both thermodynamically and kinetically. The thermodynamic instability comes from the large surface area of the foam structure which is driven to decrease by the surface energy, leading to the collapse of the liquid films and the destruction of the liquid foam structure [9, 10]. The liquid foams are kinetically unstable because of the inevitable liquid drainage under the gravity and the suction of the liquid to the Plateau borders that causes film thinning [4]. For the usual organic liquids, e.g. water, there are numerous bipolar molecular surfactants or frothers that may be used to stabilize the liquid foam structure. For metals, however, there are no such bipolar molecular or atomic surfactants, and the only known stabilizers for liquid metal foams are certain ceramic particles such as SiC, Al_2O_3 and CaO particles in Al [1, 11, 12].

Although ceramic particles have long been used as effective stabilizers for liquid metal foaming, their roles in the foam stabilization are still not well understood. One explanation attributes the improved liquid foam stability to an increase in the bulk viscosity resulting from the inclusion of the ceramic particles. The viscosity increase slows down the liquid drainage and thus mitigates the inherent kinetic instability. A piece of evidence is from the increase in the apparent viscosity found in liquid Al containing CaO particles introduced through internal oxidation [12, 13]. However, such observations have not been made in other systems, such as liquid Al containing SiC or Al_2O_3 particles. Another explanation ascribes the improved foam stability to the appropriate wetting behavior of the ceramic particles in the liquid metal [14-16]. Support for this explanation comes from experiments, on liquids containing a molecular surfactant, in which the foam life was found to be dramatically prolonged by the addition of particles falling into a special wetting angle range (40 to 70°) [14, 16]. However, these results, since they are based on organic liquids already containing a molecular surfactant, are not directly relevant to the processing of the liquid metal foams which are stabilized by ceramic particles alone without bipolar surfactants. The use of molecular surfactants, which are themselves foam stabilizers, masks the true foam stabilization effect due to the solid particles alone. Theoretical analysis incorporating several mechanisms predicted an effective wetting angle range of 20 to 90° for foaming Al [15]. There is clearly a need for experimental studies of the role of ceramic particles in stabilizing liquid metal foams and for a scheme for the selection of the metal-ceramic system for optimum foam stability.

In this work, surfactant-free liquids containing inert particles, with a continuously adjustable wetting behavior, are used to determine the optimum foaming condition. Water-ethanol solution and inert polymer particles are used to mimic liquid metal and ceramic particles respectively. The wetting behavior is adjusted by varying the concentration of the fully miscible water-ethanol solution in which the polymer's wetting behavior is varied continuously through the changing surface tensions [17]. A direct goal of this work is to determine, by the experiments, the wetting property that leads to optimum foam stabilization effect.

Procedure

Water-ethanol solutions of various concentrations were prepared using deionized water (H_2O) and research grade pure ethanol (C_2H_5OH). Pure polymer particles were obtained from Aldrich (PTFE) and Polysciences (PE and PVC). 30 ml water-ethanol solutions containing polymer particles were prepared in 200×25 mm round glass test tubes. The solution was then shaken vigorously in the closed tube until maximum foam volume was produced. The foam volume, or foam height, measured 60 seconds later, is used as a measure of foam stability. To ensure suspension, the polymer particles were blended into the pure ethanol first, before water was added to the required concentration. The viscosity of the water-ethanol solution with evenly dispersed inert polymer particles was measured using a rotational viscometer with a spindle 19 mm in diameter and 65 mm in thickness at 60 RPM. The foam structure was observed in an optical microscope by spreading the liquid foam on a glass slide. All the experiments were conducted at room temperature.

Results and Discussion

Figure 1a plots the foam stability, expressed by the foam height, against the concentration of the water-ethanol solution, for foams stabilized by PTFE (1 µm), PE (20 µm) and PVC (75 µm) particles, all at 3.3% (weight percent, as used in the rest of the paper for particle concentration). It shows that the foam stability depends strongly on the water-ethanol concentration. The particles have no foam stabilization effect on pure ethanol, as shown by the zero foam height. With increasing water concentration, the foam height increases, reaching a maximum, and then decreases with further increase of water concentration. As shown in Fig.1a, the peak foam volume occurs at different concentrations for different polymer particles, at 60%, 90% and 95% for PTFE, PE and PVC particles respectively (volume percent of water is used for liquid concentration). The effect of particle size on the foam volume will be discussed later (Figure 4).

Figure 1b plots the foam stability against the wetting angle. The wetting angle was obtained from an earlier measurement made on PTFE, PE and PVC against the water-ethanol concentration [17]. It shows that there is an optimum wetting angle for foam stability which, for the three types of polymer, falls into the narrow angular range of 75 to 85°. This shows that the foam stabilization effect by solid particles is closely correlated with the wetting behavior of the particles in the liquid. We note that the wetting angle of SiC in liquid Al is 75° [18]; the effectiveness of SiC in stabilizing liquid Al foams is hence consistent with the present result. The wetting property of ceramic particles in a given liquid metal depends on temperature and alloy composition, as well as on the type of ceramic material [18, 19]; the present work shows that these factors must be considered in the selection of the optimum foam stabilizer and processing condition.

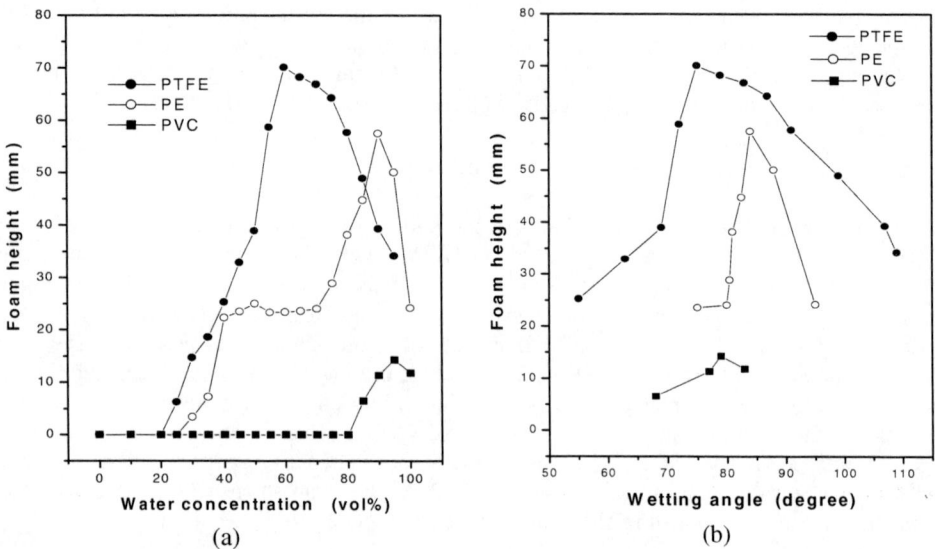

Figure 1. (a) Liquid foam height plotted against water concentration for water-ethanol solutions containing 3.3% PTFE, PE and PVC particles. The optimum foam stability for PTFE, PE and PVC occurs at 60%, 90% and 95% water respectively. (b) Foam height against the wetting angle, showing an optimum wetting angle range of 75 to 85°.

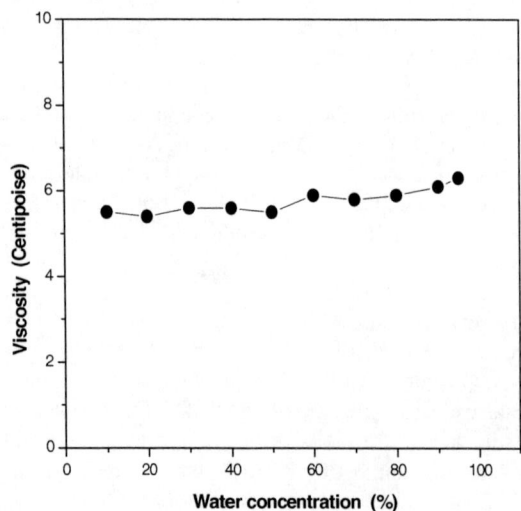

Figure 2:

The bulk viscosity of water-ethanol solution with 3.3% 1 µm Teflon particles versus water concentration. The change of viscosity is insignificant, in strong contrast to the drastic changes in the liquid foam stability.

Our results show that the foam stabilization effect by the partially wetting solid particles is unrelated with the bulk viscosity of the liquid. Figure 2 plots the bulk viscosity measured against the concentration of the liquid containing 3.3% 1 µm PTFE particles. With increasing water concentration, the viscosity changes very slightly, by less than 15%, in strong contrast to the drastic change in the liquid foam stability. The absence of a correlation between foam stability and liquid viscosity is contrary to an earlier explanation attributing the stabilization of liquid Al by CaO particles to a viscosity increase [13]. It is possible that the viscosity increase observed in liquid Al is an apparent effect due to the segregation of the ceramic particles to the liquid metal surface, increasing the surface viscosity via the Gibbs-Marangoni effect [20]. The present result is in agreement with similar findings earlier that showed the lack of viscosity change by the addition of partially wetting solid particles, for example in aqueous alkali solutions (containing surfactant dodecyl sulfonate) with hydrophobic SiO_2 particles [21].

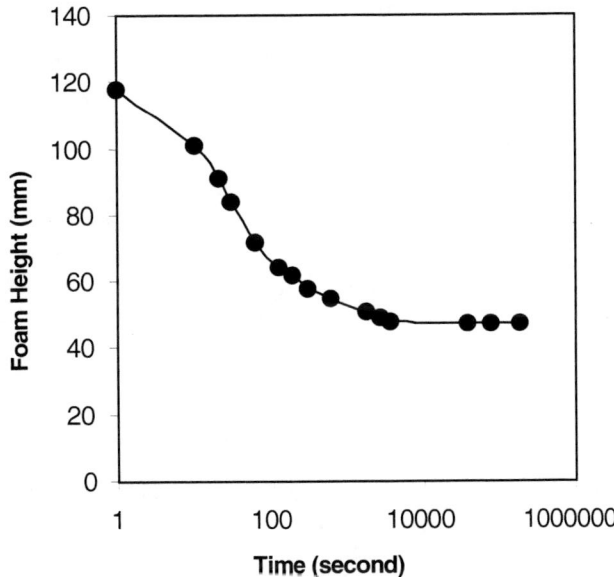

Figure 3: Foam height plotted against foam holding time for ethanol-60%water with 3.3% 1 µm Teflon particles. The foam decays rapidly initially, and then reaches a stable level that lasts for a very long time.

The addition of partially wetting solid particles changes not just the foam height but also the rate at which the foam decays. Figure 3 shows the foam height decrease with time for a foam of ethanol-60%water containing 3.3% 1 µm Teflon particles. The foam height first decreases rapidly with time, at a rate that decreases with time, and eventually stabilizes. There are two noteworthy features. One is that the initial decay rate depends on the wetting behavior of the particle used, and is lowest at the optimum wetting angle. This indicates that the inherent kinetic instability of the liquid foam is alleviated by the addition of solid particles with wetting angles falling into the 75-85° range. Another observation is that the final stable foam height at the optimum wetting angular range last for a very long time, more than a couple of weeks, in the still and closed environment of the test tube. The persistent liquid foam indicates that the final foam structure, after the initial drainage, is presumably metastable. This observation also further demonstrates that the liquid bulk viscosity is unimportant to the long-term stability of the liquid foam, and that the solid particles, if having the optimum wetting property, may be as effective as the molecular surfactants in forming highly stable liquid foams.

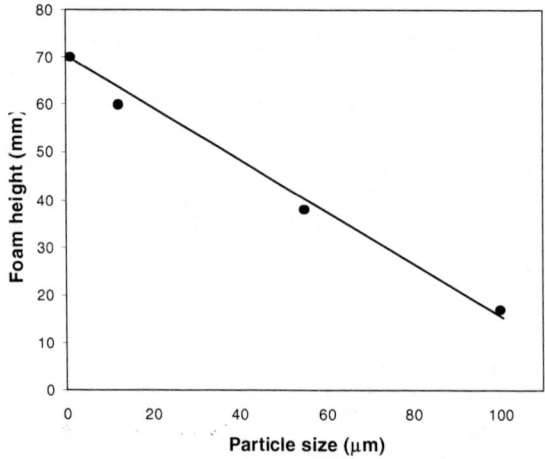

Figure 4: Foam height of ethanol-60%water solution containing 3.3wt% PTFE particles, plotted against particle size. The foam height decreases approximately linearly with particle size.

At a given particle weight concentration, the foam volume decreases with increasing particle size, as shown in Figure 4 in which the foam height is plotted against particle size for a ethanol-60%water solution containing 3.3wt% PTFE particles; this is the liquid concentration for optimum foam stability which is unchanged by particle size. The foam height decreases with particle size approximately linearly. Because, at a given weight concentration, the particle number density decreases with the particle size to the cube power, the observed linear decrease cannot be explained simply in terms of changes in the number density. The particle size affects the foam stabilization effects also because the large particles are less effective foam stabilizers due to their lack of agility and difficulty to segregate to the liquid surfaces. The ceramic particles used for stabilizing liquid metal foams (e.g. SiC in Al) are typically tens of micrometers in size [11]. The observed effectiveness of fine particles clearly points to the need to use fine ceramic particles for the optimum stabilization of liquid metal foams. The use of nanometer ceramic particles to stabilize liquid metal foams is an interesting possibility.

The liquid foams at the optimum wetting angle have a very fine structure, in contrast to the metal foams stabilized by ceramic particles for which the cell size is rarely smaller than 1 mm [11]. The structure of ethanol-60% water foam containing 3.3% 1 μm Teflon powder, observed in an optical microscope under polarized illumination, is shown in Figure 5. The average bubble size is close to 50 μm; some small bubbles are less than 10 μm in size. The foam structure in Figure 5 is typical of foams without extensive drainage [22] and is characterized by liquid bubbles separated by a thick body of liquid rather than polyhedral bubbles with thin liquid films in between.

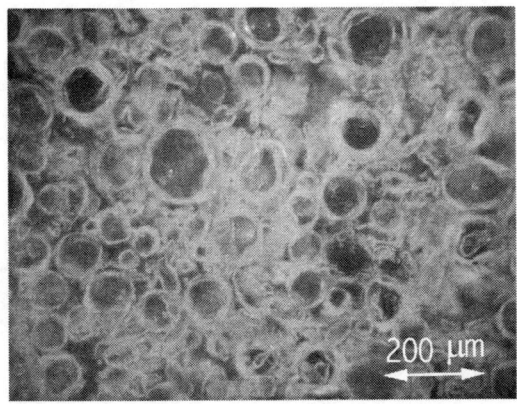

Figure5: Optical microscopy image of the liquid foam structure of ethanol-60%water with 3.3wt.% of 1 μm Teflon powder. The average bubble size is about 50 μm.

The present experimental simulation results support the view that the effectiveness of solid particles in stabilizing liquid foams depends strongly on the wetting behavior of the solid particles in the liquid [14-16]. Because our experiment used liquids that do not contain bipolar molecular surfactants, the findings are directly relevant to the stabilization of liquid metals by ceramic particles. The water-ethanol solution offers the advantage that it is fully miscible and that the wetting behavior is continuously adjustable by varying the liquid concentration. This eliminates the need to use different solid particles or the use of molecular surfactants for changing the wetting angle; the use of molecular surfactants obviously would mask the true foam stabilization effects due to the solid particles. The present experiment therefore demonstrates the role of the wetting behavior in the foam stabilization due solely to the solid particles. The 75-85° optimum wetting angle identified in our work is better defined and narrower than those given by previous experimental results [14-16].

It has long been known that the liquid foam stabilization is a complicated phenomenon because it involves a combination of kinetic and thermodynamic processes [4]; the observed 75-85° optimum wetting angle still defies a simple, quantitative explanation. There are a few qualitative explanations relating the foam stabilization effect with the wetting characteristics. One relates the foam stabilization effect with the capillary drainage of the liquid film to the Plateau borders [16]; the stabilization effect of the solid particles derives from a reduction in the liquid surface tension, hence a lessening in the suction pressure, caused by the segregation of the solid particles to the liquid surface. Another explanation is based on an increase in the rigidity of the liquid surface imparted by the mechanical strength of the solid particles adhering to the surface due to their partially wetting character [14]. A feature common to these models is the segregation of the solid particles to the liquid surface driven by the lowering of the system energy. Ross and Nishioka [23] provided a thermodynamic explanation of foam stabilization based on the original Gibbs-Marangoni effect. Once the particles have segregated to the liquid surface, a decrease in surface area or energy, as driven by the surface tension, would force some of the particles back into the liquid interior, causing the energy to increase. The liquid foam would be stable if this energy increase outweighs the surface energy decrease. This condition is expressed by

$$\left| A \frac{\delta \gamma}{\delta A} \right| > \gamma$$

where A is the surface area and γ is the surface tension. The left hand side of the above inequality is the Gibbs surface elasticity. The Gibbs surface elasticity due to a monolayer of partially wetting spherical particles can be readily calculated, and can be shown to give rise to a critical wetting angle of 90.2° below which the liquid foam is unstable. This is clearly at variance with the present experimental results that have shown stable liquid foams at much smaller wetting angles. This model also predicts a foam stability that increases monotonically with the wetting angle, also contrary to our observation of a peak foam volume at the optimum wetting angle and the decrease of foam stability with further increase of the wetting angle. The presence of a peak in the foam stability versus the wetting angle may be explained by the clustering of the solid particles, or the micelle formation, that occurs when the wetting angle is large and the energy is lowered by the replacement of solid-liquid interfaces with solid-solid interfaces. The maximum foam stability occurs at a concentration close to the critical micelle concentration (CMC) [9]. The bulky micelles do not segregate to the film surfaces as readily as the small particles, resulting in the loss of foam stability.

Conclusions

Surfactant-free water-ethanol solutions containing polymer particles with continuously adjustable wetting behavior are used to simulate the stabilization of liquid metal foams by ceramic particles. The foam stabilization effect depends strongly on the wetting angle. A wetting angular range for optimum foaming stability is identified to be 75 to 85°. The liquid bulk viscosity is unaffected by the wetting angle and, for the liquid system investigated, is unimportant to the foam stabilization effect. At a given particle weight concentration, the foam volume decreases approximately linearly with particle size. The improved stability by partially wetting particles can be generally understood as due to the Gibbs-Marangoni effect. The observed 75-85° optimum wetting angle range cannot yet be explained quantitatively.

References

1. M. F. Ashby et al., Metal Foams: A Design Guide (Boston: Butterworth-Heineman, 2000), 3-5.
2. A. G. Evans, J. W. Hutchinson, and M. F. Ashby, "Cellular Metals," Current Opinion in Solid State & Materials Science, 3 (1998), 288-303.
3. L. J. Gibson, "Mechanical behavior of metallic foams," Annual review of materials research, 30 (2000), 191-227.
4. D. Weaire and S. Hutzler, The Physics of Foams (Oxford: Clarendon Press, 1999), 126-143.
5. L. J. Gibson and M. F. Ashby, Cellular Solids: Structure and Properties (Cambridge: Cambridge University Press, 1997), 8-11.
6. G. J. Davies and S. Zhen, "Metallic foams: their production, properties and applications," Journal of Materials Science, 18 (1983), 1899-1911.

7   J. Banhart and J. Baumeister, "Production Methods for Metallic Foams," Proceedings of the Materials Research Society, 521 (1998), 121-132.
8   L. D. Zardiackas et al., "Structure, Metallurgy and Mechanical Properties of a porous Tantalum Foam," Journal of Biomedical materials Research, 58 (2001), 180-187.
9   J. J. Bikerman, Foams (New York: Springer-Verlag, 1973), 5-13.
10 F. Sebba, Foams and Biliquid Foams - Aphrons (Chichester: John Wiley & Sons, 1987), 56-62.
11 M. Thomas, D. Kenny, and H. Sang, in US Patent Database (Alcan International Limited, USA, 1997), Vol. 5622542.
12 L. Ma and Z. Song, "Cellular structure control of aluminum foams during foaming process of aluminum melt," Scripta Materialia, 39 (1998), 1523-1528.
13 J. Banhart, "Manufacturing Routes for Metallic Foams," Journal of Metals, 12 (2000), 22-27.
14 G. Johansson and R. J. Pugh, "The influence of particle size and hydrophobicity on the stability of mineralized froths," International Journal of Mineral Processing, 34 (1992), 1-21.
15 G. Kaptay, "Interfacial criteria for ceramic particle stabilised metallic foams," Metal Foams and Porous Metal Structures, ed. J. Banhart, M. F. Ashby and N. A. Fleck (Bremen, Germany: MIT Verlag, 1999), 141-148.
16 S. W. Ip, Y. Wang, and J. M. Toguri, "Aluminum foam stabilization by solid particles," Canadian Metallurgical Quarterly, 38 (1999), 81-92.
17 J. R. Dann, "Forces involved in the adhesive process. I. Critical surface tensions of polymeric solids as determined with polar liquids," Journal of Colloid and Interface Science, 32 (1970), 302-320.
18 V. Laurent, D. Chatain, and N. Eustathopoulos, "Wettability of SiC by aluminum and Al-Si alloys," Journal of Materials Science, 22 (1987), 244-250.
19 J. V. Naidich, "The Wettability of Solids by Liquid Metals," Prog. Surf. Memb. Sci., 14 (1981), 353.
20 L. E. Scriven and C. V. Sternling, "The Marangoni Effect," Nature, 187 (1960), 186-188.
21 F. Tang et al., "The effect of SiO2 particles upon stabilization of foam," Journal of Colloid and Interface Science, 131 (1989), 498-503.
22 C. Monnereau, M. Vignes-Adler, and K. Kronberg, "Influence of gravity on foams," Journal de Chimie Physique, 96 (1999), 958-967.
23 S. Ross and G. Nishioka, "Foaming behaviour of partially miscible liquids as related to their phase diagrams," Foams: Proceedings of a Symposium organized by the Society of Chemical Industry, Colloid and Surface Chemistry Group, (London: Academic Press, 1976), 17-31.

SECTION III

OXIDE-REDUCTION AND P/M PROCESSES

MULTIFUNCTIONAL METALLIC HONEYCOMBS BY THERMAL CHEMICAL PROCESSING

Joe K. Cochran[*], K. J. Lee[*], Dave McDowell[*,1], Tom Sanders[*]

School of Materials Science and Engineering[*] & GWW School of Mechanical Engineering[1]
Georgia Institute of Technology
771 Ferst Dr. N.W., Atlanta, GA 30332-0245 USA

Abstract

Thin-walled metallic honeycombs with highly functional cell geometries and microstructure are being developed at Georgia Tech. Using conventional powder processing techniques, precise shapes are formed with non-metallic precursors and subsequently converted to the metal state by a direct reduction process. Forming is at room temperature using techniques including extrusion, spray coating on sacrificial cores, slurry casting, and dry pressing. When direct reduction is coupled with extrusion to form cellular articles of constant profile, it is termed Linear Cellular Alloy (LCA) technology. To produce fine geometry with good precision, fine powder with sizes about 1/50th of the feature dimension are necessary. By using fine powders in the micron and sub-micron range, articles that have small dimensions (i.e., micro-channels) and low density are routinely made. Through a controlled conversion process, the articles formed from brittle powder precursors (principally oxides) are chemically converted into a metal article with structural integrity and substantially the original shape. Demonstrated alloys include stainless steels, maraging steels, Inconel 617 and 718, Super Invar, and copper alloys. Strength and specific energy absorption of a maraging steel honeycomb at 1.5 gm/cc were approximately 400 MPa (at 2.1 gm/cc were approximately 600 MPa) and 100 Joules/gm, respectively. Potential applications are lightweight structural components with high energy absorption capability in defense and civil transportation areas. Copper based honeycombs have been tested to provide a high heat transfer rate at low pressure loss. Thermal management applications such as forced air heat sinks, energy absorption applications such as high pressure/high strain rate blast/impact abatement device, and electrochemical applications such as solid oxide fuel cells (SOFCs) using a hybrid metal/electrolyte honeycomb structure are discussed

Introduction

Affordable Ordered Cellular Metals

The objective of this article is to give an overview of the LCA program, in areas focused on advancing the technology, measurement and modeling results, and efforts in interfacing with potential end users to achieve viable products in emerging technologies.

For higher performances and weight efficiency, light weight cellular metal/alloy components are becoming more prevalent in structural designs. From a multi-year study program of ultralightweight metallic foams, Evans,et al. [1] concluded: "Methods have been developed to create cellular metals with a wide range of topologies. Stochastic materials are inexpensive but place material in locations where it contributes little to material properties (other than density). Periodic materials can be made by several (for the most part) expensive techniques. They can be designed to optimize multifunctionality by placing material at locations where mechanical and other performance indices are simultaneously maximized. Inexpensive manufacturing methods that enable control of the topology are needed." LCA technology has demonstrated the potential to become a viable, inexpensive manufacturing method of high quality cellular metals with ordered geometries in production quantity. The viability of the process needs to be tested in at least three areas: design flexibility, manufacturability, and affordability. Design flexibility is discussed in terms of cell geometry/size variety and a solid phase composition that can be tailored to applications. Manufacturability is judged by shape forming capability and ease of conversion of precursor to final composition. Affordability is related to LCA production cost, as well as application specific cost-performance ratio.

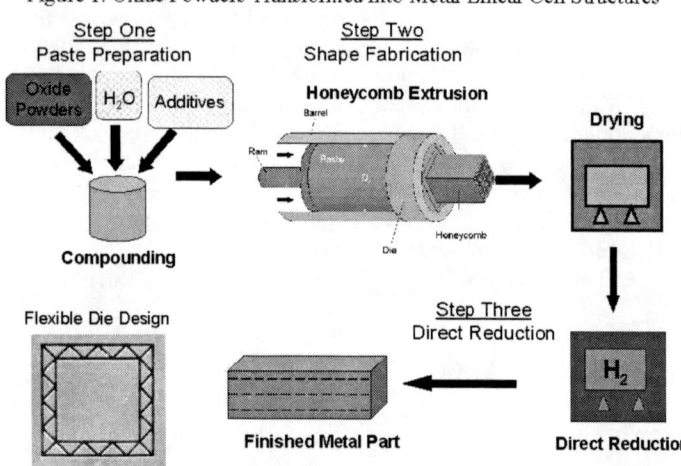

Figure 1. Oxide Powders Transformed into Metal Linear Cell Structures

LCAs are produced by powder paste extrusion (a proven mass production technique) at or near room temperature, followed by a heat treatment that converts the oxide to metal and simultaneously sinters the metal to a useful density (Fig. 1) [2]. Unlike other processing technologies for cellular metals (e.g., foaming and melt casting), LCA technology de-couples cell geometry and metal/alloy composition. Composition is controlled only by formulation of the oxide powder mixture according to chemical specification of the intended product. For example, commercially available metal oxides of Co, Ni, Cu, Zn, Mo, W, Mn, and Nb in any combination can readily be reduced by hydrogen. Shape forming uses a paste containing the oxide powder mixture in an organic-water system to impart the plasticity for extrusion. For non-plastic powder mixtures, paste rheology is dominated by the viscous liquid phase [3]. Extrudability of the paste is primarily controlled by water/binder content, solids loading, and powder size distribution. The material choice is independent of cell geometry and the cell geometry is optimized by design. As an example, a collection of extrusion dies is shown in Fig. 2 and products extruded through these dies are shown in Fig. 3. This process allows for significantly more complex distributions of cell

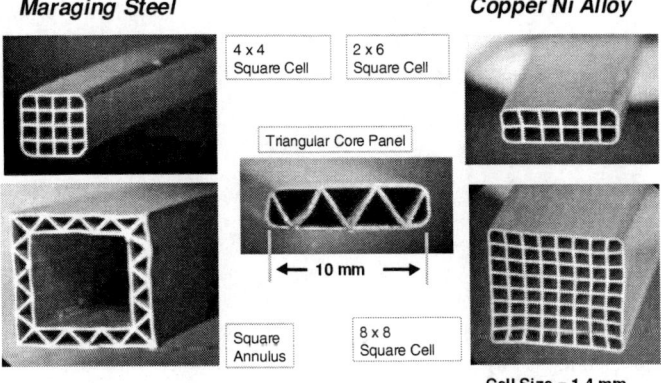

Figure 2. Extrusion Dies for Thin Wall Metallic Honeycomb

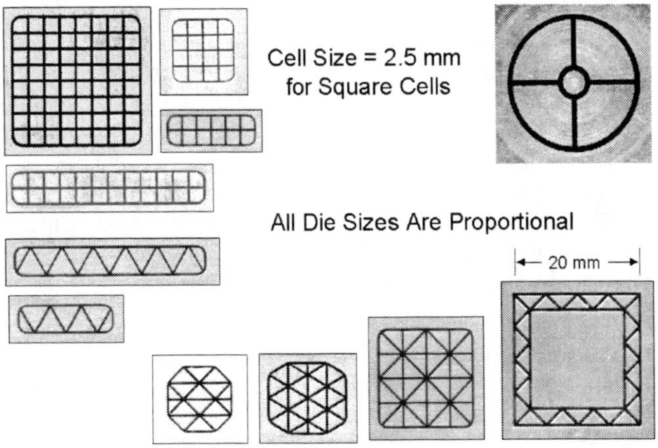

Cell Size = 2.5 mm for Square Cells

All Die Sizes Are Proportional

|← 20 mm →|

Fig. 3 Examples of Linear Cellular Alloys by Direct Reduction

Maraging Steel   *Copper Ni Alloy*

4 x 4 Square Cell   2 x 6 Square Cell

Triangular Core Panel

← 10 mm →

Square Annulus   8 x 8 Square Cell

Cell Size = 1.4 mm

shapes and sizes, as well as precision of cell alignment and wall thickness compared to the more conventional stamped and expanded sheet metal honeycomb process. Complexity of cell geometry is limited only by extrusion die manufacturability and powder paste rheology.

LCA Manufacturing Economics

Honeycomb fabrication by paste extrusions is a precision, high volume, inexpensive manufacturing technique. This is demonstrated by automotive catalytic converter production, which over the past 30 years has an established low cost structure.
Laboratory scale process information for direct reduction of LCA is available for pilot-full capacity plant design and associated cost modeling. Using copper as a low cost material option

for this technology, a preliminary production cost analysis was performed. Copper LCA is low cost because the oxide powder is abundant and inexpensive, and also because of the relative low reduction and sintering temperature provides for economical heat treatment. Production cost of LCA honeycomb largely depends on volume, type of raw material and thermal treatment, product size (weight, density), and yield. Volume production is a universal key for lowering cost. Production yield, however, is a complex function of product design, density and size, user specified tolerance, and above all, the quality assurance procedures that would be implemented in the manufacturing operation. For Cu LCAs used as heat sinks for CPU cooling, a production cost of $ 3.00 to 7.00 per lb of product is projected. The heat sink's simple design and small size (250 CC) contributed to a competitive price for this part. This cost is approximately the same as the cost for stochastic foam made from molten aluminum. The assumption for the heat sink cost projection used medium scale production volumes at a plant size capable of producing 0.5 million pounds per year and up. Figure 4 shows the location of the Cu LCA heat sink in cost vs. thermal resistance, as compared to all other competing solutions.

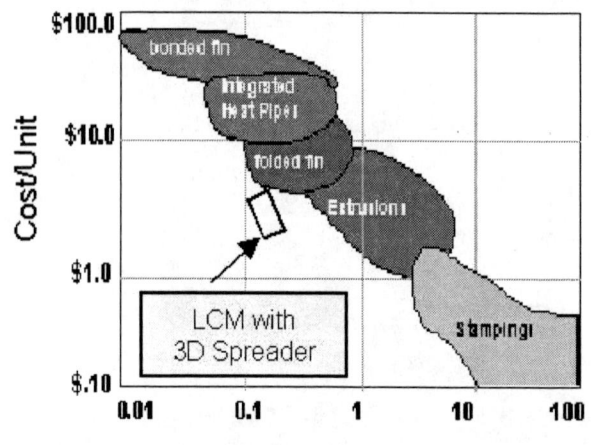

Figure 4. Comparison to Commercial Heat Exchangers

LCA Development Thrust Areas

LCA technology has been under development since early 2000, pushed by the process improvements in extrusion precision and metallurgical quality, and pulled by industrial needs for higher performance structures at lower density. Currently thin-walled LCA extrusions have cell sizes in the range of 0.5 to 2.0 mm with a web thickness of 0.10 to 0.20 mm, at a relative density from 0.12 to 0.30. Alloys that have been demonstrated and characterized to date include several types of steel, Inconels (nickel based), and copper and copper alloys. High thermal conductivity has been measured for copper from the reduction process (360 W/mK) [4] and heat transfer rates of 90 W/cm^2 have been calculated based on optimized copper cellular design models. High strength (600 MPA at 2.1 g/cc) and energy absorption (100 kJ/kg at 2.1g/cc) have been demonstrated for maraging steel cellular structures [5]. No other lightweight structures approach these values in properties.

Material systems are under development for specific application areas including, maraging steel LCAs for high strength and fracture toughness [6], a nickel based superalloy for high temperature properties, and certain copper based alloys for superior thermal management [4]. Processing improvements for these alloy honeycombs have resulted from extrusion paste formulation and modeling [7], in-house die design, and reduction/sintering heat treatment.

Metallurgical characterization has demonstrated that the direct reduction metal is comparable to conventionally processed counterparts. [8,9] These developmental successes are supported by concurrent mechanical property measurements and finite element analyses. Highlights that underscore this two year old program are the superior strength and energy absorption capability demonstrated by maraging honeycombs (Fig. 5), the exceptional dimensional stability of Super Invar (ultra low CTE of 0.4 ppm/C, Fig. 6) and the low predicted thermal resistance of a forced convection heat sink based on Cu LCA (see Application Case Example #1).

Application Case Example 1
Monolithic Forced Convection Heat Exchanger (in prototyping)
The primary objective of a series of engineering designs for a "Monolithic Forced Convection Heat Exchanger with Integrated Heat Spreader" has been to establish a set of performance criteria for thermal resistance and pressure drop (Fig. 7). The term "monolithic" refers to the integration of high-density webs (the heat exchanging surface) with a wrap-around

Figure 5. Comparison of LCA Energy Absorption to Other Lightweight Metals.

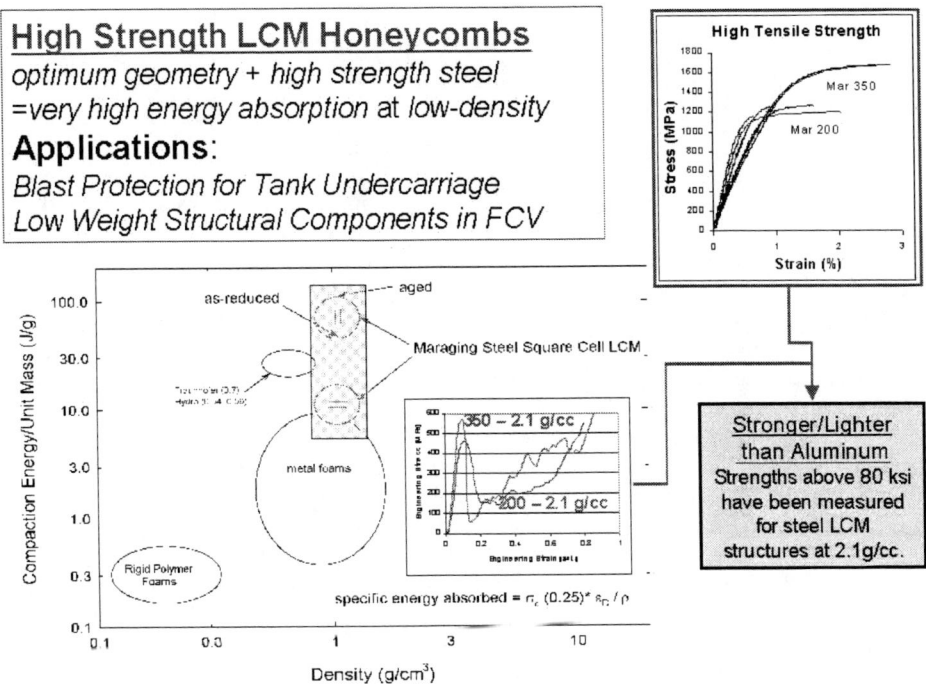

solid base (the 3D heat spreader). Extrusion of this monolith will both reduce manufacturing cost and insure low contact thermal impedance between the two elements. The cell size of honeycomb is in the optimum mesoscale range (2 to 4 mm) for a balanced trade-off between heat transfer rate and pressure loss. Low cost and thermal impedance are the enabled features of the underlying technology (honeycomb extrusion coupled with direct reduction).

Solutions have been calculated for the lower bound of total heat transfer rate (upper bound thermal resistance) and upper bound of total heat transfer rate (lower bound thermal resistance). The lower bound total heat transfer rate solutions are based on assuming that the temperature gradient associated with the fin heat transfer solution at the inlet is applied along

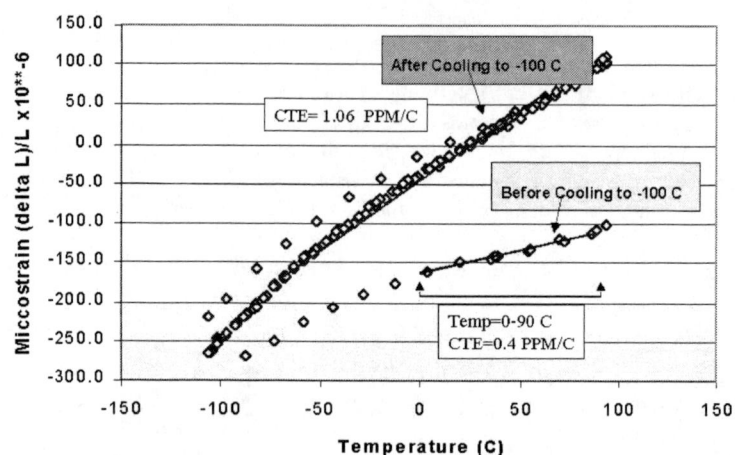

Fig. 6 Thermal Expansion of Super Invar
Before & After Cooling to -100 C

the length. A characteristic length is assumed for the fins to reach adiabatic tip conditions, which is an approximate correction on fin efficiency. The upper bound solution is based on assuming an isothermal heat exchanger. Generally, the effective Biot number is low enough that for a 3D heat spreader, the heat exchanger efficiency (compared to the isothermal case) is in the 85% to 95% range, with the former representative of a peripheral 3D heat spreader. The term "3D heat spreader" refers to the co-extrusion of the heat spreader with the cross-section, effectively embedding a three-dimensional heat spreading capability within a mesoscale forced convection heat exchanger.

The use of a 3D heat spreader as skeletal network of LCA extrusion is a key to driving up efficiency and driving down thermal resistance, along with the capability to extrude thin walls relative to cell size. The ideal upper bound total heat transfer for these specifications is 365.8 W for T_{base} at 100 oC (209.0 W for T_{base} 70 °C) with an associated heat sink thermal resistance of $R_{sa} = 0.191$ °C/W; this represents a heat flux removal from the chip of 101.3 W/cm^2 for $T_{base} = 100$ °C (57.9 W/cm^2 for $T_{base} = 70$ °C). These numbers for thermal resistance are at the low end of expectations for extrusions and appear suitable for advanced technologies such as heat pipes and integrated vapor chambers.

Application Case Example 2
Hybrid Solid Oxide Fuel Cells (currently under development)
The objective of this effort is to develop SOFCs based on extruding honeycomb having alternating layers of a metallic interconnects and a solid oxide electrolyte (e.g., ceria or zirconia). The monolithic honeycomb will be formed by simultaneously extruding the addressing projected cooling needs in the industry at potentially significantly lower cost than electrolyte and metal component, followed by heat treating in a controlled atmosphere. After adding gas manifolds and electrodes via slurry deposition, an operational SOFC can be constructed with revolutionary cost savings. The approach can provide the fuel cell industry with a device in meeting a cost constraint of $500/kilowatt required for commercialization.

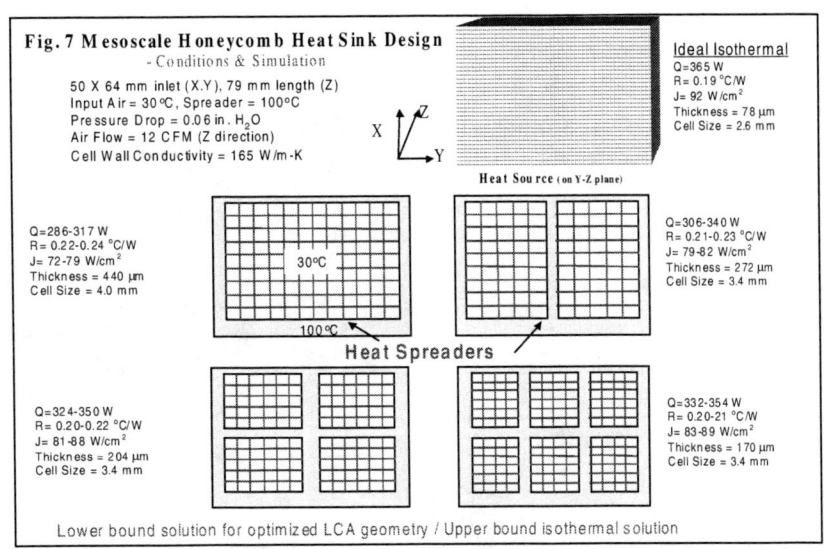

Figure 8. Hybrid Metal/Electrolyte Monolithic SOFC.

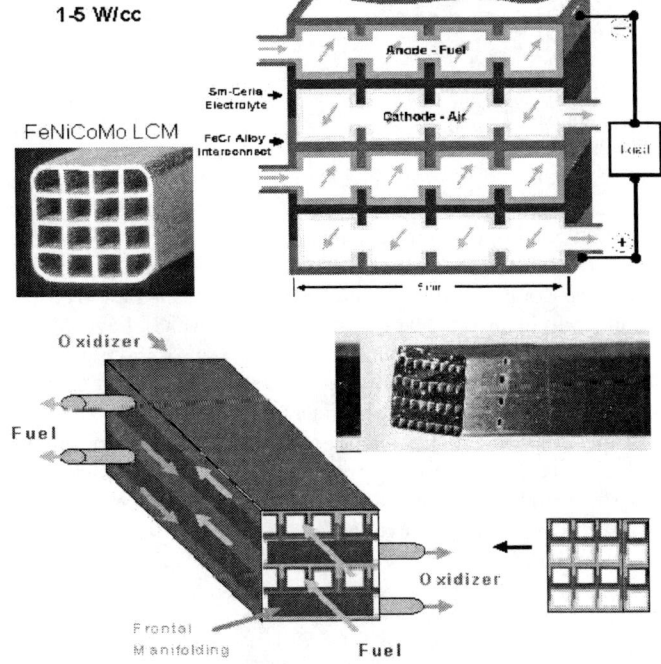

Volumetric power density is projected to be at one W/cc, based on producing solid electrolyte honeycomb with dimensions currently being fabricated at Georgia Tech. Further increase in power density with corresponding decrease in cost can be made via dimensional scale-down.

Power density in fuel cells is proportional to the electrode surface area and the surface area increases inversely to reduced dimensions. By adopting state-of-the-art extrusion dimensions at a factor of five smaller, a five-fold increase in power density to 5 W/cc is possible.

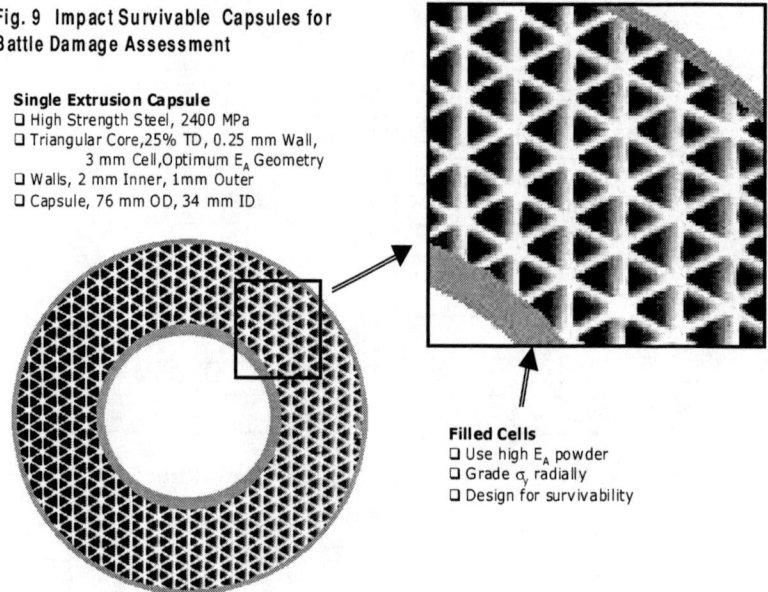

Fig. 9 Impact Survivable Capsules for Battle Damage Assessment

Single Extrusion Capsule
- High Strength Steel, 2400 MPa
- Triangular Core, 25% TD, 0.25 mm Wall, 3 mm Cell, Optimum E_A Geometry
- Walls, 2 mm Inner, 1mm Outer
- Capsule, 76 mm OD, 34 mm ID

Filled Cells
- Use high E_A powder
- Grade σ_y radially
- Design for survivability

Application Case Example 3
Reactive Powder Filled Blast Protection Capsule (at concept stage)

The objective of this effort is to develop a process for fabricating alloy honeycomb structures with spatially variable cell geometry and integral face sheets. These linear cellular alloys may be in a multitude of geometrical cross sections and are intended for high strength and high strain rate applications. It is well established that foam filled tubes have synergistic energy absorption; meaning the combined performance is greater than the performance sum of the tubes and the foam individually. Starting with LCAs, which have demonstrated energy absorption at the upper bound of a tube bundle, an energy absorption level unmatched by any other means appears achievable by loose-filling cells with a carefully formulated mixture of reactive powders. Insertion of appropriate materials into cells of linear structures adds multiple energy absorbing capabilities to the structure; namely contribution from plastic deformation of the high strength honeycomb, energy dissipation due to powder compaction, and energy dissipation during shock induced chemical reactions. A realistic estimation of energy level capable of being absorbed with high strength LCA is on the order of 5×10^9 J/M^3 at densification and reaction pressure of ~2000 MPa (Fig. 10).

Acknowledgements

This work is sponsored by DSO of DARPA (N00014-99-1-1016) under Dr. Leo Christodoulou and by ONR (N0014-99-1-0852) under Dr. Steven Fishman. The "Hybrid SOFC" is supported by DSO of DARPA under Dr. Bob Nowak and is a product of collaboration with Professor M. Liu of MSE at Georgia Tech. The "Reactive Powder Filled Blast Protection Capsule Concept" is a product of collaboration with Professor N. Thadhani of MSE at Georgia Tech. We deeply appreciate their support and guidance.

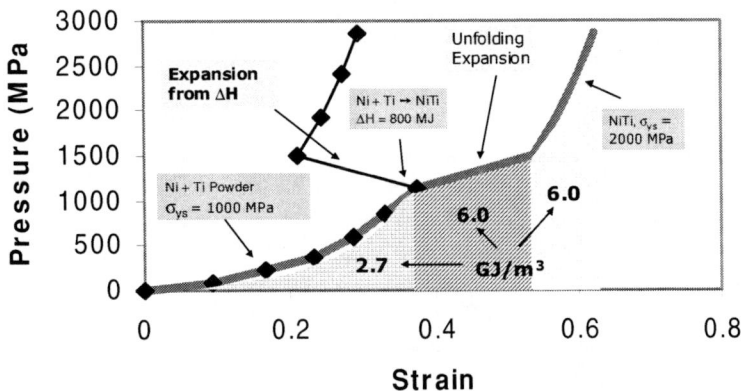

Fig. 10 Pressure versus Strain with Reaction for Powder Compaction

References

1. Evans, A. G., Hutchinson, J. W., Fleck, N. A., Ashby, M. F., Wadley, H.N. G., "The Topological Design Of Multifunctional Cellular Metals", Proceedings of the 2nd International Conference on Cellular Metals and Metal Foaming Technology (MetFoam 2001), Bremen (Germany), 18.-20. June

2. Cochran, J., Lee, K.J., McDowell, D., Sanders, T.H., et al., 2000, "Low Density Monolithic Metal Honeycombs by Thermal Chemical Processing," Proceedings of the 4th Conference on Aerospace Materials, Processes, and Environmental Technology, September, Huntsville, AL.

3. K.M. Hursyz, "Paste Mechanics of Fine Extrusion", Ph.D. Dissertation, MSE, Georgia Institute of Technology, Nov. 2001.

4. Church, B.C., Dempsey, B.M., Clark, J.L., Sanders, T.H., Jr., and Cochran, J.K., "Copper Alloys from Oxide Reduction for High Conductivity Applications," Proceedings of IMECE 2001: 2001 International Mechanical Engineering Congress and Exposition, November 11–16, 2001 -New York, NY.

5. Hayes, A. M., "Compression Behavior of Linear Cellular Steel", M.S. Thesis, School of Materials Science and Engineering, Georgia Institute of Technology, Atlanta, GA, August, 2001.

6. Hayes, A., Dempsey, B., McDowell, D. L. Isley, S., "Mechanics of Linear Cellular Alloys", Proceedings of IMECE 2001: 2001 International Mechanical Engineering Congress and Exposition, November 11–16, 2001 -New York, NY.

7. A.M. Hayes, D.L. McDowell, and J.K. Cochran, "Properties of Maraging Steel Honeycomb under Compressive Quasistatic and Dynamic Loading," In: Proceedings of the 2002 TMS

Annual Meeting; Processing and Properties of Lightweight Cellular Metals and Structures: Structural and Multifunctional Applications, February 17-21, 2002, Seattle, Washington.

8. K.M. Hurysz, R. Oh, J.K. Cochran, T.H. Sanders, Jr., and K.J. Lee, "Modeling Powder Extrusion Pastes for Forming Light Weight Multifunctional Structures," In: Proceedings of the 2002 TMS Annual Meeting; Processing and Properties of Lightweight Cellular Metals and Structures: Structural and Multifunctional Applications, February 17-21, 2002, Seattle, Washington.

9. J.L. Clark, J.K. Cochran, T.H. Sanders, K.J. Lee, "Metal Honeycomb from Oxide Paste: Maraging Steel and Super Invar Structure and Properties," In: Proceedings of the 2002 TMS Annual Meeting; Processing and Properties of Lightweight Cellular Metals and Structures: Structural and Multifunctional Applications, February 17-21, 2002, Seattle, Washington.

10. J.H. Nadler, T.H. Sanders, Jr., and J.K. Cochran, "Reduction of Iron and Chromium Sesquioxide Powder Mixtures for Metal Honeycomb Structures," In: Proceedings of the 2002 TMS Annual Meeting; Processing and Properties of Lightweight Cellular Metals and Structures: Structural and Multifunctional Applications, February 17-21, 2002, Seattle, Washington.

11. B.C. Church, J.K. Cochran, and T.H. Sanders, Jr, "Copper Extrusions from Oxide Reduction for High Conductivity Applications," In: Proceedings of the 2002 TMS Annual Meeting; Processing and Properties of Lightweight Cellular Metals and Structures: Structural and Multifunctional Applications, February 17-21, 2002, Seattle, Washington.

METAL HONEYCOMB FROM OXIDE PASTE: MARAGING STEEL AND SUPER INVAR STRUCTURE AND PROPERTIES

Justin L. Clark, Joe K. Cochran, Thomas H. Sanders, Kon J. Lee

School of Materials Science and Engineering, Georgia Institute of Technology
771 Ferst Dr. N.W., Atlanta, GA 30332-0245 USA

Abstract

A study is under way to produce complex metal honeycombs through a powder oxide extrusion process. The extruded oxide pastes are subsequently reduced in a hydrogen atmosphere to a metallic body. One of the alloy regions being investigated is the FeNiCo system including 18% Ni maraging steels and Super Invar. Maraging steels are well known for their excellent combination of high yield strength and fracture toughness and Super Invar combines ultra low thermal expansion with reasonable strength. A focus of this study is the characterization of these bulk alloys produced through the reduction process. As in traditional powder processing methods, porosity, homogeneity, shrinkage, and microstructure must be understood with respect to the various processing steps. Elastic modulus, yield strength, and elongation of the bulk material are required for predicting the performance of the honeycomb. To date, densities approaching 100% theoretical and yield strengths approaching 70% of values have been observed. The relationships between processing, microstructure and properties have been analyzed.

Introduction

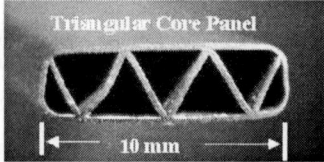

Figure 1: Maraging steel extrusions.

Powder processing is a conventional fabrication technique used to refine, shape, and consolidate raw materials into technologically useful products. The use of extrusion to produce honeycomb structures similar to those utilized as catalytic substrates in automobiles is the focus of a program at the Georgia Institute of Technology. In this program several metal oxide powders are combined into a paste and extruded through any of several dies of various designs, sizes, and honeycomb densities (Figure 1). The extruded material is dried and heated in a hydrogen atmosphere to 1200-1300°C to burnout binder, reduce the oxide powders, and sinter the wall material to near theoretical density. The result is a thin-walled (~150 µm) metal honeycomb. The density of structures ranges from 10 to 50 %.

Alloys Investigated

Numerous alloy systems may be pursued using the powder extrusion technique including, but not limited to, copper, nickel, and iron alloys. Several limitations arise when considering alloying components that are difficult to reduce such as titanium and aluminum. However, the use of hydrides and intermetallics are being investigated as an alternative for incorporating such elements. The alloys that are the focus of this study include two maraging steel grades and Super Invar.

Maraging Steel The maraging steel compositions studied in this program were chosen due to their combination of high strength and toughness. Based on the Fe-Ni system, these steels derive their strength from two microstructural phenomena: a martensitic transformation on cooling from solutionizing temperatures and a precipitation event during the maraging heat treatment. Additionally, the primary alloying components are readily reduced in hydrogen with the exception of titanium and aluminum. Two grades were chosen for study; maraging 350 (M350) and maraging 200 (M200). The nominal compositions and mechanical properties of these alloys are given in Table 1. Due to the significant strengthening of the titanium alloy addition, titanium hydride was used to alloy the maraging batches.

Super Invar The Super Invar alloy was investigated in order to extend the range of mechanical properties being investigated. This alloy is not considered for typical engineering structure applications. It is chiefly known for its low coefficient of thermal expansion over a temperature range of approximately 0-100°C.(3) However, the alloy possesses strengths much lower than the maraging steel, but with a much higher elongation. This allows processing effects of this form of chemo-powder metallurgy to be better validated. The choice of this alloy also fortuitously does not deviate significantly from the composition of the maraging steels. This effectively means that paste rheology and the impurities from raw materials will be common to both systems. The composition and mechanical properties of wrought Super Invar can be found in Table 1.

Table 1: Composition and mechanical properties of wrought alloys studied.

	Composition (wt%)					
	Fe	Ni	Co	Mo	Ti	Al
M200*	69.9	18	8.5	3.3	0.2	0.1
M350*	63.2	18	12.5	4.8	1.4	0.1
Super Invar**	62.89	31.75	5.36	--	--	--

	Mechanical Properties			* - See reference (1)
	UTS (MPa)	YTS (MPa)	El (%)	** - See reference (2)
M200*	1500	1400	10	
M350*	2450	2400	6	
Super Invar**	483	276	40	

Focus of Study

The focus of this study was to characterize the properties of the bulk material produced by oxide paste extrusion and reduction. This data is necessary to sufficiently model the behavior of the honeycomb structures. Mechanical properties, densities, and microstructural details are presented in this paper.

Experimental

The fabrication of metal honeycomb is a multi-stage process that includes batching, compounding, extruding, drying, reduction, and, optionally, heat treatment. All these steps are common regardless of the alloy system involved.

Processing

A general description of the extrusion process is given below. A detailed study of the factors involved in paste development and extrudability is presented by Hurysz. (4)

<u>Raw Materials</u> The raw materials used in the production of the metal honeycombs were chiefly metal oxide powders with average particle sizes below 10 μm. Molybdenum metal powder and titanium hydride powder were exceptions. Specifically, titanium oxide is not readily reducible and must be incorporated into the alloy by alternative means. Molybdenum metal powder, on the other hand, was used because the acicular morphology of the molybdenum oxide particles was a complicating factor in the rheological behavior of the pastes.

Paste Batching and Compounding The pastes were prepared through conventional means. The precursor powders were dry blended with a methylcellulose binder in the appropriate proportions. The proportion of oxide powders was determined by the metal yield upon reduction. The rheological behavior of the pastes was controlled by the amount of liquid phase present. This included the binder, water, and lubricant. The proportions were chosen to give the best possible extrusion. The compositions of the M200, M350, and Super Invar powder processed alloys, and the raw materials used are presented in Table 2. After all these components were mixed, the batch was processed through a Buss kneader to ensure paste homogeneity.

Table 2: Composition of powder-processed alloys

	Composition (wt%)				
	Fe	Ni	Co	Mo	Ti
M200	69.9	18	8.5	3.3	0.2
M350*	63.2	18	12.5	4.8	1.4
Super Invar	62.89	31.75	5.36	--	--
Raw Materials	Fe_2O_3	NiO	Co_3O_4	Mo metal	TiH_2
* - Composition w/o Ti had Fe balance adjusted					

Extrusion The homogeneous pastes were extruded through a hydraulic ram extruder. For the purpose of this study no honeycomb dies were utilized. Instead, a flat strip was extruded and collected for testing.

Drying The paste was dried slowly in a constant humidity chamber to facilitate drying. This helped to uniformly remove a significant amount of water without cracking and allowed the extrusion to be handled with much less difficulty. The extrusions were then placed into a convection dryer to complete the water removal process.

Reduction The heating schedules for the alloys involved in the study were identical. The dried extrusions were heated in the presence of a hydrogen/argon atmosphere to 1350°C, soaked for 10-12 hrs., and allowed to cool in the furnace. The majority of the reduction process is complete around 700°C, and a high temperature soak is required for maximum densification.

Heat Treating The maraging steel alloys require a heat treatment to reach their optimal mechanical properties via a precipitation event. The alloys were heated to 480°C and soaked for 5 hours to attain peak hardness.(5) The Super Invar composition was tested as reduced at room temperature and after a soak at liquid nitrogen temperatures. As shown below, for this composition, the martensite start temperature is apparently near –50°C and hold at -195°C completes about 80% of the transformation.

Characterization

Characterization of the powder-processed alloys involved density measurements, mechanical testing, and optical and scanning electron microscopy. Thermal expansion data were also collected on the Super Invar alloy for comparison to published data. The data are presented in the subsequent section.

Density Measurements The Archimedes method was used to determine the apparent density of the extruded metal strips.

Mechanical Testing Mechanical testing was performed on the metal strips to determine yield strength and elongation. This was accomplished through tensile tests of strips that were EDM machined to a dog bone profile. A representative specimen is shown in Figure 2 with approximate dimensions based on the shrinkage of a maraging

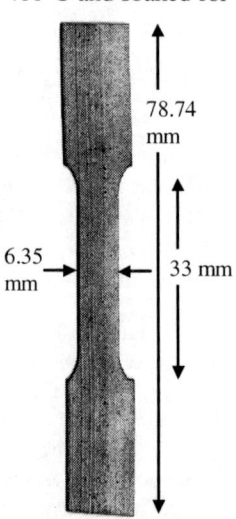

Figure 2: Tensile test specimen, thickness ~0.5mm

steel alloy. Specimens were tested on a screw type test frame with a 1000 lb load cell. Crosshead displacement and extensometer data were used to calculate strain.

Microscopy   Optical and scanning electron microscopy were used to investigate features of polished surfaces as well as several fracture surfaces. Polished samples were etched with a nital solution to reveal the microstructure.

Results and Discussion

Maraging Alloys

The reduction of the powder-processed alloys requires that the samples be processed in a hydrogen atmosphere at high temperatures. Due to the furnace mass, quenching the samples from the soak temperature was not practical. However, due the sluggish kinetics of the Fe-Ni austenite transformation, furnace cooling is sufficient to yield the necessary martensite structure (Fig. 3). Typical apparent densities measured on the test specimens averaged 95% of theoretical density. The source of the porosity was due to the inherent porosity associated with powder packing and sintering, and to the presence of extrusion defects. The homogeneity of the polished specimens was confirmed using the x-ray mapping capabilities of the SEM.

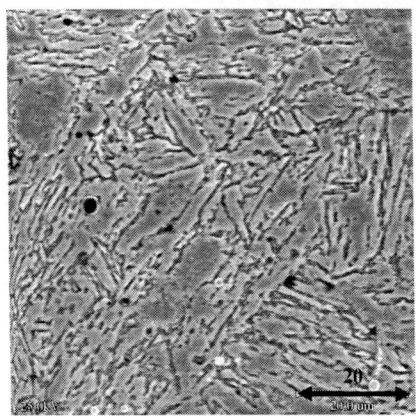

Figure 3: 200 grade maraging steel etched with nital to show martensitic structure.

Table 3: Properties of powder processed alloys.

	Mechanical Properties			
	UTS (MPa)	YTS (MPa)	Elongation (%)	5% porosity UTS prediction*
M200	1195	1110	2	1210
M350	1675	1530	3	1975
Super Invar	415	285	25	389
Super Invar/N$_2$**	600	450	7	--
*Values based on equation (1) and wrought UTS values from Table 1				
**Superinvar quenched in liquid nitrogen.				

Mechanical Data   The mechanical data for the fully alloyed specimens is tabulated in Table 3. The first composition to be tested was a M350 alloy without the titanium addition. The stress-strain data for this alloy is shown in Figure 4. The yield strength data corresponds reasonably well with extrapolated data from the ASM Handbook, Vol 1.(6) The figure suggests that the yield strength should double with the addition of titanium at the levels given by the wrought composition of the same grade. Subsequent M350 compositions were alloyed with titanium through the use of titanium hydride. It was hoped that the hydride would decompose to titanium and not oxidize during the reduction process. Samples with the Ti addition were prepared and tested under the same conditions. The mechanical behavior can be compared in Figure 4. The data shows a significant increase the yield strength of the titanium-alloyed composition. However, the increase does not correspond to a doubling of the yield strength, nor does it reach the strengths given by the wrought alloy properties. The ultimate and yield strengths are approximately two-thirds the wrought value. Bocchini (7) reports a relationship for sintered materials in which the strength is a exponential function of the porosity,

$$\sigma = \sigma_o e^{-4.3P} \qquad (1)$$

where, σ is strength at a fractional porosity of P, σ_o at zero porosity, and 4.3 is an empirical constant that fits a variety of materials. Using this relationship, predicted strengths could be compared to experimental results (Table 3). The experimental results are reasonably close to strength values predicted for 5% porosity.

The M200 grade maraging steel was also prepared with titanium hydride and tested. The results are shown in Figure 5. The data shows the susceptibility of the mechanical behavior to extrusion defects owing to failure prior to yield. The successful test yields strengths very close to those predicted for 5% porosity.

While the strengths for the maraging steel grades are promising, the percent elongations at failure appear to suffer greatly from the presence of porosity. The maximum elongation for all M350 specimens tested did not exceed 3%. It was hoped that investigation of the fracture surface would provide clues to the low tensile elongation values.

Fracture Surface Scanning electron microscopy was used to investigate the details of the fracture surfaces. The fracture surface of an M200 specimen is shown in Figure 6. The surface was representative of the M350 surface as well. Features of note in this micrograph are the dimpled surface, and the presence of large pores and second phase particles. The dimpled surface suggests that the specimen failed in a ductile manner. Intergranular failure was not observed in any of the maraging steel specimens. The large pores are likely the result of air trapped in the paste during extrusion. These pores cannot be removed by sintering and effectively reduce the loading area causing the specimen to fail at lower loads. The second phase particles are, in this study, common to both the maraging alloys and the Super Invar alloy processed in this study. The particles are associated with dimples and range up to several microns in diameter. Energy Dispersive Spectroscopy (EDS) analysis of the larger particles indicates concentrations of silicon, titanium, and aluminum. Since titanium was an intended alloying element in the maraging alloys, its presence is justifiable. The presence of silicon and aluminum is linked to impurities in the raw materials. The supplier's assay of the hematite powder lists several impurities including silica, alumina, and titania. Based on the EDS information and the powders chemical makeup, it can be assumed that the particles are an oxide complex of several raw material impurities. The particles appear to play a part in the failure of the specimens since they appear throughout the fracture surface. German and Smugeresky report Ti-rich precipitates segregating at prior particle boundaries to be responsible for the loss of ductility in HIP samples of 250 grade maraging steel.(8) The size of these precipitates was in the 0.5-1.0 μm range and

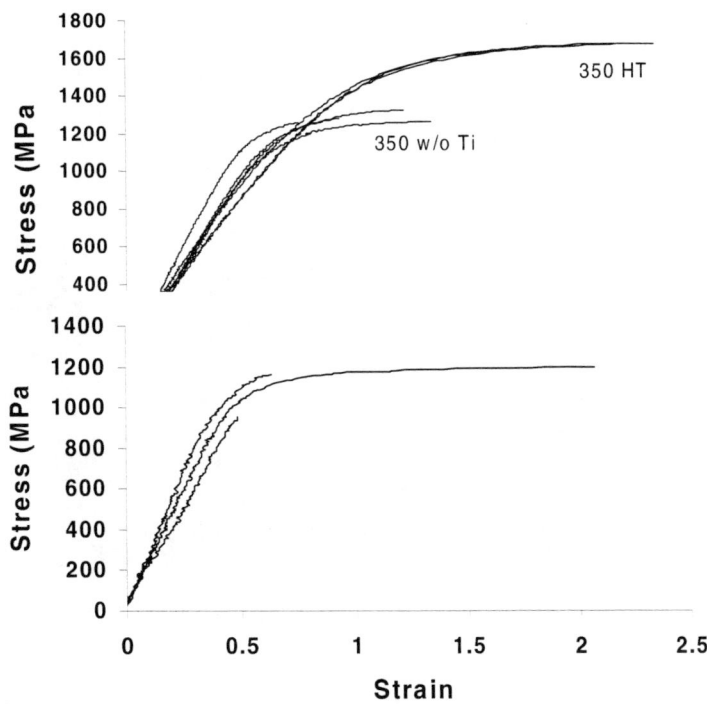

Figure 5: Mechanical behavior of M200 powder-processed alloy.

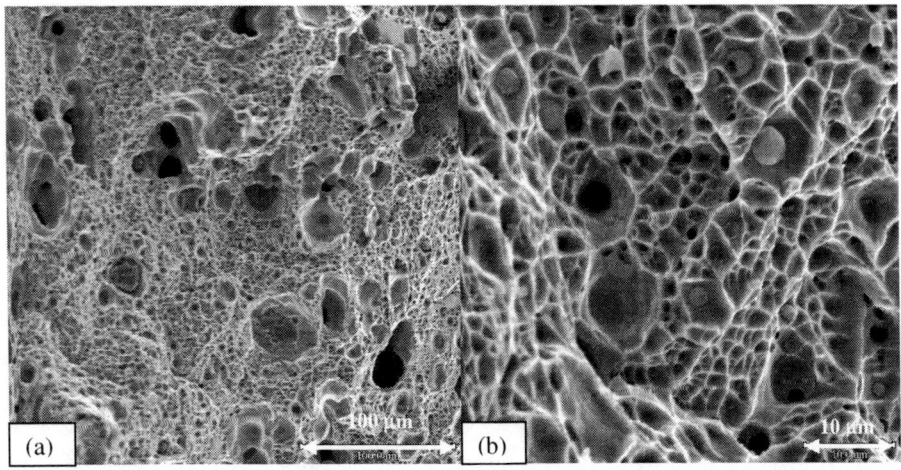

Figure 6: Fracture surface of M200 tensile specimen at two magnifications.

they were also associated with dimples. To what degree the larger particles found in this study degrade mechanical performance beyond the effects of porosity alone is unknown.

Super Invar Alloy

The Super Invar alloy was processed in a similar manner to the maraging alloys and the resultant 95% average apparent density was expected. Similar low-expansion Fe-Ni alloys have been processed by more traditional powder metallurgy means to yield similar densities. Thomas and Jones report 5% residual porosity for Fe-Ni powder metal alloys compacted to 30 tons/in^2 and heated to 1350°C for 3 hrs.(9) The Super Invar microstructure after processing is shown in Figure 7a. The pores range in size from 1-10 µm. The larger pores associated with extrusion defects do not appear in this micrograph.

Figure 7: Super Invar, (a) showing as processed structure, and (b) showing martensite after liquid nitrogen quench, nital etch.

Mechanical Data The tensile data for the Super Invar alloy is shown in Figure 8. The data is tabulated in Table 3. Mechanical properties appear to meet published data. However, this may be due to dissimilar heat treatments, i.e. as reduced rather than annealed.

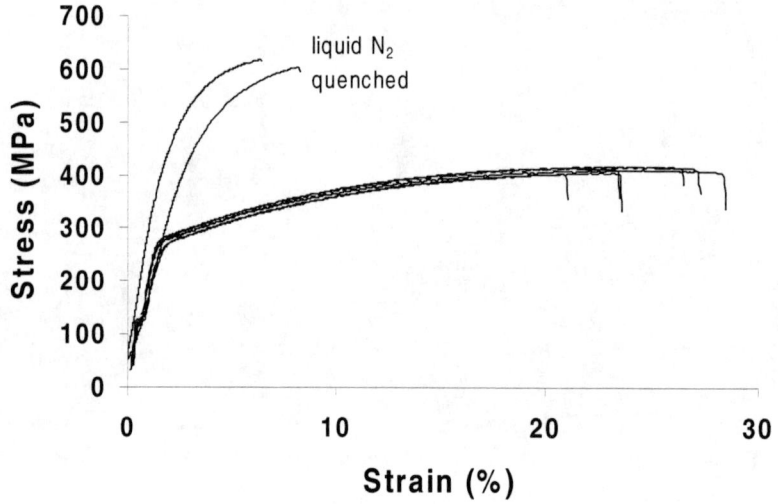

Figure 8: Mechanical behavior of powder processed Super Invar.

Fracture Surface The micrograph in Figure 9 shows a representative fracture surface of the Super Invar alloy. Again dimples, large pores, and second phase particles characterize the fracture surface. The dimples suggest ductile failure and the large pores are associated with extrusion defects. The second phase particles were discussed previously with respect to the maraging steels. Their presence in the Super Invar composition, which does not require titanium additions, further amplifies the conclusion that impurities in the raw materials are playing a role in the fracture of these samples.

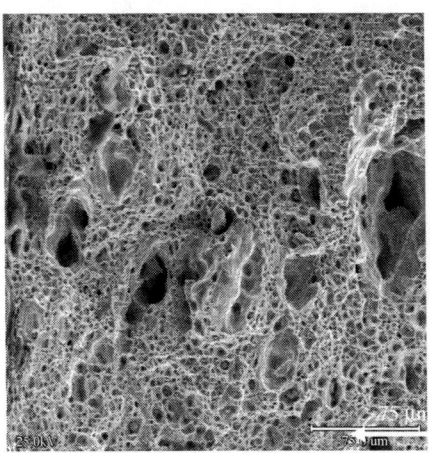

Figure 9: Fracture surface of as-reduced Super Invar tensile specimen.

Thermal Expansion: The invar type alloys have been investigated chiefly for their low expansion properties. Consequently, the study of the powder-processed Super Invar included a thermal expansion analysis. The results are reported in Figure 10. The analysis shows two distinct expansion curves that have been attributed to a martensitic phase change occurring at sub-zero temperatures. The presence of martensite was confirmed using optical microscopy and tensile tests on liquid nitrogen quenched samples. The partial transformation of the austenite to martensite is depicted in Figure 7b. The volume change of the Super Invar sample at -100°C,

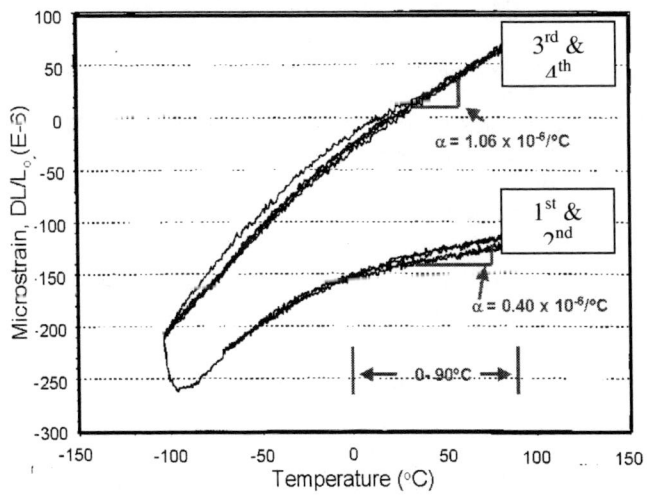

Figure 10. Thermal expansion of direct reduction super invar in the as-reduced state (cycles 1&2) and after cooling below −100°C to partially transform to martensite (cycles 3&4).

shown in the thermal analysis, suggests that the martensite start temperature for this alloy lies in a range between –50 and -100°C. Further, the partial martensite transformation at liquid nitrogen temperatures (-196°C) indicates that the martensite finish temperature lies below these temperatures. The mechanical behavior for the liquid nitrogen quenched Super Invar can be compared in Figure 8. The degree of martensite transformation during thermal analysis is unclear. However, the 0.40 x 10-6/°C value determined from the 1st and 2nd cycles, prior to the phase change, correspond well to published data (2).

Conclusions

Ferrous alloys are being produced using a powder processing technique that yields intricate shapes, near dense materials, and good mechanical properties. Strength and ductility of the maraging steel alloys show need for improvement. Investigations of fracture surfaces suggest that low ductility is due in part to the presence of second phase particles. These particles are largely the result of impurities in the raw material powders. Large pores associated with extrusion defects were also identified at the fracture surfaces. Powder-processed Super Invar shows good overall properties, but suffers the degrading effects of large pores and second phase impurity particles. Improvements in raw material purity and extrusion quality should lead to increased strength and ductility.

References

1. Joseph Davis et al., eds., Metals Handbook, vol. 1 (Materials Park, OH: ASM International, 1990) 793.
2. High Temp Metals, Inc., Product Literature, 12910 San Fernando Rd, Sylmar, CA 91342-3601.
3. Saito, H et al., eds., Physics and Applications of Invar Alloys, (Tokyo, Japan: Maruzen Company, Ltd., 1978), pp. 530-1.
4. K.M. Hursyz, "Paste Mechanics of Fine Extrusion" (Ph.D. thesis, Georgia Institute of Technology, 2001).
5. Raymond F. Decker, ed., Source Book on Maraging Steels (Metals Park, OH: American Society for Metals, 1979), 12.
6. V.D. Eisenhuttenleute, ed., Steel: A Handbook for Materials Research and Engineering, vol. 2 (Springer-Verlag, Dusseldorf, 1993) 216.
7. G.F. Bocchini, "The Influence of Porosity on the Characteristics of Sintered Materials," The International Journal of Powder Metallurgy, 22 (3) (1986), 185-202.
8. V. Thomas and D.J. Jones, "Low-Expansion Nickel-Iron Alloys Prepared by Powder Metallurgy," Symposium on Powder Metallurgy, 1954, 200-203.

REDUCTION OF IRON AND CHROMIUM SESQUIOXIDE POWDER MIXTURES FOR METAL HONEYCOMB STRUCTURES

J.H. Nadler, T.H. Sanders, Jr., and J.K. Cochran

School of Materials Science and Engineering
Georgia Institute of Technology
771 Ferst Drive, Atlanta, Georgia 30332-0245

Abstract

The reduction of Fe_2O_3 and Cr_2O_3 powder mixtures has been studied to understand the microstructural evolution of chromium-containing alloys for metal honeycomb structures. The effects of sintering, porosity and sample composition are discussed. All samples were reduced with hydrogen in an infrared gold image furnace at 1573K. None of the samples exhibited complete reduction of Cr_2O_3. However, the amount of reduced chromium increased with higher Cr_2O_3 compositions. The results, in agreement with previous work, suggest that gaseous diffusion is the most prevalent limitation to Cr_2O_3 reduction. Sample preparation was also observed to have an effect on the amount of Cr_2O_3 that will reduce. Sintering the pellets prior to reduction yielded higher chromium concentrations when compared to direct reduction in the powder compacts. Thermodynamics and kinetics of Cr_2O_3 reduction will be discussed in terms of free energy of formation of reduced solid solutions as well as the rate limitations of gaseous diffusion and interfacial chemical reactions.

Processing and Properties of Lightweight Cellular Metals and Structures
Edited by Amit Ghosh, Tom Sanders and Dennis Claar
TMS, 2002

Metallic honeycomb structures feature important properties such as light weight and comparable strength with respect to monolithic geometries, making them extremely useful in a wide range of applications. Some of these applications include load bearing structures, thermal management, sound absorption, buoyancy, and other areas where the beneficial properties of metals are needed to coincide with low-density structures.

A new process for making metal honeycomb has been developed at the Georgia Institute of Technology (1). Metal honeycomb fabrication begins with a plastic paste composed of ceramic powder, binders and lubricants, extruded through a die of the desired cross sectional geometry. The extruded shape is subsequently sintered to improve mechanical strength. These extruded ceramic honeycombs are then converted to metal using direct reduction in a hydrogen atmosphere.

Several metallic oxide mixtures are currently being investigated to determine their feasibility for directly reduced metal honeycombs. Among these alloys are compositions that reduce to stainless steels, nickel-based superalloys, maraging steels, and copper alloys. Many of these alloy systems contain iron and chromium. This combination has significant scientific and practical importance in terms of its oxides behavior during hydrogen reduction. Iron oxide (Fe_2O_3) reduces at a significantly lower temperature and higher rate when compared to chromium oxide (Cr_2O_3). In addition, previous work has shown that Cr_2O_3 is much more susceptible to reduction when the resulting metal is allowed to alloy with an iron metal matrix (1,9). The microstructures of these Cr_2O_3-containing mixtures are often plagued by porosity and unreduced particles of oxide, artifacts that can be deleterious to the desired mechanical properties of the target alloy.

When considering direct hydrogen reduction of a metal oxide for metal alloy honeycombs, there are three major considerations: the reducibility of the individual oxides, reduction rate, and the reduced structure-property relationships. Each factor affects the other in ways that are not always intuitive. Whether a single metal oxide is reduced under a given set of thermal and atmospheric conditions is determined by the thermodynamics of the reduction reaction as elucidated by the Ellingham diagram. However, for a mixture of metal oxides it is not sufficient to consider the reducibility of the individual components, nor is it necessarily appropriate to consider a homogeneous mixture of the initial constituents. As reduction does not occur instantaneously, the relative amounts of components as well as their thermodynamic stability with respect to each other will determine what *should* reduce and what *will* not.

The goal of this study is to contribute to the constitutive practices for evaluating and predicting the microstructural evolution of directly reduced iron-chromium alloys. The Fe_2O_3 - Cr_2O_3 system was chosen as the initial focus of this study for two reasons: 1) There exists a wide range of literature with which to compare observations and 2) There is extensive thermodynamic data for both the Fe_2O_3-Cr_2O_3 and Fe-Cr systems. Future work will use similar techniques with nickel-chromium alloys as a basis for the nickel-based superalloys Inconel 617 and 718.

Background

Reduction Models

Several models have been proposed for gas – solid reactions where the reactants and products both consist of a solid and a gas. Strangeway performed an exhaustive study on the kinetics of Fe_2O_3 reduction (3). The simplest treatment is the shrinking core model, which is based on a solid consisting of an unreacted core surrounded by a reacted layer, with sharp interface between the two regions. This interface is where the reaction takes place (4). There are three rate limiting factors in this model: gas phase mass transfer, diffusion of products and reactants through the reaction phase boundary, and chemical reactions occurring at the interface. The rates of pore diffusion can be calculated, but the reaction rate must be measured experimentally. In addition, these reactions rarely occur topochemically, that is, with a well defined interface. During the reduction of hematite, there is a transition zone separating the reacted and unreacted regions (3). In addition, gas - solid reaction rates are affected by structural factors such as grain size, porosity and the extent of sintering. These effects are included in the rate constant of a particular reaction, but do not appear explicitly in the model, so each model must be material-specific. Szekely and Evans developed a structural mathematical model in order to define specific physical criteria that affect the reactivity of porous solids (5). This model, however, was restrictive as well, as it assumes the original structure is maintained in spite of the reaction.

The Practicality of Direct Chromium Oxide Reduction

Provided that the equilibrium partial pressure of H_2O is not exceeded by oxygen from sources other than Cr_2O_3 and a very low metal yield is acceptable, it is possible to reduce Cr_2O_3 with hydrogen at elevated temperatures (2). Very small concentrations of H_2O exist in equilibrium with Cr_2O_3 in the presence of H_2. At increasing temperatures, the partial pressure of H_2O changes very slowly. In a U.S. patent, Rohn produced chromium by passing hydrogen over Cr_2O_3 at 1773K with a maximum flowrate of 50 l/min. In 24 hrs, this process yielded 50 –100 grams of chromium: 14 to 18% of the theoretically possible amount of chromium metal (6). In another U.S. patent, Alexander heated Cr_2O_3 in the presence of CaH_2, relying on the thermal decomposition of the hydride to reduce the Cr_2O_3 (7). There was, however, no evidence supporting the role of evolved hydrogen in reduction with respect to the calcium metal which was in contact with the Cr_2O_3. Newbury and colleagues used hydrogen at elevated temperatures and pressures using sodium to maintain a low H_2O partial pressure to reduce Cr_2O_3 (8). Other researchers have produced chromium metal by alloying Cr_2O_3 with iron metal powder (2). Cr_2O_3 powder was mixed with iron metal and reduced. Under these conditions, at 1200°C, complete reduction of the Cr_2O_3 was reported possible in a strongly flowing stream of hydrogen.

Iron Oxide – Chromium Oxide Reduction

Much research has been carried out on several aspects of the effect of chromium on the reduction of Fe_2O_3. Gurevich et al. (9) studied the hydrogen reduction of Cr_2O_3 – FeO solutions at several temperatures. To effectively reduce Cr_2O_3, it needs to be in contact with iron. An iron - chromium solid solution with compositions as high as 20 wt.% was detected under these conditions. In addition, if $FeCr_2O_4$ forms during reduction, the system will first reduce to Fe + Cr_2O_3, with some chromium passing into solution. At 1200°C, an Fe-Cr solid solution with a maximum 1.2 wt.% chromium content exists in equilibrium with $FeCr_2O_4$. Chinje and Jeffes performed several investigations of Fe_2O_3 - Cr_2O_3 reduction with compositions as high as Fe_2O_3 – 5 wt.% Cr_2O_3 (10). Over the stages of reduction, four phases were observed: $(Fe,Cr)_2O_3$ solid solution, Fe_3O_4 forming a stable spinel with Cr_2O_3, FeO with a small amount of dissolved Cr_2O_3, and Fe alloys with Cr. Forming mixed crystals reduces the activities of iron oxides (e.g. Fe_3O_4-Cr_2O_3 or FeO-Cr_2O_3, for example). As the wustite lattice dissolves Cr_2O_3, the diffusion coefficient is lowered, increasing the oxygen potential. This reduces the reaction rate for a given H_2/H_2O ratio.

Several authors have observed a decrease in total reduction accompanied by both increases and decreases in the overall reduction rate with respect to the stage of reduction (10,11). During the initial stages of Fe_2O_3 reduction, the reordering to Fe_3O_4 should be easiest when the Fe_2O_3 is pure since part of the ordering is due to electron transfer, which would be shorter in pure Fe_3O_4. In addition, iron ion diffusivities are higher in magnetite with respect to substitutional ferrites such as $FeCr_2O_4$. Changes in rate of reduction were also attributed to predominant reduction rate limitations such as interfacial reaction rate and mass transport. For example, the presence of iron metal can act as an intergranular diffusion barrier for the reducing gas. When this barrier is activated, the only way reduction can proceed is by oxygen diffusion through the metal matrix. (10) Kedr et al. observed a decrease in the degree of reduction with increased Cr_2O_3 additions (11). The hindering of fractional reduction was attributed to unreduced Cr_2O_3. More interesting was the effect of Cr_2O_3 on reduction rate. A decrease in reduction rate was observed up to 2.5 wt% Cr_2O_3, while an increase occurred up to 10 wt.%.

Experimental

Sample Preparation

The experiments were perfomed to determine the effects of processing and composition on Cr_2O_3 reduction. Oxide powders were mechanically mixed before being pressed into a rectangular die at 45kN. Approximate weight and dimensions of each pellet were respectively 4g and 0.25 x 0.65 x 0.65 cm. Sintered samples were placed on an alumina tray and placed in a bottom loading furnace. Sintering consisted of heating to 1300°C at 5°C/min. in air for 24 hours and cooling at 5°C/min to avoid cracking. Sample target compositions (in weight percent) were Fe-1.25Cr, Fe-2.50Cr, Fe-5.00Cr, Fe-10.00Cr, and Fe-25.00Cr. The three preparations of pressed pellets were investigated: Fe_2O_3- Cr_2O_3 powder, sintered Fe_2O_3- Cr_2O_3 powder, and Fe_2O_3 – Cr metal powder.

Reduction Apparatus

The sample to be reduced sits on a thin alumina plate, which in turn sits inside a fused silica tube. An alumina sheathed C-type thermocouple is in physical contact with one of the samples. To ensure that the emissivity of the alumina sheath was close to that of the samples, it was coated with an oxide powder of the same composition as the samples. The entire assembly is placed in an infrared tube furnace. A mixture of 20% hydrogen with an argon balance flowing at 1000 cc/min was used as the reducing agent.

Composition Evaluation

Samples were sectioned with an abrasive cutter and ground and polished flat using conventional metallurgical techniques. Compositions were evaluated using energy dispersive x-ray spectroscopy (EDS) on an Hitachi S-800 scanning electron microscope (SEM). The microscope accelerating voltage was 20 kV. Sample EDS data was attained by first magnifying the image to approximately 10kX, and scanning the metal matrix using the reduced area scan method. For each sample, scans were performed for a total of 150 seconds, with pauses and random matrix area selection every 30 seconds. As a result, each scan is actually an average of five metal matrix area scans. Sample weight loss was used as a means of confirming EDS results.

Density Calculations

Density was determined using the Archimedes method. Porosity was determined by comparing density measurements to the theoretical density of each sample. The theoretical density was calculated using the rule of mixtures with respect to composition. Density calculations for reduced samples included consideration for the fraction of any unreduced Cr_2O_3. Surface or open porosity was calculated using the weights of samples whose surfaces were saturated with water.

Results and Discussion

Simplified Thermodynamic Model

The purpose of thermodynamic modeling of the reduction process in this work is to determine how much chromium sesquioxide will reduce in the presence of Fe_2O_3. The simplified model presented here is an initial step in developing a more comprehensive understanding of the extent of Cr_2O_3 reduction. The general reaction that occurs during the reduction of a powder mixture of Fe_2O_3 and Cr_2O_3 is

$$xFe_2O_3 + (1-x)Cr_2O_3 + 3H_2 = 2[xFe,(1-x)Cr] + 3H_2O, \qquad (1)$$

where x is the mole fraction of Fe_2O_3. Preliminary calculations made the assumption that both Fe_2O_3 - Cr_2O_3 and Fe – Cr formed ideal solutions, obeying Rauolt's law. Further refinement was accomplished using published Fe – Cr activity data (12). Additional refinement can be done using an optimized Fe_2O_3 - Cr_2O_3 phase diagram to calculate activity. Several other reactions can occur before and during reduction. Before reduction starts, and as long as hematite is present, the powder compact sinters, forming a sesquioxide solid solution. It has been previously reported that during the reduction of iron and chromium sesquioxide, intermediate phases such as $FeCr_2O_4$, , FeO, and Fe_3O_4 occur in different stages of reduction (13). These interactions that may occur during these stages have not been considered in this simplified model. Instead, an assumption is made that iron and chromium exist either as sesquioxide or metal solid solutions.

Free energy of formation of the single oxides was calculated from available $C_p(T)$ data. By combining the free energies of formation of the sesquioxide solid solutions, the iron – chromium solid solutions and the oxidation of hydrogen into water vapor, the free energy of reduction with respect to composition can be calculated. It is assumed that the amount of chromium that reduces to form a pure metal is negligible. Rather, all reduced chromium immediately becomes part of the iron-chromium solid solution. The maximum amount of chromium that can dissolve into iron can now be determined by varying the p_{H_2}/p_{H_2O} ratio. Figure 1 is a plot of free energy of reduction for an iron – chromium solid solution at 1300°C. The two plots shown indicate two extreme solutions where no chromium can reduce into iron solution when p_{H_2}/p_{H_2O} values are 10^3 and higher while $p_{H_2}/p_{H_2O} = 10^8$ permits reduction over the entire range of compositions.

Iron Oxide – Chromium Oxide Reduction

All samples in the Fe_2O_3-Cr_2O_3 studies were reduced for 32 hours at 1300°C. Figure 2 exhibits two compositions of Fe_2O_3 -Cr_2O_3 pressed powder compacts that were reduced for 32 hours at 1300°C. It can be clearly seen here that in none of the cases did all of the Cr_2O_3 reduce to metal.

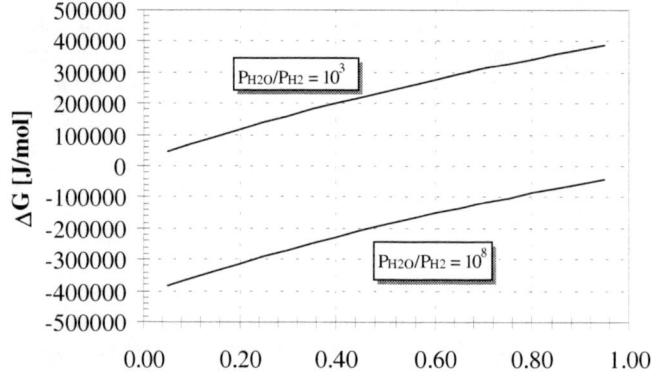

Figure 1: Free energy of formation of reduced Fe-Cr solid solutions with respect to composition at reduction temperatures of 1300°C.

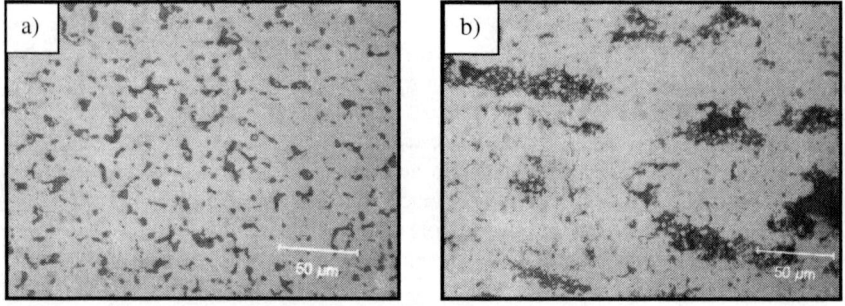

Figure 2: Optical micrographs of microstructures of reduced Fe_2O_3 -Cr_2O_3 powder compacts. Compositions: a) Fe-10Cr, b) Fe-25Cr.

Figure 3 demonstrates the fraction of chromium that entered the iron metal solution with respect to the total amount of chromium in the powder compact. Three treatments are considered: Fe_2O_3- Cr_2O_3 powder, sintered Fe_2O_3- Cr_2O_3 powder, and Fe_2O_3 – Cr metal powder. The greater reduction values associated with the sintered samples appear to be a function of the density, illustrated by the density – composition graphs in figure 4 and SEM micrographs in figure 5. The addition of Cr_2O_3 retards sintering and therefore increases the porosity of sintered samples (13). This porosity probably facilitates mass transport, which keeps the p_{H_2}/p_{H_2O} ratio high, allowing reduction to proceed at a higher rate when compared to the other samples. The dramatic drop in density is probably due to the volume changes associated with the hematite – magnetite step of reduction which caused extensive cracking in the sintered powder compact. Chromium metal powder additions exhibit similar behavior to Cr_2O_3 powder at higher concentrations.

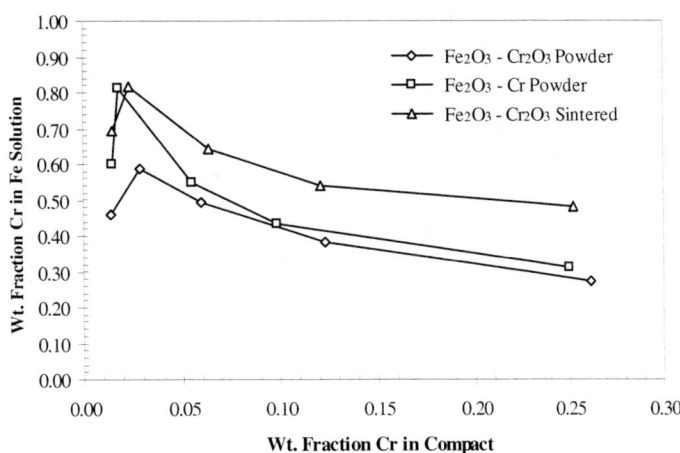

Figure 3: Fraction of chromium in iron solution versus total chromium in

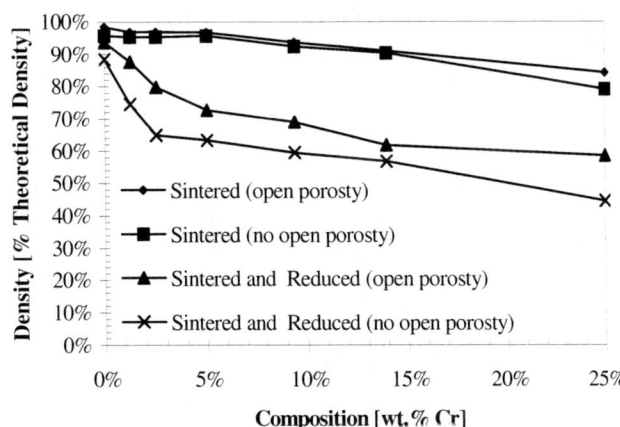

Figure 4: Density Fe_2O_3 -Cr_2O_3 powder compacts sintered in air and reduced. Density was measured after sintering and then again after reduction.

Due to the aforementioned thermodynamic barrier to reducing Cr_2O_3, one might assume that under all circumstances the less Cr_2O_3 added to the compact, the greater the fraction of chromium that will enter solution. If this were always the case, then the presence of iron metal is a thermodynamically sufficient condition for the reduction of a given amount of chromium. These ideas are consistent with the curves in figure 3 until the chromium concentration reached approximately 2.5 wt.%. At this point, it would appear that another barrier is present below this composition. Since only a small amount of H_2O is in equilibrium with chromium metal at a given temperature, the reduction process is extremely sensitive to mass transport. These results

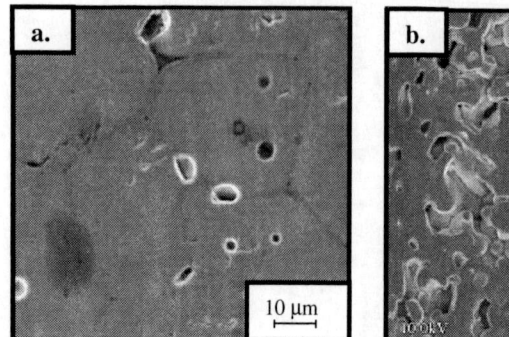 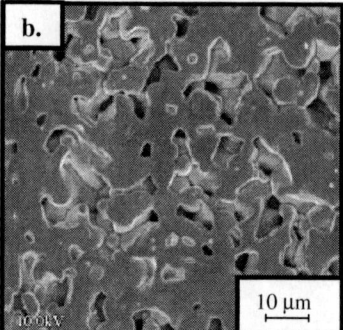

Figure 5: SEM micrographs of sintered Fe_2O_3-Cr_2O_3 microstructures with compositions: a)Fe and b) Fe-25Cr

are consistent with the work of Kedr who observed a reduction rate maximum at approximately 1 wt.% chromium (11). It is likely that for the reducing conditions in this experiment, concentrations of chromium below 1.5% encounter some kind of diffusion barrier, which prevent further reduction for the duration of the experiment. It is possible that the number density of the Cr_2O_3 particles is small enough and that Fe_2O_3 reduction proceeds so rapidly that it significantly prevents the gaseous diffusion of reducing gas to these particles, slowing Cr_2O_3 reduction.

Conclusions

The reduction of Fe_2O_3- Cr_2O_3 mixtures has been investigated. Thermodynamic calculations were used to determine the free energy of reduction associated with a given iron-chromium composition and H_2/H_2O partial pressure ratio. The maximum amount of fractional chromium reduction occurs near 2.5wt.% chromium, indicating that for the reduction conditions used, rate limitation though gaseous diffusion was activated. This result is in agreement with previous work which observed a rate maximum near 1 wt.% chromium. Sintering the oxide samples in air resulted in greater reduction, which appears to be directly related to the increased porosity associated with Cr_2O_3 additions.

Acknowledgements

This work is sponsored by DSO of DARPA (N00014-99-1-1016) under Dr. Leo Christodoulou and by ONR (N0014-99-1-0582) under Dr. Steven Fishman.

References

1. J.K. Cochran et al., "Low Density Monolithic Metal Honeycombs by Thermal Chemical Processing," <u>Fourth Conference on Aerospace Materials, Processes, and Environmental Technology</u>, in press.

2. C.G. Maier, "Sponge Chromium," US Bur. Mines Bull., 436 (1942).

3. P.K. Strangway, "The Kinetics of Iron Oxide Reduction,"(Ph.D. Thesis, University of Toronto, 1966).

4. J. Szekely, J.W. Evans, and H.Y. Sohn, Gas-Solid Reactions (New York, NY: Acedemic Press, 1976), 65-107.

5. J. Szekely and J.W. Evans, "Studies in Gas-Solid Reactions: Part I. A Structural Model for the Reaction of Porous Oxides with a Reducing Gas," Met. Trans., 2 (1971), 1691-1698.

6. W. Rohn, "Reducing Chromium with Hydrogen," U.S. Patent 1,915,243, (1933).

7. P.P. Alexander, "Reducing Refractory Oxides Such as Those of Chromium," U.S. Patent 2,038,402, (1936).

8. E. Newbery and J.N. Pring, "Reduction of Metallic Oxides with Hydrogen at High Pressures," Proc. Roy. Soc. London, 92 (1916), 276.

9. Y.G. Gurevich, I.D. Radomysel'skii, L.F. Barschevskaya, N.R. Frage, Y.I. Pozhidaev, "Thermodynamic Analysis of the Reduction of a Mixture of Iron and Chromium Oxides by Hydrogen," Sov. Powder Metall. Met. Ceram. 14 (1) (1975), 13-16.

10. U.F.Chinje and J.H.E.Jeffes, "Effects of chemical composition of iron oxides on their rates of reduction: Part 2: Effect of trivalent metal oxides on reduction of hematite to iron", Ironmaking and Steelmaking, 17 (5) (1990), 317-324.

11. M.H.Kedr, "Isothermal Reduction Kinetics of Fe_2O_3 Mixed with 1-10% Cr_2O_3 at 1173-1473K," ISIJ Int., 40 (4) (2000), 309-314.

12. R.B. Reese, R.A. Rapp, and G.R. St.Pierre, "The Chemical Activities of Iron and Chromium in Binary Fe-Cr Alloys," Trans. Met. Soc. AIME, 242 (1968) 1719-1726.

13. U.F.Chinje and J.H.E.Jeffes, "Effects of chemical composition of iron oxides on their rates of reduction: Part 1: Effect of trivalent metal oxides on reduction of hematite to lower iron oxides," Ironmaking and Steelmaking, 13 (2) (1989) 90-95.

COPPER ALLOYS FROM OXIDE REDUCTION FOR HIGH CONDUCTIVITY APPLICATIONS

Benjamin C. Church, Joe K. Cochran, and Thomas H. Sanders, Jr.
School of Materials Science and Engineering
Georgia Institute of Technology
771 Ferst Drive
Atlanta, GA 30332-0245

Abstract

Extrusion of oxide powders allows fabrication of thin-walled metal articles to produce controlled geometry, low-density copper alloy architectures. Shapes formed with copper oxide powders mixed with alloying oxides are reduced and sintered to produce high relative densities in the thin walls. This technology has produced square cell honeycomb extrusions, which are being characterized for heat exchange applications. This effort is to determine the bulk properties of alloys produced by this type of thermo-chemical powder processing and to explain behavior based on the final chemistry and microstructure of the alloys. Compositions investigated include Cu, Cu-Ni, and Cu-Ag alloys. Alloys have been characterized for relative density, porosity, pore size distribution, and grain size. Mechanical properties, tensile strength and elongation, and thermal conductivity were measured and results were analyzed based on porosity and chemistry of the alloys. Properties were compared to alloys from conventional processing and powder metallurgy.

Processing and Properties of Lightweight Cellular Metals and Structures
Edited by Amit Ghosh, Tom Sanders and Dennis Claar
TMS, 2002

Introduction

A technology developed in MSE at Georgia Tech allows fabrication of thin-walled metal articles. Using conventional powder extrusion, honeycomb shapes of a variety of cell structures are formed with oxide powders and subsequently converted to the metal state by direct reduction. By using fine powders in the micron and sub-micron range, articles that have small dimensions (i.e., micro- to milli-meter) may be formed. Practical experience has shown that maximum particle sizes of 1/50 of the feature dimension are necessary for producing features with good precision. After reduction, sintering consolidates the honeycomb to dense wall, lightweight geometries with structural integrity and substantially the original shape. Demonstrated alloys include stainless steel, maraging steel, Inconel 617 and 718, Super Invar, and copper alloys [1].

Extruded square cell honeycomb in copper-1wt% nickel is presented in Figure 1. This thin-walled honeycomb, at ~20% relative density, RD, shows the geometrical precision available for property design. Shapes are formed by the extrusion of pastes comprised of metal oxides at room temperature. After forming, the articles are converted to metal by means of hydrogen reduction. The thin walls of the honeycomb allow for the transport of water vapor out of the material during reduction. Using this method, it is possible to create cell walls that are roughly 30 µm in thickness after reduction. Depending on cell size and geometry, the extrusions can approach 8% RD.

The combination of a high conductivity alloy with the linear cellular structure of the honeycomb makes these types of structures applicable to systems such as heat sinks for next generation CPUs. With the continuing advance of processor speeds, the need for higher performance heat dissipation systems has grown. To serve this need, square cell honeycomb copper and copper alloys are being characterized as heat sinks. Copper honeycomb is lightweight (<20-25% RD) with 10:1 cell size/wall thickness ratio, as shown in Figure 1. This translates to high heat-transfer-surface-area to volume ratio of 2400 m^2/m^3, compared to 30-150 m^2/m^3 for shell-and-tubes, and 400-600 m^2/m^3 for folded fin units. Calculations for Cu honeycomb at 25% RD (Figure 1) indicates low-pressure drop (51.2 Pascal) and high heat flux (542.6 W/cm^2) using water as the medium and assuming 400 W/mK for copper conductivity. When air is used as the cooling fluid, the heat flux removal is approximately 25 W/cm^2. This compares well to the best heat sinks available for current CPUs. High efficiency heat sinks appear feasible for LCAs in electronic areas due to low-pressure drops and thin walls.

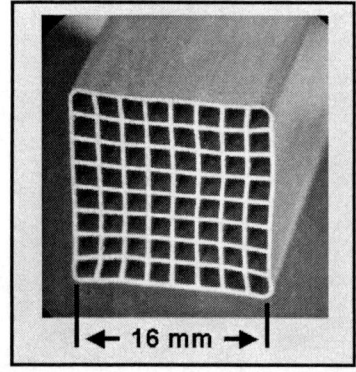

Figure 1: Copper square cell honeycomb for heat sink applications; honeycomb density = 2.0 g/cc, Cell size = 2.0mm.

The object of this paper is to explore the physical properties of copper and copper alloys reduced and sintered in hydrogen to determine if they meet mechanical and thermal requirements for the next generation of forced air heat sinks.

Experimental Procedure

Metal Oxide Constituents

Copper oxide (Cu_2O) (-325 mesh) was obtained from American Chemet Corporation. Nickel oxide (NiO) was obtained from Ceramic Color and had an average particle size of 6.5 µm. Silver oxide (Ag_2O) was obtained from Alfa Aesar with a reported particle size of –325 mesh.

Pellet and Extrusion Sample Preparation

Powders were prepared for compaction into pellets by first weighing the proper amounts of constituent metal oxides. A solution of 10wt% polyvinylbuterol (PVB) dissolved in methanol and acetone was added to the powder to act as a binder. Pellet compaction was performed using a standard twin piston steel die with a 1.905 cm diameter and a hydraulic press at a load of 44,600 N. Stearic acid dissolved in acetone was used to lubricate the die wall.

The process used to produce extrusions began with blending the proper amounts of constituent metal oxide powders with Methocel A4M methocellulose powder from Dow Chemical in a commercially available food processor. A solution of water and pegosperse, a lubricant, were then added to the powder mixture in the processor in order to produce a granulated mixture. The hydrated powder was then processed through a Buss kneader type compounder to form a highly sheared, homogeneous paste. The paste was then loaded into a 5.08 cm diameter Loomis extruder and extruded at various speeds and pressures depending on paste properties and the geometry of the extrusion die being used. Samples were then placed in a humidity controlled drying oven at a slightly elevated temperature (~40 °C) for approximately 48 hours. The drying process produces a rigid structure that can be cut to length and handled easily.

Several dies have been designed to produce extrusions of various geometries. A flat strip die was designed for the characterization of the bulk materials. This die produces a flat strip from which tensile specimens can be machined for mechanical testing and metallographic analysis using either optical, scanning electron microscopy (SEM), or transmission electron microscopy (TEM) techniques. Honeycomb dies were designed to create linear cellular geometries of square, triangular, and circular cells.

Reduction and Sintering

Pellets and extruded honeycomb samples were reduced using the same procedure. Furnace atmosphere was comprised of ultra high purity hydrogen for all samples. Sintering temperatures depended on the chemistry of the system. Copper, Cu-Ni, and dilute Cu-Ag samples were sintered at 1050 °C. Higher silver containing Cu-Ag samples were sintered at 750 °C due to the eutectic temperature of that system.

Characterization

Density measurements were made using the Archimedes method. Samples for microstructural analysis were mounted in Bakelite and prepared using standard metallographic techniques. Etching of all copper base samples was with a mixture of one part 3% H_2O_2, one part ammonium hydroxide, and one part water. The etchant was prepared fresh prior to use. Etching times varied with the composition of each sample. Optical analysis was carried out on a Leica BM IMR microscope. Scanning electron microscopy and energy dispersive spectrometry were utilized for the analysis of microstructure and fracture surfaces. SEM analysis was performed on a Hitachi S-800 SEM operating at voltages in the range of 5 to 25 kV.

Conductivity calculations were made from thermal diffusivity data taken from pellet samples. Thermal diffusivity was measured using a pulsed laser technique at Oak Ridge National Laboratory. The thermal diffusivity (D_{TH}) was used to calculate thermal conductivity (K) by the following formula:

$$D_{TH} = K / (C_P \times \rho) \qquad (1)$$

where C_P is the heat capacity of the material at constant pressure and ρ is the measured density of the pellet. Pellets were sanded to 320 grit and sprayed with a graphite powder prior to testing. The graphite was used to ensure the complete absorption of the laser pulse by the otherwise reflective metal samples

Tensile bars were machined by an electrical discharge machining technique starting with as- reduced strip to produce a standard dog-bone shape. Tensile testing was done on a Satec screw driven frame at a strain rate of 0.1 inches per minute. Tensile bar gage area cross sections were 0.6 mm (typical thickness of reduced strip) by 6.4 mm with a gage length of 50 mm. Strain was measured dynamically with an extensometer.

Results and Discussion

Copper-Nickel

The copper nickel system is well suited for application to the metal oxide paste extrusion/ reduction process described earlier. Both constituent oxides are easily reduced using hydrogen gas as can be seen in an Ellingham diagram [2]. In addition, the elements form a complete solid solution across the entire composition range facilitating the analysis of various compositions. Additions of nickel to copper result in increased strength and corrosion resistance but decrease conductivity. As the current study places a focus on high thermal conductivity, the nickel content of the samples tested was restricted to a maximum of 8 wt% nickel.

Post reduction density data for copper and Cu-Ni samples, shown in Figure 2, indicate that

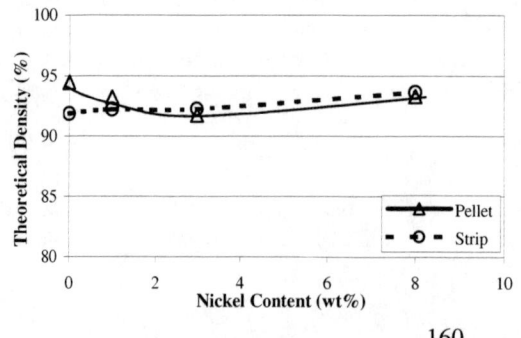

Figure 2: Post reduction density of Cu-Ni pellets and strip made from reduced metal oxides.

both pressed oxide pellets and extruded oxide pastes sinter to a similar degree. In addition, the microstructures of pellets and strip, as shown in Figure 3, are also of a similar nature. Grain size after reduction was approximately 25 µm and did not vary significantly with nickel content or between pellet, strip, and honeycomb extrusion. As shown in the micrographs in Figure 3, the post reduction structure is homogeneous and pores tend to be widely dispersed. However, additional testing must be performed to determine the extent of any density gradients in the pellets and strip. Such gradients have been shown to have an effect on properties as a function of location within a pressed pellet [3].

Bocchini [4] states that the conductivity of a sintered P/M material can be expected to fall into a particular range depending on the porosity level and the morphology of the pores. As seen in Figure 4, the data for Cu and Cu-Ni pellets are generally within this predicted range. Relative conductivity as shown in Figure 4 was calculated using published data [5] for the Cu-Ni system.

It is well known that the type of solute element present in copper has a strong effect on the conductivity of the material. Some elements, such as iron, phosphorous and silicon, are more detrimental than others and can have a significant effect on conductivity even in parts per million concentrations. Certain impurities, such as silicon, would likely be present in the reduced material as an oxide since the reduction of SiO_2 by hydrogen is not spontaneous. Thus, the impact of such non-reducable impurities would be small since the atoms would in effect be prevented from entering the copper matrix. Iron oxide, on the other hand, would reduce in a hydrogen atmosphere and would therefore have a significant effect on conductivity. Another potential impurity would be oxygen.

Since this type of impurity would likely take the form of an isolated metal oxide particle as opposed to a solute addition, it would be expected that any remaining oxygen would have a negligible effect on conductivity. To date, unreduced metal oxides have not been observed during SEM or optical analysis. Impurities have the potential to be far more detrimental than

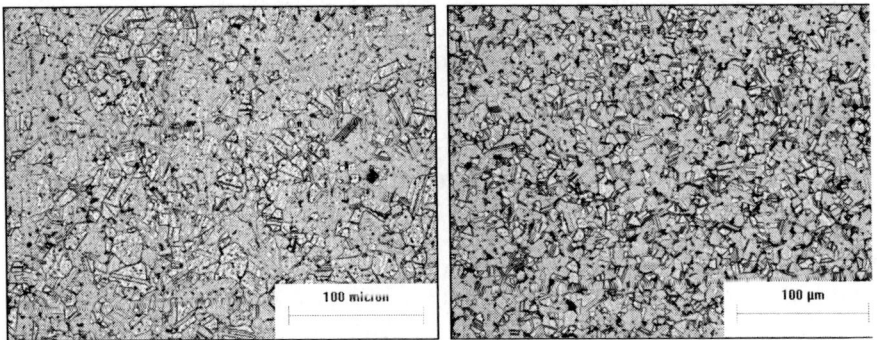

Figure 3: Optical micrographs of copper 3 wt% nickel; pellet (left) and extruded strip.

porosity as is illustrated in Figure 5. An iron impurity level of approximately 100 ppm has a similar effect on conductivity as a pure matrix with roughly eight percent porosity. Future work will look to measure the actual amount of impurities present in the as-reduced samples. It has been shown that the type of binder/lubricant

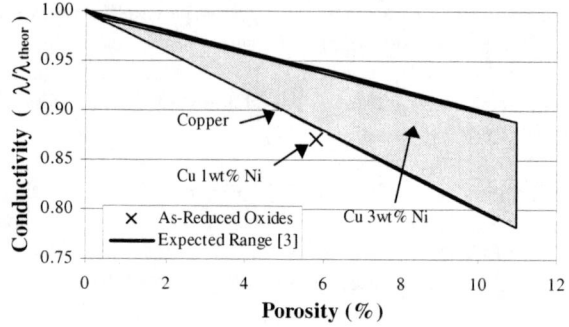

Figure 4: Effect of porosity on thermal conductivity of Cu-Ni

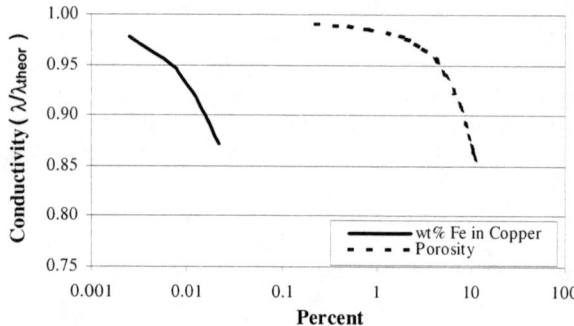

Figure 5: Relative effect of iron impurity and porosity on the conductivity of a copper material.

used in P/M copper materials has an effect on post-processing conductivity [3, 6]. The binder used for pellet samples was PVB while Methocel was used for extrusions. The effect of the binder type on post-reduction properties of strip and pellets has not been clearly identified though it is believed that the binder effect may be less pronounced than that for traditional P/M processes due to the presences of oxygen and water vapor produced during the reduction of the metal oxides. In addition, the initial structure is very porous which facilitates transport of gas in and out of the structure. The nature of this process also allows for samples to be fired in air prior to the reduction step. A pre-reduction firing would allow for the removal of any organics present such as the methocel binder.

Mechanical properties of extruded strip are shown in Table 1. It is well known that mechanical properties of powder metal materials have a large dependence on sintered density / porosity level [4,7]. The elongation and tensile strength of the pure copper extruded strip falls below expected values [6] for pure copper P/M materials of similar porosity levels. Since the Cu-Ni samples are not standard alloy compositions, comparison between the reduced oxide strip and published values is difficult. One interesting trend in the mechanical property results is the tendency of small alloying additions of nickel having a noticeable improvement in both strength and elongation. Small alloy additions may introduce additional driving force for sintering which then results in a more dense structure and hence better mechanical properties than pure material without any alloy additions. The plot of density versus nickel content shown in Figure 2 does in fact show a slight increase in relative density with increasing nickel content for extruded strip. A fracture surface from a Cu 3 Ni tensile specimen is shown for reference in Figure 6.

Table 1: Mechanical properties of extruded Cu, Cu-Ni, and Cu-Ag strip made from reduced metal oxides.

Sample	Theoretical Density (%)	Yield Strength (MPa)	Tensile Strength (MPa)	Elongation (%)
Copper	91.9	77	174	8.9
Cu 1 Ni	92.2	112	234	15.5
Cu 3 Ni	92.2	108	193	9.1
Cu 8 Ni	93.6	115	242	12.8
Cu 1 Ag	93.0	85	184	9.2

Copper-Silver

Silver is used as an alloying element for high conductivity copper alloys to improve resistance to softening at elevated temperatures [7]. Silver is an attractive alloying element in that it has a very small impact on conductivity both when in solution in copper and when present as a second phase in a copper matrix. A plot showing the electrical conductivity of the copper-silver and copper-nickel systems is shown in Figure 7.

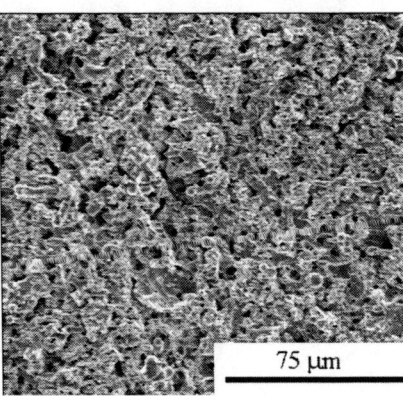

Figure 6: SEM micrograph showing the fracture surface from a Cu 3 wt% Ni extruded strip tensile specimen.

Pellets made from metal oxides of copper and silver were pressed and reduced at different temperatures. Samples containing more than 1 wt% silver were not reduced at high (1050 °C) temperatures in order to avoid the formation of a liquid phase during the sintering process. As shown in Figure 8, the post reduction densities of samples reduced at 750 °C were well below those of samples reduced at higher temperatures. Cu 1 Ag pellet samples reduced at 1050 °C were on average less dense than pure copper samples though they were still comparable to other systems tested.

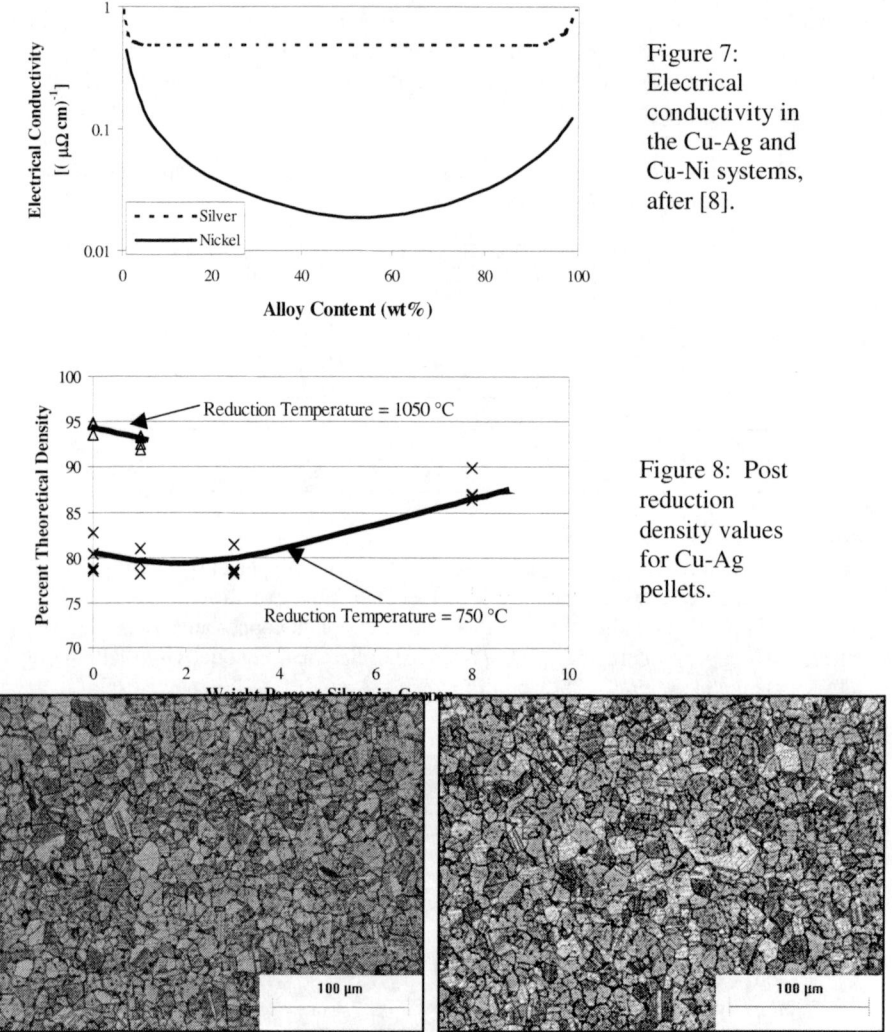

Figure 7: Electrical conductivity in the Cu-Ag and Cu-Ni systems, after [8].

Figure 8: Post reduction density values for Cu-Ag pellets.

Figure 9: Optical micrographs of Cu 1 wt% Ag pellet (left) and extruded strip.

Optical micrographs of Cu 1 Ag pellets and strip reduced at 1050 °C, Figure 9, show that the as-reduced structures of the different sample types are quite similar. Grain size for the different samples are on the order of 20 μm.

Mechanical property results given in Table 1 indicate that the Cu-Ag material may be slightly stronger than pure copper strip produced from reduced oxides. This is likely due to the higher density of the Cu 1 Ag strip relative to the pure copper. The difference in strength is slight and is not be statistically significant. Additional testing would be required to fully define any potential room temperature strengthening of the material due to the silver addition.

As with the Cu-Ni samples, conductivity values for the Cu-Ag pellets agree with the predicted range for porous materials as shown in Figure 10. Tabulated data for the thermal conductivity of various Cu-Ag alloys was not available as in the case of the Cu-Ni system. Therefore, the plot of relative conductivity of the Cu-Ag pellets was created using a theoretical conductivity of 400 W/mK for all samples. It can be seen in Figure 7 that the conductivity of Cu 1 Ag would be lower than pure copper. Thus, the data shown in Figure 10 would be shifted slightly toward higher relative conductivity values if more precise reference data for Cu 1 Ag were available.

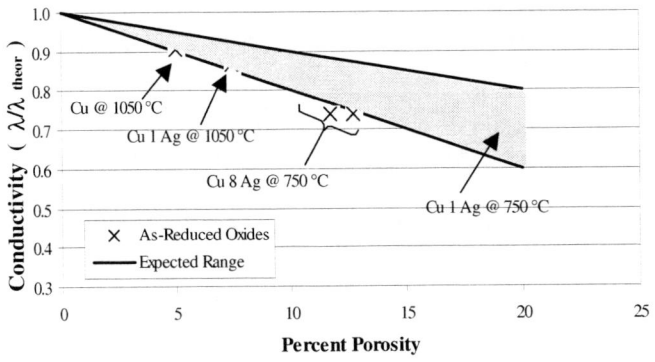

Figure 10: Relative conductivity of Cu-Ag pellets reduced at 1050 and 750 °C.

The use of silver as an alloying addition to be used in the honeycomb extrusion process appears to be viable for low silver additions only. As silver content increases, porosity becomes a significant problem due to the lower sintering temperatures required at those compositions.

Conclusions

Metal oxide powders can be used to form metallic structures with bulk properties similar to traditional P/M materials. Extrusion of a paste comprised of metal oxides and other constituents can be utilized to fabricate thin wall honeycomb type structures. After reduction and sintering, these honeycomb structures have a high relative density in the cell walls. The bulk material displays thermal and mechanical properties similar to traditional P/M materials of similar porosity levels. These metallic honeycombs are well suited for applications demanding high performance per unit mass.

Acknowledgments

This work is sponsored by DSO of DARPA (N00014-99-1-1016) under Dr. Leo Christodoulou and by ONR (N0014-99-1-0852) under Dr. Steven Fishman.

References

[1] J.K. Cochran et al., "Low Density Monolithic Metal Honeycombs by Thermal Chemical Processing," Fourth Conference on Aerospace Materials, Processes, and Environmental Technology, *in press*.
[2] David R. Gaskell, Introduction to Metallurgical Thermodynamics, Second Edition (Hemisphere Publishing Corporation, 1981), 287
[3] M. Dombroski et al., "Electrical Conductivity Gradients in Press and Sinter Copper," 1987 Annual Powder Metallurgy Conference Proceedings, 43 (1987), 303-322.
[4] G.F. Bocchini, "The Influence of Porosity on the Characteristics of Sintered Materials," International Journal of Powder Metallurgy, 22 (3) (1986), 185-202.
[5] C.Y. Ho et al., "Thermal Conductivity of Binary Alloy Systems," Journal of Physical and Chemical Reference Data, 7 (1978), 959-1177.
[6] P.W. Taubenblat, ed., New Perspectives in Powder Metallurgy: Vol. 7, Copper Base Powder Metallurgy, (Princeton, NJ: Metal Powder Industries Federation, 1980).
[7] J.R. Davis, ed., Metals Handbook, Desk Edition, Second Addition (Materials Park, OH: ASM International, 1998).
[8] K. Schroder, ed., CRC Handbook of Electrical Resistivities of Binary Metallic Alloys, (Boca Raton, FL: CRC Press, 1983) 50-52, 215, 216, 224-228.

MODELING POWDER EXTRUSION PASTES FOR FORMING LIGHT WEIGHT MULTIFUNCTIONAL STRUCTURES

K. M. Hurysz, R. Oh, J. K. Cochran, T. H. Sanders, Jr., and K. J. Lee

School of Materials Science and Engineering
Georgia Institute of Technology
771 Ferst Drive
Atlanta, GA 30332-0245

Abstract

The extrusion of pastes of oxide powder mixtures through a thin-wall honeycomb die results in a low-density metallic structure following reduction. The pastes are a combination of two phases: a solid phase composed of the particular oxide or oxide mixture carried by a fluid solution of water, binder, and additives. The key to forming high quality, defect free extrudate lies in the optimization of paste properties and is contingent on solids loading and fluid-phase rheology. To extrude efficiently, the paste must be sufficiently compliant to allow flow through the die, yet provide a high enough yield stress at low shear rates to avoid deformation following extrusion. To develop a model that predicts paste bulk shear stress, capillary rheometry was used to characterize aqueous binder gels and powder pastes using the binder gels as the fluid phase. Rheology models based on solids loading were successfully applied to permit rapid optimization of solids content and binder gel solutions for extrusion of articles having complex geometry.

Introduction

Extrusion and paste composition

Extrusion is a material forming process that permits the shaping of paste into a linear form having constant cross section. The paste is a homogeneous mixture of two phases: a solid phase composed of the particular oxide or oxide mixture carried by a fluid solution of water, binder, and additives. To maximize extrusion efficiency, the paste must be sufficiently compliant to flow through a die yet have a high enough bulk shear stress at low shear rates to avoid deformation following extrusion.

A novel technology has been developed at Georgia Tech for the extrusion of thin walled (< 150 µm) ceramic honeycomb and subsequent thermochemical processing of the green structure to metal alloy [1]. A wide variety of alloy compositions are possible, limited by thermodynamic stability of the oxide as determined by the appropriate Ellingham diagram.

Loose, dry powders naturally contain voids and provide low resistance to shear mixing. The fluid phase is composed of the solvent, usually water, and additives that may include binders, lubricants, and surfactants. As it is introduced, inter-particle spaces are gradually filled. Particles with wet surfaces agglomerate into pellet-like aggregates. When fluid content approaches pore volume, incremental additions of liquid cause the aggregates to agglomerate into a single, non-tacky, cohesive body. This state typically persists when pore saturation is between 90 and 100%. The bulk shear stress remains infinite until a critical fluid fraction is reached at 100% saturation. The solids content at full pore saturation is termed ϕ_{max}. Increasing fluid content beyond full pore saturation causes particles to arrange in a less closely packed configuration and results in a tacky cohesive body that moderately resists shearing. Further fluid additions cause discontinuities in the paste structure and make it no longer able to support itself [2].

Each solute component influences the rheology of the fluid phase. As paste rheology is controlled by solid-fluid interactions [3], behavior of the fluid phase exerts a strong influence on the properties of the final paste. The binder provides rheological control of the paste after the addition of the fluid phase and promotes green strength of the extrudate [4]. Chemical composition, molecular weight, molecular weight distribution, and concentration of binders can be varied to control paste properties [5-7]. The most successful binder solutions exhibit extreme shear thinning behavior. They are highly viscous at the low shear stresses present in the extruder barrel and immediately following extrusion, and fluid under the high stresses experienced near the die.

Benbow and Bridgewater [8] developed an analysis technique using flow through a die of circular cross section to measure the rheological properties of pastes. The four parameter model in equation 1 assumes a linear relationship between the extrusion pressure, P and extrudate velocity, V, and has the form:

$$P = 2(\sigma_o + \alpha V)\ln\left(\frac{D_o}{D}\right) + \frac{4L(\tau_o + \beta V)}{D} \tag{1}$$

In this equation, D_o is the diameter of the barrel, D is the diameter of the die land, and L is the length of the die land. Variables σ_o and α are associated with flow into the die land, τ_o and β with flow through the die land. Here, σ_o is the bulk shear stress, α is a parameter characterizing the effects of velocity on σ_o, τ_o is the die wall shear stress, and β is a parameter characterizing the effects of velocity on τ_o. A six parameter model that assumes an exponential relationship between P and V provides a more exact fit to the data but can be less desirable due to the difficulty encountered in comparing similar pastes.

Extension of fluid phase properties to pastes

Einstein [9] developed a relationship between solids content and the viscosity of an infinitely dilute suspension of rigid spheres. A variety of extensions to this theory have since been presented in the attempt to describe the rheological behavior of higher concentration suspensions. Several of the most commonly used models are the Mooney [10], Dougherty-Krieger [11], and Chong [12] equations, 2-4, respectively. These typically make use of the solids loading (ϕ), the maximum solids loading (ϕ_{max}), and empirically derived hydrodynamic and crowding factors (K_H). The Chong equation must first be generalized to allow direct comparison of K_H [13]. Several methods can be used to arrive at an experimental value for ϕ_{max} whereas K_H is completely empirical. Einstein [9] calculated $K_H = 2.5$, but positive and negative deviations from this value have been noted experimentally [14-15].

$$\sigma_r = \exp\left(\frac{K_H \phi}{1 - \frac{\phi}{\phi_{max}}}\right) \tag{2}$$

$$\sigma_r = \left(1 - \frac{\phi}{\phi_{max}}\right)^{-K_H \phi_{max}} \tag{3}$$

$$\sigma_r = \left[1 + \frac{K_H \cdot \phi_{max}}{2}\left(\frac{\frac{\phi}{\phi_{max}}}{1 - \frac{\phi}{\phi_{max}}}\right)\right]^2 \tag{4}$$

This study concentrated on the application of high solids content suspension viscosity models to pastes. Use of these models requires the substitution of relative bulk shear stress for relative viscosity. Relative property values are given by the ratio of the solid phase to fluid phase property. A simplified method is presented that allows rapid prediction of paste properties using limited data and the appropriate model equation.

Experimental Procedure

Paste preparation

The pastes were made by weighing individual components into the appropriate fractions to result in 1500 g batches. The solid components were dry blended for 15 min in a commercial blender fitted with an open mixing paddle. The lubricant, previously dissolved in water, was added to the powder-binder blend. Mixing continued for 2 min following the fluid addition.

The granulated powder was introduced into a Buss kneader for pugging. The Buss kneader is a compounding extruder that uses a single reciprocating screw for continuous high shear mixing of a variety of solid and liquid components [16]. Highly reproducible homogenization and compounding are realized within the processing length of the kneader. Dwell time in the kneader was approximately 5 min.

Measurement of maximum solids loading

Two methods were employed to measure ϕ_{max}. The oil drop test involves the wet mixing of powder raw materials with oil by hand using a small stirrer. Use of oil prevents evaporation and provides good wetting of the powder. The liquid fraction is increased drop wise until a cohesive, workable paste having smooth texture and shape retaining character is formed. Maximum solids content is then calculated from the weight of oil needed for pore saturation, the powder weight, and respective densities.

Experimentally, ϕ_{max} can be determined from rheological measurements by considering where the pressure to extrude becomes infinite ($1/\sigma_o = 0$). The relationship between the reciprocal of bulk shear stress and the ratio of fluid volume, V_f to solid volume, V_s approximates a straight line. The critical value of V_s at the intercept is the solids content at full pore saturation.

Capillary rheometry

Extrusion testing was conducted using a custom, stainless steel piston extruder fitted to a Satec universal testing machine. A schematic of the testing device can be found in Figure 1. It was established for these pastes that wall stresses between the paste and the barrel were negligible. The measurement procedure developed by Benbow and Bridgewater for this case was followed.

The extrusion pressures at several speeds were measured with a single barrel of paste. The barrel was charged with paste to a depth of 100 mm and tamped to remove large air pockets. The piston was inserted into the barrel and driven downward by the testing machine at each of six constant extrudate velocities: 68.54, 27.42, 13.71, 6.85, 2.74, and 1.37 mm/s. Testing proceeded from fastest to slowest followed by a return to the fastest velocity. Repeat

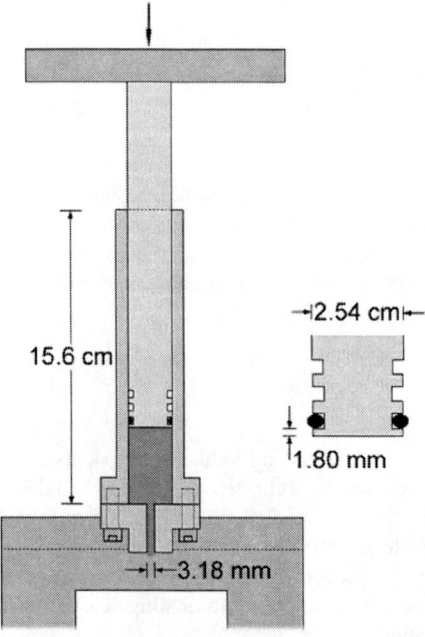

Figure 1 Schematic of the capillary rheometer.

Table I Characterization and batch composition of 350 Maraging steel pastes.

Raw Material	Source	Particle size (μm) at finer than			BET Surface (m²/g)	Amount in Batch
		10 wt.%	50 wt.%	90 wt.%		
Fe_2O_3	Pea Ridge Iron Ore Sullivan, Mo	0.81	4.48	10.04	1.513	66.7%
NiO	Ceramic Color New Brighton, PA	3.26	6.27	11.76	0.482	16.7%
Co_3O_4	Ceramic Color New Brighton, PA	3.02	3.65	4.47	4.036	11.9%
Mo	Atlantic Equipment Bridgeville, PA	2.67	6.98	15.60	0.543	3.5%
TiH_2	Reading Alloys Reading, PA	1.96	6.82	15.35	0.478	1.1%

Fluid Component	Source					Amount in Fluid
A4M Methocel	Dow Chemical Co. Midland, MI					19.6%
100S Pegosperse	Lonza Fair Lawn, NJ					2.0%
H_2O						balance

measurements did not deviate from each other by more than 10% and no corrections were applied to results. Statistical analysis of error in property measurements was not undertaken because of the length of time required to prepare and characterize each sample.

The data obtained for each paste are the ram force at each velocity for each die. Three interchangeable dies of circular cross section, $1/8$ inch (3.18 mm) diameter, and L/D ratios of 1, 8, and 16 are fitted to the rheometer. Extrapolating the plot of the L/D ratio against P to zero enables the calculation of bulk shear stress.

Six pastes of 350 grade maraging steel composition and varying solids fractions were prepared with a constant water to binder weight ratio of 4:1 and a constant lubricant concentration of 0.5% of the powder weight. The solids fractions (ϕ) tested were 0, 30, 40, 50, 55, and 60 vol.%. The test at 0 vol.% solids permits calculation of relative properties.

Material

A4M Methocel, a moderate molecular weight methylcellulose from Dow Chemical was used as a binder. Pegosperse 100S, an ethoxylated stearic acid from Lonza was used in small concentrations as a lubricant. These additives burn out cleanly in the reducing atmosphere used during firing. Distilled, deionized water was used as a solvent in all experiments.

For paste material, a mixture of oxides that when reduced forms 350 grade maraging steel was used as the solid phase. A series of identically prepared pastes were formulated having batch compositions as described in Table I.

Results and Discussion

Properties of the paste

The data from each set of paste experiments are plotted on a graph of die length to diameter ratio against extrusion pressure. From (1), as L/D approaches 0, $P = 2(\sigma_o + \alpha V)\ln(D_o/D)$.

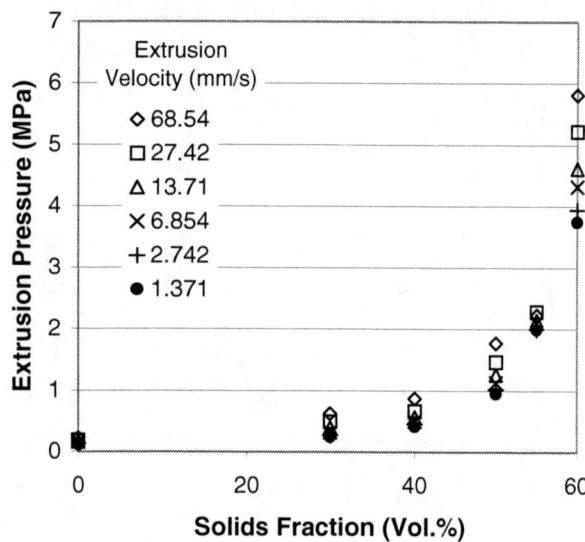

Figure 2. Summary of the extrapolations of $L/D \rightarrow 0$ for each extrudate velocity and paste composition.

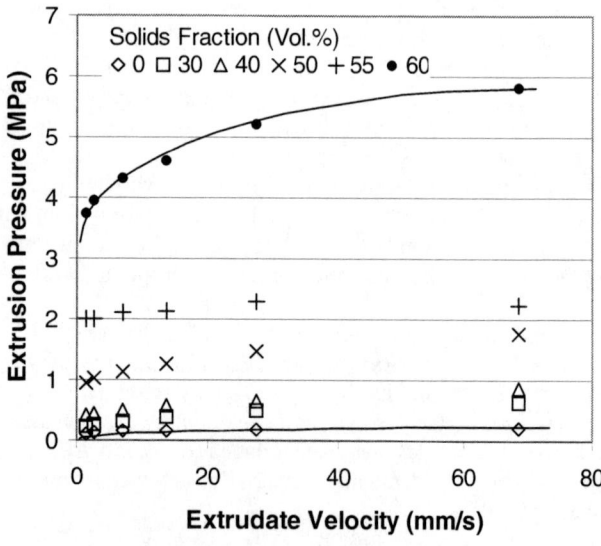

Figure 3. Crossplot of extrudate velocity against extrusion pressure for each paste composition.

Figure 2 is a summary of the extrapolations of $L/D \to 0$ for each extrudate velocity and paste composition. Increasing the solids content results in exponential increases in extrusion pressure. Extrusion pressure is especially sensitive to solids content as ϕ approaches ϕ_{max}. Higher solids contents result in an increased frequency of particle-particle and particle-surface interactions. The dimensionally thinner fluid phase separating and lubricating the particles translates directly to rapid increases in paste bulk shear stress at higher solids contents.

Figure 3 is a crossplot of extrudate velocity against extrusion pressure. Extrudate velocity is proportional to shear rate. The observed behavior remains similar to that seen in dispersed suspensions. At low solids content, the pastes behave as Bingham plastic bodies. The Benbow and Bridgwater models used in this investigation assume that fluid phase and paste samples show Bingham plastic behavior.

Each sample tested shows a distinct yield point and a near linear pressure increase with a corresponding increase in velocity. The range of solids contents that produced the best honeycomb extrusions showed minimal curvature. As ϕ approaches ϕ_{max}, there is a transition in behavior from Bingham plastic to pseudoplastic (shear thinning). This transition suggests that property prediction will be prone to error as ϕ_{max} is approached.

Determination of maximum solids loading

The oil drop test tends to underestimate ϕ_{max} because hand mixing is a low shear operation and can be ineffective in breaking up agglomerates. The utility of the oil drop test arises from the fact that a single, small sample can be used to determine the maximum solids loading without the difficulty of extensive processing and rheometric investigation. The amount of liquid present at ϕ_{max} is typically assumed to be 90% of the oil drop test value [17]. By applying this criterion to the test results, ϕ_{max} was estimated to 62.1%.

By plotting the inverse of the bulk shear stress against the ratio of fluid to solid volume for the six pastes as shown in Figure 4, ϕ_{max} was measured to be 64.1 vol.%.

Application of suspension models to pastes

Non-linear optimization can be used to determine the best fit ϕ_{max} and K_H values associated with each model equation. The results of these calculations are presented in Table II and will be discussed in a future publication.

Figure 5 shows the relative properties of the pastes and reference plots of the Dougherty-Krieger, Mooney, and Chong equations at the oil drop test estimate of $\phi_{max} = 62.1\%$ and

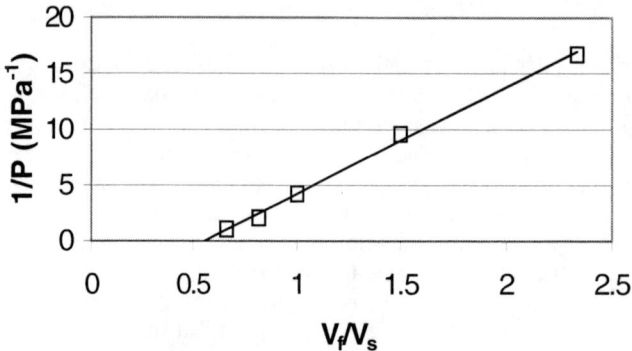

Figure 4   Determination of ϕ_{max} from measurements of paste bulk shear stress.

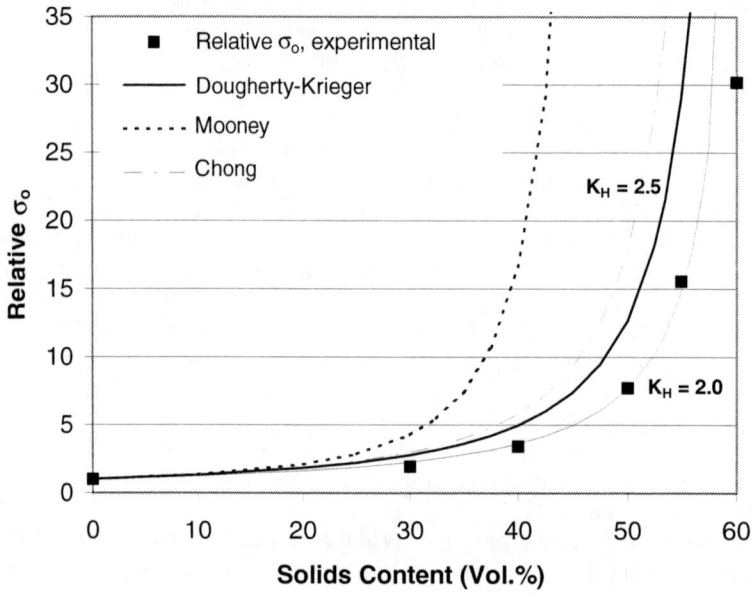

Figure 5   Relative bulk shear stress of the 350 maraging steel pastes compared to plots of the Mooney, and Chong equations at ϕ_{max} = 62.1% and K_H = 2.5 and the Dougherty-Krieger equation at ϕ_{max} = 62.1% and K_H = 2.0 and 2.5.

Einstein's calculated K_H = 2.5. The Dougherty-Krieger equation is also solved with K_H = 2.0. The improvement in fit at this value of K_H stands in agreement with the high viscosity suspension work of J. Lee, *et al*.

With a minimum amount of effort and material, paste properties can be characterized by measuring ϕ_{max} with the oil drop test and using Einstein's calculated K_H. The Dougherty-Krieger equation is the best of the three models under these conditions and remains within a factor of 2 of experimental results up to a solids content of 90% of ϕ_{max}. If K_H is fixed at 2.0, the Dougherty-

Krieger equation remains within a factor of 2 of experimental results up to a solids content of 97% of ϕ_{max}.

The testing of pastes with multiple solids contents allows for a more precise determination of ϕ_{max} and the fitting of a value of K_H to the data. Non-linear optimization techniques result in

Table II Values of ϕ_{max} and K_H resulting in minimum error.

Equation	ϕ_{max} (Vol.%)	K_H
Dougherty-Krieger	67	2.3
Chong	70	2.4
Mooney	85	1.7

values of ϕ_{max} and K_H most consistent with a legitimate physical interpretation when the Dougherty-Krieger equation is used.

Conclusions

The oil drop test estimates ϕ_{max} to be 62.1%. A plot of $1/P$ against the ratio of fluid volume to solid volume is approximately linear. The value of V_s at the x-intercept is the maximum solids loading, 64.1%.

With a minimum amount of effort and material, paste properties can be predicted by measuring ϕ_{max} with the oil drop test and using Einstein's calculated K_H. The Dougherty-Krieger equation is the best of the three models considered under these conditions. When significant data for a paste system exist, the Dougherty-Krieger equation minimizes error at values of ϕ_{max} and K_H most consistent with a legitimate physical interpretation.

Acknowledgements

This work is sponsored by DSO of DARPA (N00014-99-1-1016) under Dr. Leo Christodoulou and by ONR (N0014-99-1-0852) under Dr. Steven Fishman. We deeply appreciate their support and guidance.

References

[1] J. K. Cochran, et al., "Low density monolithic metal honeycombs by thermal chemical processing," Fourth Conference on Aerospace Materials, Processes and Environment Technolgy, in press.

[2] J. S. Reed, Introduction to the Principles of Ceramic Processing, (New York, NY: John Wiley & Sons, 1988), 200-202.

[3] K. M. Hurysz, et al., "Optimum iron oxide/chromium oxide paste formulation for complex geometry extrusion," Ceramic Transactions (USA), 115 (2000), 173-181.

[4] A. Ekonomakou, A. Tsetsekou, and C. Stournaras, "The influence of binder properties on the plasticity and properties of raw extruded ceramics," Key Engineering Materials, 132-136 (1997), 420-423.

[5] Dow Chemical Corporation, METHOCEL Cellulose Ethers Technical Handbook.

[6] J. J. Benbow, *et al.*, "Prediction of paste extrusion pressure," Ceram. Bull., 68 (10) (1989), 1821-1824.

[7] J. E. Schuetz, "Methylcellulose polymers as binders for extrusion of ceramics," Ceram. Bull., 65 (12) (1986), 1556-1559.

[8] J. Benbow and J. Bridgwater, Paste Flow and Extrusion (Oxford, Eng.: Clarendon Press, 1993), 25-44.

[9] A. Einstein, Ann. Phys., "Eine neue Bestimmung der Molekuldimensionen," 289 (19) (1906).

[10] M. Mooney, "The viscosity of a concentrated suspension of spherical particles", J. Colloid Sci., 6 (1957), 162-170.

[11] I. M. Krieger and T. J. Dougherty, "A Mechanism for Non-Newtonian Flow in Suspensions of Rigid Spheres," Trans. Soc. Rheol., 3 (1959) 137-152.

[12] J. S. Chong, E. B. Christiansen, and A. D. Baer, "Rheology of Concentrated Suspensions," J. Appl. Poly. Sci., 15 (1971) 2007-2021.

[13] K. M. Hurysz, "Paste Mechanics for Fine Extrusion," (Ph.D. dissertation, Georgia Institute of Technology, 2001).

[14] J. Lee, J. So, and S. Yang, "Rheological Behavior and Stability of Concentrated Silica Suspensions," J. Rheol., 43 (5) (1999) 1117-41.

[15] D. J. Jeffrey and A. Acrivos, "The Rheological Properties of Suspensions of Rigid Particles," A. I. Chem. E., 417 (22) (1976).

[16] Buss America, Buss kneader operating principle and system advantages.

[17] J. Benbow, J. Bridgwater, op.cit., 34-36, 126.

EFFECT OF PROCESSING VARIABLES ON SOLID-STATE FOAMING OF TITANIUM BY SUPERPLASTICITY

N.G. Davis, D.C. Dunand

Department of Materials Science and Engineering
Northwestern University
2225 North Campus Dr.
Evanston, IL 60208

Abstract

Solid-state foaming of titanium and titanium alloys can be achieved by hot-isostatic pressing powders with argon gas, followed by high temperature expansion of the resulting high-pressure argon bubbles. This foaming technique, which is currently under industrial development at Boeing Corp. for Ti-6Al-4V as their LDC material, is limited in terms of maximum porosity by the low creep rate and ductility of the metal, which lead to slow pore growth and early cell wall fracture limiting. We address these issues by performing the foaming step under transformation superplastic conditions where the foam is thermally cycled around the α/β allotropic temperature of titanium. This induces superplasticity due to the complex superposition of internal transformation stresses and the biasing stress from the pore pressure. Variations in processing conditions such as the initial powder size, backfill pressure and initial powder aspect ratio form are studied and discussed with respect to the foaming kinetics, the terminal porosity and the open-to-closed porosity ratio.

Introduction

Due to the outstanding mechanical properties, low density and high chemical resistance of titanium, titanium-based foams have many potential applications including load-bearing applications as sandwich cores in the aerospace and ship building industries [1, 2] and as porous implants in the biomedical industry, for which the excellent biocompatibility and fatigue properties of titanium are also essential [1-5]. However, due to the high melting temperature of titanium and its propensity for contamination at elevated temperatures, liquid-state foaming techniques have proven challenging for titanium, leaving solid-state foaming as a promising foaming technique. Solid-state foaming of titanium was first patented by Kearns et al., [6], and later by Martin for Ti-6Al-4V [7]. Other studies concerning the solid-state foaming of Ti-6Al-4V [8-10] and unalloyed Ti [11] have followed. The foaming process, as described in Refs. [6-12], and schematically depicted in Figure 1, consists of Hot Isostatic Pressing (HIPing) titanium powder in the presence of argon gas. The resulting compacted material consists of a continuous titanium matrix containing discreet micron-sized high-pressure argon pores. On exposure to elevated temperatures, the bubbles expand by creep of the surrounding metal due to the high argon pressure. Kearns et al. [6] reported porosities of up to ~40 % after heat treatments at very high temperatures (1250 °C) and long hold times (several days). The growth rate of the pores was controlled by the slow creep deformation of the metal and the backfill argon gas pressure.

In a previous publication [11], we have demonstrated that deformation of the titanium matrix surrounding the high-pressure argon pores could be enhanced using transformation superplasticity (TSP). Transformation superplasticity is found in polymorphic materials (e.g., titanium [13-15], Ti-6Al-4V [16, 17] and many other polymorphic metals and alloys and relies on internal mismatch strains produced by the density difference between the two allotropic phases coexisting during a phase transformation. These mismatch strains are biased in the direction of an externally applied stress, resulting in a net strain increment in the biasing direction accumulated during the phase transformation. By repeatedly cycling through the phase transformation, tensile elongations well in excess of 100 % can be accumulated without fracture or cavitation. Using thermal cycling to repeatedly induce the phase transformation and the argon pressure inside the pores as the biasing stress, we recently reported that foaming titanium under superplastic conditions led to faster foaming rates and higher terminal porosities [11].

In the present paper we study the effect on superplastic foaming of CP-Ti of processing parameters: initial titanium form (powders versus wires (Figure 1)), argon backfill pressure and initial powder size and size distribution.

Experimental Procedure

Spherical powders of commercial-purity titanium (CP-Ti) were packed, encapsulated in evacuated steel canisters, which were back-filled with argon. To achieve various initial pore size and spacing, both sieved and unsieved powders were used. In the unsieved case, powders (from Starmet, Concord, MA) with a median size of 130 µm and a mesh size of –100 (<177 µm) were used. Also used were sieved compacts with CP-Ti powder sieved to –35/+40 (355–500 µm with an average size of 423 µm) and –400/+500 (25-37 µm with a 31 µm average size from Ti Science and Technology Co., Xian, China). Powders were densified by hot isostatic pressing (HIPing) at 890°C with 100 MPa for 120 minutes. The unsieved powders were HIPed by UltraClad (Andover, MA) with 3.3 atm argon backfill while the coarse and the fine powders (with argon backfill pressure of either 3.3 atm or 7 atm) were HIPed at Bodycote, IMT, Inc. (Andover, MA). Furthermore, a canister was filled with 250 µm diameter CP-Ti wires from Alfa Aesar (Ward Hill, MA) arranged length-wise (Figure 1) and HIPed with 7 atm argon backfill under the same conditions as above by Bodycote, IMT, Inc.

Cubic samples with 9 mm edges were machined by EDM from the billets of all spherical powder types, and two 5 mm x 5 mm x 10 mm parallelepipeds samples were similarly machined from the HIPed wire compact. Specimens were foamed by either thermal cycling or isothermal anneal. Experiments were performed in a rapid thermal cycling furnace (Materials Research Furnaces, Allenstown, NH). Unsieved specimens were foamed in this furnace with graphite radiant heaters either encapsulated in quartz tubes (evacuated to 0.01 mtorr) together with a tantalum gettering foil, or within the vacuum furnace evacuated to 10-30 mtorr residual argon atmspohere. All other specimens were foamed in the vacuum furnace using Mo radiant heaters, evacuated to 1-5mtorr residual argon atmosphere. For superplastic foaming experiments, specimens were thermally cycled between 830°C and 980°C, spanning the allotropic temperature range of CP-Ti (nominally at 882°C, but expected to be higher and wider due to oxygen content [18]) with a 4 min. cycle period. Isothermal foaming experiments were

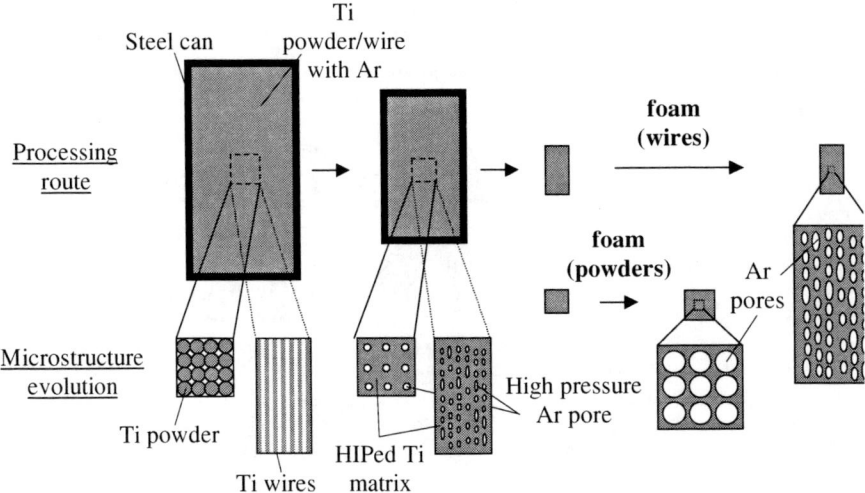

Figure 1. Schematic of foaming process, where metal in the form of powders or wires is first compacted by HIP in the presence of argon and the resulting high-pressure argon bubbles are then expanded under high pressure inert atmosphere.

performed at 960 °C or 903 °C. The latter is the effective temperature of the above-mentioned thermal cycle, defined as the temperature at which the isothermal creep rate is the same as the time-averaged creep rate during thermal cycling in the absence of TSP [15, 16], found by using lattice activation energies given in Ref. [19] for α-Ti and β-Ti. Thus, by comparing the thermal cycling foaming kinetics with those at the effective temperature, the effect of TSP on foaming is directly revealed. Sieved specimens were foamed either with 1 atm Ar cover gas or encapsulated in an evacuated quartz tube while sieved specimens were foamed at 1-5 mtorr.

Frequent interruptions by excursions to room temperature occurred to determine specimen porosity. Archimedes density measurements were performed on all specimens in distilled water. A thin layer of vacuum grease was used to seal the surface of each specimen in order to prevent infiltration of water into the open porosity, yielding a direct measurement of each specimen's total porosity (sum of open and closed porosity). Density measurements were also performed using helium pycnometry with an Accupyc 1330 (Micromeritics, Norcross, GA) on unsealed specimens, allowing for measurement of closed porosity.

Cubic specimens were cut in half along one of the cube face diagonals and the wire samples were cut both parallel and perpendicular to the wire length with a low-speed diamond saw, mounted in acrylic resin, and polished to 0.05 μm alumina. Foamed specimens were vacuum-infiltrated with acrylic resin at regular intervals during polishing, filling open porosity to retain the original pore shape during polishing.

Results

Figure 2 shows the as-HIPed pore structure of various initial compacts. Initial pores in the HIPed, sieved and unsieved compacts retained the general shape of the gap between powders in the original preforms (Figure 2 (a)-(e)), and were thus non-spherical. Figure 2 (a) and (b) are composite micrographs showing the initial pore structure for specimens made from the coarse

Table I. Summary of initial and final conditions for materials used in the present study.

	Powders							Wires	
	Unsieved		25-37 µm		355-500 µm			longitudinal	perpendicular
	isothermal	TSP	TSP		TSP			TSP	
backfill pressure (atm)	3.3	3.3	3.3	7	3.3	7		7	
initial porosity (%)	0.3	0.3	1.3	2.1	1.7	3.42		0.26	
initial pore diameter (µm)	25	25	5	10	100	100	300 *		25
final porosity (%)	26	42	35	30	14	13		16	
final pore diameter (µm)	200	250	50	50	1000*	500*	2500*		100

* Maximum pore size

powder with 3.3 atm and 7 atm argon backfill, respectively. Figure 2 (c) shows the pore morphology for as-HIPed material made from the fine powders with 3.3 atm argon backfill. Figure 2 (d) shows that for an argon backfill of 7 atm for the same type of powder. The pore morphology for the unsieved specimens is shown in Figure 2 (e). These figures show that initial average pore size and initial porosity varied with initial powder particle size argon backfill pressure (Table I). Preforms backfilled with 7 atm argon generally show larger initial pore size than the same particle sized compact backfilled with only 3.3 atm argon, as expected from the larger amount of gas present. Additionally, coarse powder particles led to initial pore size larger than either the unsieved powders or the fine powders. Also, unsieved powder specimens have a larger inter-pore spacing, and a lower initial porosity than the specimens made from fine powders. As-HIPed wire compacts show numerous pores that are elongated and partially aligned in the direction of the wire alignment (Figure 2 (f)). The pores perpendicular to the direction of the wire alignment show a triangular morphology and prior wire boundaries are visible.

Figure 3 (a)-(c) show specimen porosity as a function of time for various initial particle size distributions and argon backfill pressures. Figure 3 (a) shows foaming of unsieved powders for thermal cycling (830-980°C), for isothermal foaming at the effective temperature of the thermal cycle (903°C), and isothermal foaming at 960°C. Foaming using TSP achieved higher porosities than isothermal foaming. Both isothermal and thermally cycled specimens foamed for 11.8 h. had only small amounts of open porosity, as observed by the difference between the total and closed porosity: an eighth of the total porosity was open for the isothermally foamed specimen (960°C) while a tenth of the total porosity was open for the thermally cycled foam. Reproducibility was tested by foaming three specimens at 903°C for 24h yielding similar porosities for each specimen: 26%, 22%, and 21%. Since foaming kinetics and terminal pore fractions are dependent on the initial pore fraction (which can vary slightly with position in the HIPed canister due to powder settling), the reported amount of variation is considered reasonable. Figure 3 (b) depicts the foaming behavior under thermal cycling conditions for compacts with unsieved powders and sieved powders (fine and coarse) with 3.3 atm or 7 atm argon backfill. Figure 3 (c) shows only the first five hours of the foaming curves shown in Figure 3 (b). The fine powders with the highest argon backfill pressure shows the fastest initial foaming rate; this specimen was at maximum porosity (33%) after only 1 h of foaming.

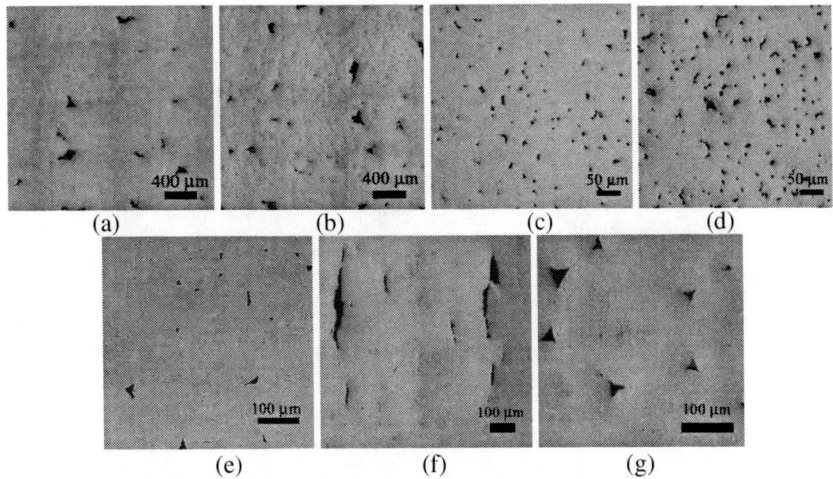

Figure 2. Optical micrographs showing as-HIPed microstructures: (a,b) irregularly shaped argon-filled pores for 355-500 μm powders HIPed with 3.3 atm and 7 atm argon backfill, respectively; (c,d) 25-37 μm powders HIPed with 3.3 atm and 7 atm argon backfill, respectively; (e) unsieved powders HIPed with 3.3 atm argon backfill; (f,g) 250 μm wires HIPed with 7 atm argon backfill showing initial pores along and perpendicular to the wire length, respectively.

However, this specimen did not achieve the highest final porosity. The unsieved powders achieved the highest total porosity at 41% and had reached close to maximum final porosity after two hours of foaming. Foams made from the fine powders and backfilled with 3.3 atm argon, achieved a maximum total porosity of 37% after 2h of thermal cycling, which is higher than those of the same particle size backfilled with 7 atm, but slightly lower than that for the unsieved powders. For both backfill pressures of the fine powder foams the total porosity decreased with further thermal cycling after reaching a maximum level of porosity.

Specimens made with coarse powders foamed much more slowly and to much lower total porosities than either the unsieved or fine powder particles; the specimen backfilled with 3.3 atm argon achieved a final porosity of 14% while that backfilled with 7 atm argon foamed to 13% porosity.

The pore structures of the above specimens are shown in Figure 4 (a)–(h). A comparison of foams processed by isothermal foaming with those foamed by TSP (thermal cycling) is shown in Figures 4 (a) and (b) for unsieved powders. In both cases, pores are rounded and have a diameter of up to 200 μm (many of the smaller pores result from the metallographic plane intersecting far from the equatorial plane of the roughly spherical pore). Pore coalescence is much more pronounced in the isothermally foamed specimen (arrows in Figure 4 (a)) than in the thermally cycled specimen, which exhibits many examples of pore walls as thins as 10 μm (arrows in Figure 4 (b)). The etched matrix shows colonies of serrated α grains typical of CP-Ti annealed in the β field [20]. Figure 4 (c) and (d) show the pore structure of the foamed specimens made from the fine powder; again, in both cases, the pores are rounded. Pores in these foams are the smallest of all foamed specimens in this study with, a maximum pore size of about 50 μm. Composite micrographs for the coarse foams with 3.3 atm and 7 atm argon backfill are shown in Figure 4 (e) and (f), respectively. These pores appear spherical after foaming. The coarse foam with the higher backfill has a much larger maximum pore size (1 mm) than for the 3.3 atm backfill (0.5 mm). However, the average pore is smaller in the 7 atm foam than when compared to the 3.3 at foam with coarse powder as is seen when comparing

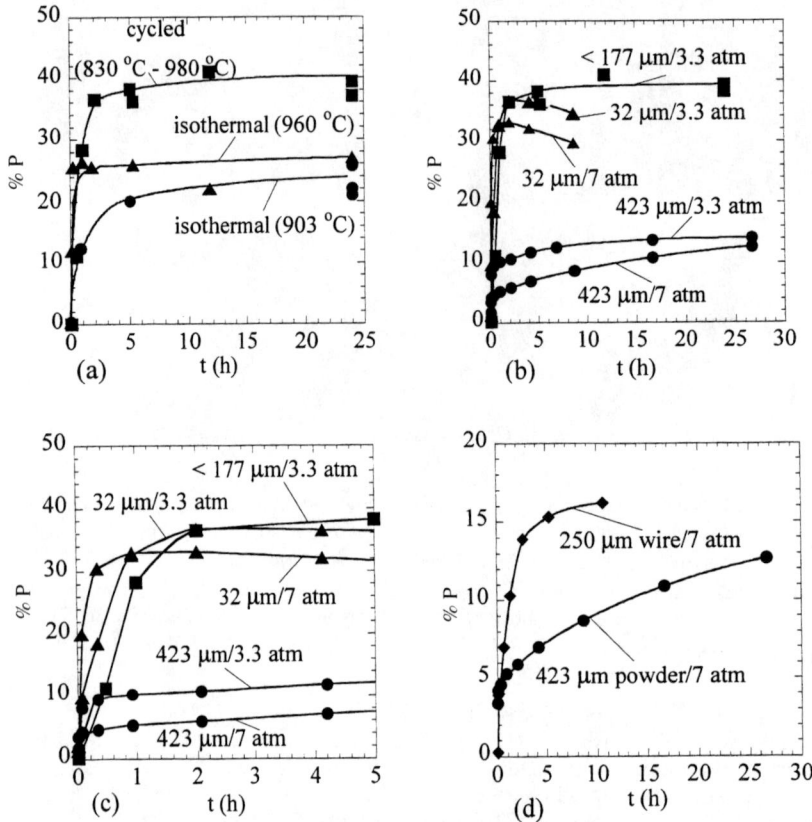

Figure 3. Total porosity as a function of time for performs with: (a) unsieved powders foamed by thermal cycling and isothermal experiments; (b) and (c) unsieved powders (filled squares), powders sieved to 25-37 μm (triangles) and 355–500 μm (circles) all thermally cycled; and (d) 355-500 μm powders and 250 μm wires, both backfilled with 7 atm argon and thermally cycled.

Figure 4 (e) with Figure 4 (f). In Figure 4 (g) and (h), the pore morphology of the wires is shown parallel and perpendicular to the length of the wires, respectively. Parallel to the length of the wires, the pores are elongated, and generally aligned along the wire length. In addition, some pore coalescence is observed. In the perpendicular direction, the pores are spherical in nature and fairly well spaced.

Discussion

TSP foaming can be directly compared to isothermal foaming carried out at the effective temperature (neglecting the small pressure variation due to thermal cycling). Figure 3 (a) shows that the overall foaming rate and terminal porosity are significantly higher under TSP conditions than under creep conditions at the corresponding effective temperature (903 °C) for unsieved powder foams. This confirms that foaming under TSP conditions leads to faster deformation rates, as is also observed for bulk CP-Ti and Ti-6Al-4V subjected to uniaxial and multiaxial

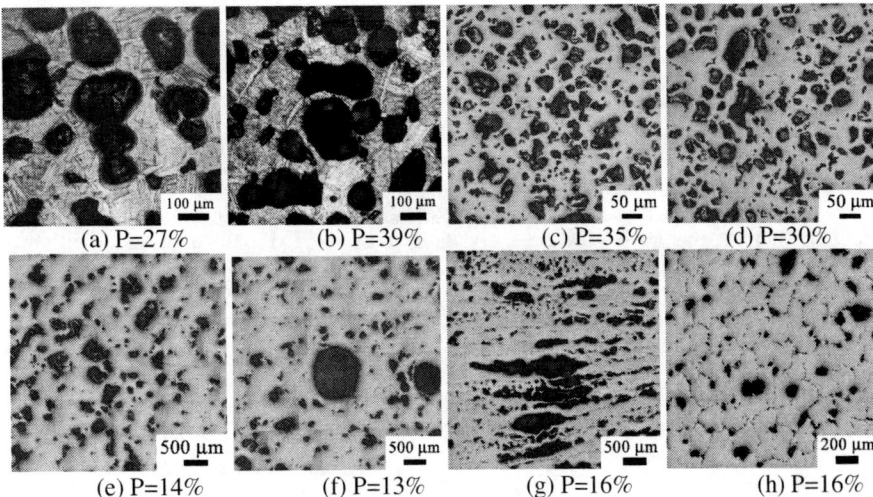

Figure 4. Optical micrographs of pore microstructures for different initial condition: (a,b) unsieved specimen foamed isothermally at 960 °C for 5.25 h and by thermal cycling for 5 h, respectively; (c,d) 25-37 μm powders backfilled with 3.3 atm and 7 atm foamed for 8.7 h by thermal cycling, respectively; (e,f) 355-500 μm powder specimens foamed by thermal cycling for 26.7 h with 3.3 and 7 atm argon backfill, respectively; (g,h) wire specimen foamed by thermal cycling showing pore morphology in the directions parallel and perpendicular to the wire length, respectively.

tests [15, 16]. Even when the cycling experiments are compared to isothermal experiments at 960 °C, terminal total porosities are higher for the cycled specimens. In other materials, if no allotropic transformation exists, foaming enhancement due to internal mismatch strains may be induced in other ways, such as by thermal cycling in composites due to the difference in the thermal expansion between matrix and reinforcing phases as has been demonstrated for the case of uniaxial deformation [21].

The foaming characteristics for the unsieved specimens include a plateau, after which there is no appreciable change in porosity. In the present study for the unsieved materials, time to reach final porosity is shortest for the isothermal foaming at 960 °C (0.5 h.) and longest for foaming isothermally at 903 °C (5 h.), with cyclic foaming being intermediate (2 h.). The average foaming rate, calculated as the ratio of the final porosity to the corresponding time, is roughly an order of magnitude slower at 903 °C than for thermal cycling. For foaming at 960 °C, the foaming rate was found to be roughly half an order of magnitude faster than for cyclic foaming but the terminal porosity was much lower. Foams made from fine powder particles backfilled with 7 atm argon backfill have the highest foaming rate. Both backfill pressures for the fine powder foams, as well as the unsieved foams exhibit much higher foaming rates than the coarse powder foams (either backfill pressure) or the foams made from the wires. A foaming body is sometimes modeled as a series of small pressure vessels where the pressurized argon in each pore is causes creep deformation of the surrounding matrix similarly to a creeping pressure vessel, a problem that is described by Finnie and Heller [22]. Assuming equi-sized pores on a cubic lattice, one pore, and its associated matrix can be equated to this spherical pressure vessel.

In a such a case, the stress in the matrix due to a pressurized spherical pore is dependent on the pore size, pressure and inter-pore spacing as follows [22]:

$$\sigma_r = -p \frac{(R_o/r)^{3/n} - 1}{(R_o/R_i)^{3/n} - 1} \quad (1)$$

where σ_r is the stress in the radial direction, p is the pore pressure, R_o and R_i are the outer and inner radii of the spherical pressure vessel, respectively, r is the radial distance from the center of the pore and n is the power-law creep constant (unity for TSP deformation). In either the fine powder foams for either backfill pressure or the unsieved foams, the average Ro/Ri is much smaller than for the coarse or wire foams. For the fine and unsieved powder foams, the relatively thin "shells" (inter-pore spacing) likely allows for the stress fields from neighboring pores to overlap significantly more, thus enhancing foaming rate.

The total, closed and open porosity for foams made from fine and coarse powders are shown in Figure 5 (a) and (b) and for the 250 μm wires in Figure 5 (c). For both backfill pressures in the specimens made from the fine powder (Figure 5 (a)), there is more open

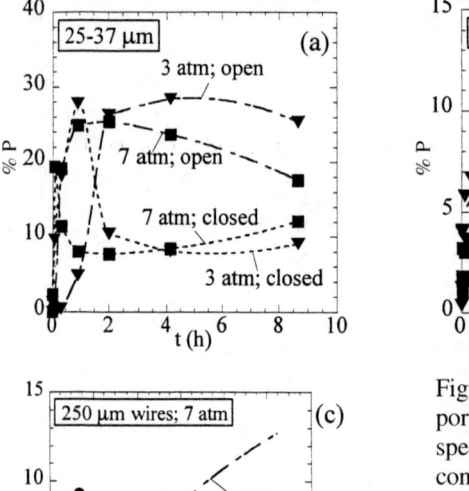

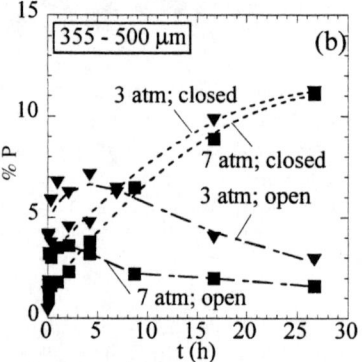

Figure 5. Closed and open porosity as a function of time for specimens with differing initial conditions foamed by thermal cycling: (a) 62–88 μm powders, (b) 355–500 μm powders and (c) 250 μm wires. Argon backfill of 3.3 and 7 atm are shown in both (a) and (b) while only 7 atm argon backfill is represented in (c).

porosity at the end of foaming than closed porosity; at maximal porosity close to three-quarters of the total porosity is open for both backfill pressures, although the total porosity is greater for the 3.3 atm backfilled specimen. Figure 5 (a) also shows that the porosity begins to open early in the foaming process for these materials with the porosity opening more rapidly for the 7 atm (squares in Figure 5 (a)) than for the 3.3 atm (triangles) backfilled foams. In fact, more than half of the porosity is open before the first half hour of foaming for the 7 atm backfilled foam. Due to this early pore wall bursting, which opens porosity to the surface of the specimen the gas escapes early during the foaming process. This removes the driving force for foaming and keeps to this early pore wall bursting, which opens porosity to the surface of the specimen the gas escapes early during the foaming process. This removes the driving force for foaming and keeps the small powder particles from achieving a greater porosity level than the unsieved powders. These remain almost entirely closed during the entire foaming process and cease to foam due to the decrease in pore pressure [11] rather pore-wall rupture.

Despite different argon backfill pressures for the fine powder foams (3.3 atm for Figure 4 (c) and 7 atm for Figure 4 (d)), slightly differing total porosities and closed-to-open porosity ratio, the pore morphology (shape, size, and distribution) look strikingly similar. The final pore size is smaller for these foams than any other foam in the study due to the small initial pore size created from the small space between the fine powders. For fine powder foams with from both backfill pressures, the total porosity decreases after reaching a maximum after ca. 1-2 h. (Figure 3 (b) and Figure 5 (a)) and the open porosity decreases (with increasing closed porosity) suggesting that the decrease in total porosity is due to sintering process.

Foams made from the coarse powders show slightly different open and closed porosity characteristics (Figure 5 (b)). For both backfill pressures, there is less open than closed porosity upon completion of pore growth. Foams with 3.3 atm argon backfill pressure have roughly a fifth of their porosity open and the specimen with 7 atm argon backfill only have an eighth open porosity. However, in the early stages of foaming there is more open porosity than closed porosity. The increase in closed porosity can be due to neighboring expanding pores causing a broken neck to pinch off, closing off the pore to the specimen surface. The final average pore size in the coarse foam with 3.3 atm is larger than the average pore size for the 7 atm argon backfilled foam. However, the maximum pore size is much larger in the 7 atm foam. The foams backfilled with 3.3 atm argon have a more uniform pore size but the general shape of the pores in both foams is similar. Some pore coalescence is observed, but due to the low amount of porosity, the inter-pore spacing is still relatively large in these foams and most pores remain discrete. Figure 5 (c) shows the total, open and closed porosity as a function of time for the CP-Ti wires foamed by thermal cycling. Terminal porosity for the wires is 16 % with four-fifths of the total porosity as open porosity. In addition, the foaming kinetics for the wires is much slower than for the unsieved foams or the fine powder foams. In Figure 4 (g) and (h), a large amount of pore coalescence is also observed. In addition, individual wires are still visible both before and after foaming (Figure 4 (g) and (h)). This evidence, along with the clear presence of wire individuality in the HIPed material (Figure 2 (g)) strongly suggests that these preforms were not HIPed under optimal conditions. This likely caused premature foaming cessation as the bond between the wires may be weak allowing the pores to easily open to the surface of the specimen. In fact, as seen in Figure 4 (c), at least half of the porosity in the wire specimen is open after about 5 hours of foaming. On continued thermal cycling, the amount of open porosity increases dramatically, but the total porosity increases very little.

Conclusions

1. Solid-state foaming by isothermal creep expansion of high pressure argon bubbles entrapped by HIPing of unsieved powders, as first demonstrated by Kearns et al. [6] in Ti-6Al-4V, was studied for CP-Ti at 903°C and 960°C. Foaming rates are significantly faster at the higher temperature but terminal porosities are equal within error (21%-27%). In both cases, pores became spherical after foaming.

2. Foaming of CP-Ti was also performed by thermal cycling between 830°C–980°C (straddling the allotropic temperature), to activate transformation superplasticity as the main deformation mechanism. During thermal cycling, the volume mismatch between allotropic phases produces mismatch strains, which are biased by the deviatoric stress from the internal pore pressure. It was demonstrated that thermal cycling leads to (i) accelerated foaming kinetics and (ii) increased maximum attainable porosities, because of the reduced strength under TSP conditions.

3. Changes in the initial conditions (particle size and backfill pressure) cause the pore size, shape and distribution of the resulting foams to differ. Small powder particles lead to a small initial pore size, which, in turn, lead to a much smaller final pore size than for unsieved powders, despite having similar total porosity. In general, higher argon backfill pressure leads to a larger initial porosity, but, for the case of the small particle size specimens, the resulting foam structure was similar to the 3.3 atm backfilled specimen. Using CP-Ti wires as the starting material leads to pores elongated in the direction of the wires axis.
4. Varying initial conditions also give rise to various foaming characteristics. Unsieved powders, foamed both by isothermal anneal and by thermal cycling, foam rapidly to a maximum porosity where a plateau is reached. The porosity remains almost entirely closed. Foaming rate is fastest for the 25-37 μm powder foams with a backfill pressure of 7 atm, which resulted in 33 % porosity after only 1 h of foaming. However, the higher porosity achieved was from the unsieved powders (41 %). Foaming of either the 355-500 μm powder foams or the foams made of the wires is much slower than for the small or unsieved powders.

Acknowledgements

This project was supported by NSF through grants DMR-0108342 and CMS-0074921 (L.C. Brinson, Principal Investigator). Also, the authors thank TST (Xian, China) and Starmet (Andover, MA) for donating Ti powders, and Erik Spoerke for the preparation of the wire performs. N.G.D. acknowledges the support of the U.S. Department of Defense, through a National Defense Science and Engineering Graduate Fellowship, the American Association of University Women through a Selected Professions Dissertation Year Fellowship and the Zonta International Foundation through the Amelia Earhart Fellowship.

References

1. J. Banhart, "Manufacture, Characterization and Application of Cellular Metals and Metal Foams," Progr. Mater. Sci. 46 (2001), 559-632.
2. M.F. Ashby, et al., Metal Foams-A Design Guide. (Woburn, MA: Butterworth-Heinemann, 2000).
3. M.J. Donachie, "Biomedical Alloys," Adv. Mater. Proc. 7 (1998), 63-65.
4. M.J. Donachie, "Introduction to Titanium and Titanium Alloys," in Titanium and Titanium Alloys Source Book, M. J. Donachie, Editor. (Metals Park, OH: ASM, 1982), 3-19.
5. S.J. Simske, R.A. Ayers, and T.A. Bateman, "Porous Materials for Bone Engineering," Mat. Sci. Forum. 250 (1997), 151-182.
6. M.W. Kearns, et al., "Manufacture of a Novel Porous Metal," Int. J. Powder Metall. 24 (1) (1988), 59-64.
7. R.L. Martin, "Integral Porous-Core metal Bodies and in situ Method of Manufacture Thereof". McDonnell Douglas Corp., USA, Patent No. 5564064, (1996).
8. D.T. Queheillalt, et al., "Creep Expansion of Porous Ti-6Al-4V Sandwich Structures," Metall. Mater. Trans. 31A (1) (2000), 261-273.
9. D.M. Elzey and H.N.G. Wadley, "The Limits of Solid State Foaming," Acta Mater. 49 (5) (2001), 849-859.
10. D.T. Queheillalt, K. A. Gable, and H. N.G. Wadley, "Temperature Dependent Creep Expansion of Ti-6Al-4V Low Density Core Sandwich Structures," Scripta Mater. 44 (3) (2001), 409-414.
11. N.G. Davis, et al., "Solid-State Foaming of Titanium by Superplastic Expansion of Argon-Filled Pores," J. Mater. Res. 16 (5) (2001), 1508-1519.

12. R. Vancheeswaran, et al., "Simulation of the Creep Expansion of Porous Sandwich Structures," Metall. Mater. Trans. 32 (2001), 1813-1821.

13. G.W. Greenwood and R.H. Johnson, "The Deformation of Metals Under Small Stresses During Phase Transformations," Proc. Roy. Soc. London. 283A (1965), 403-422.

14. C. Chaix and A. Lasalmonie, "Transformation Induced Plasticity in Titanium," Res Mech. 2 (1981), 241-249.

15. D.C. Dunand and C.M. Bedell, "Transformation-Mismatch Superplasticity in Reinforced and Unreinforced Titanium," Acta Mater. 44 (3) (1996), 1063-1076.

16. D.C. Dunand and S. Myojin, "Biaxial Deformation of Ti-6Al-4V and Ti-6Al-4V/TiC Composites by Transformation-Mismatch Superplasticity," Mat. Sci. Engng. 230A (1997), 25-32.

17. C. Schuh and D.C. Dunand, "Non-Isothermal Transformation-Mismatch Plasticity: Modeling and Experiments on Ti-6Al-4V," Acta Mater. 49 (2) (2001), 199-210.

18. J.W. Elmer, J. Wong, and T. Ressler, "Spatially Resolved X-Ray Diffraction Phase Mapping and $\alpha \rightarrow \beta \rightarrow \alpha$ Transformation Kinetics in the Heat-Affected Zone of Commercially Pure Titanium Arc Welds," Metall. Mater. Trans. 29 A (1998), 2761 - 2773.

19. H.J. Frost and M.F. Ashby, Deformation-Mechanism Maps: The Plasticity and Creep of Metals and Ceramics. (Oxford, UK: Pergamon Press, 1982), 44.

20. Metals Handbook: Metallography and Microstructures. Desk Edition. (Metals Park, OH: ASM, 1985), 464-468.

21. T.G. Nieh, J. Wadsworth, and O.D. Sherby, Superplasticity in Metals and Ceramics. Cambridge Solid State Science Series, ed. D. R. Clarke, S. Suresh, and I. M. Ward. (Cambridge, UK: Cambridge University Press, 1997), .

22. L. Finnie and W.R. Heller, Creep of Engineering Materials: McGraw-Hill Book Company, Inc., 1959).

AN INVESTIGATION OF THE EFFECTS OF SINTERING DURATION AND POWDER SIZES ON THE POROSITY AND COMPRESSION STRENGTH OF POROUS TI-6AL-4V

R. Cirincione, R. Anderson, J. Zhou, D. Mumm and W.O. Soboyejo

Princeton Materials Institute and Department of Mechanical and Aerospace Engineering
Princeton University, Olden Street, Princeton, NJ 08544

Abstract

This paper presents the results of a study of the effects of sintering duration at 1000°C on the structure and compressive strength of porous Ti-6Al-4V. The porous Ti-6Al-4V materials with 40-50% relative density were produced by sintering in an inert dry argon environment at 1000°C. The sieved powders were of diameters between 125 and 250 µm. It was found that the porosity was not significantly affected by initial powder size. However, the compression strength increaseed with increasing sintering duration (up to 24 hours at 1000°C). The strengths of the compacts produced from the smaller powders (125-150 µm in diameter) were greater than those produced from larger (180-250 µm) powders. The larger powders were also shown to be more susceptible to surface nitride formation during sintering in the presence of flowing inert dry argon atmospheres.

Introduction

Significant efforts have been made to produce ultra-lightweight titanium/titanium alloy foams in the recent years [1-8]. This has been due largely to the attractive combinations of strength (10-800 MPa) [1], density (1.0-3.5g/cm^3) [1,6] and corrosion resistance [9] that can be engineered by the use of porous Ti/Ti alloy structures. Unlike low melting point metals (Al, Cu, Zn), for which a range of techniques have been developed for the production of ultra-light foams [11], progress in the development of porous foams from higher melting point materials has been relatively slow. This is due largely to the decomposition of common foaming agents at temperatures that are significantly below those required for the processing of foam structures from higher melting point materials such as Ti-6Al-4V.

In the case of Ti-6Al-4V, some prior work has attempted to develop porous structures by hot-isostatic-pressing (HIPing) [5,6]. These have relied on the use of argon pressure in the development of porous structures. However, it has been difficult to obtain ultra-lightweight porous structures via HIPing, which is generally limited to relatively small structures with ~ 50% relative density.

In contrast, sintering has been used to produce titanium foams with porosities of ~80% [1]. Such structures have been produced by the use of organic space fillers such as carbamide (urea) and ammonium hydrogen carbonate. These are removed at relatively low temperatures (~200°C) during sintering to produce structures with 60-80% porosity. A number of researchers [7,8] have also recently explored the feasibility of producing ultra-lightweight Ti-6Al-4V structures by the sintering of hollow spheres/powders.

This paper presents the results of a preliminary study of the effects of sintering duration on the macrostructure and pore structure of Ti-6Al-4V foams produced by sintering of rapidly solidified T-6Al-4V powder at 1000 °C. Sieved powder geometries with particle sizes between 125-250 μm are used to produce controlled porous structures with relative densities between 0.4 and 0.5. The evolution of powder bonding associated with sintering is explored before investigating the effects of pore/powder size and sintering duration on room-temperature compressive strength and deformation behavior. The implications of the results are then discussed for the design of ultra-lightweight Ti-6Al-4V structures.

Material Processing and Macrostructure/Density

Processing
The porous T-6Al-4V structures were produced by sintering of – 60 mesh powder that was procured from Crucible Research, Pittsburgh, PA. In an effort to control the porosity in the sintered compacts, the powder sizes were varied by controlling the size distributions in each compact. This was accomplished by sifting with (76.2 mm diameter) standard stacks of sieves in sizes ranging from 125 to 250 μm. The size distributions were limited by the sieves that were used to achieve four powder distributions with particle diameters of 125-150 μm, 150-180 μm, 180-212 μm and 212-250 μm.
After sieving, each size distribution was placed in a covered alumina crucible. The crucible was then heated to 1000°C at a ramp rate of 20 °C/min. The heating was carried out in a flowing dry argon atmosphere. Upon reaching 1000°C, the flowing dry argon atmosphere was maintained for

0.5 hour, 1 hour, 2 hours, and 24 hours. All the samples had cylindrical cross-sections, with diameters of ~ 20 mm and heights of ~ 10 mm. After sintering, the furnace temperature was decreased at a ramp rate of 20 °C/min. The flowing argon was maintained to limit the extent of possible oxidation during sintering.

Macrostructure and Density
After processing, the sintered samples were tapped out of the crucibles. The structure of the resulting compacts was then examined under a scanning electron microscope (SEM). The density of the compacts was also estimated by dividing the measured weight by the volume, which was inferred from the dimensions of the cylindrical compacts. The results of the density measurements are shown in Table I.

Table I. Densities of Compacts Sintered for Different Durations, t, at 1000°C

Powder Size Range (microns)	Density (g/cm^3)			
	t = 30 minutes	t = 1 hour	t = 2 hours	t = 24 hours
125-150	2.18	2.25	2.22	2.62
150-180	2.12	2.16	2.07	2.26
180-212	2.06	2.08	2.20	2.17
212-250	2.06	2.04	2.01	2.19

Although varying the size of the Ti-6Al-4V particles affected the porosity, different sintering durations did not significantly alter the porosity/relative densities of the samples. Note that the relative densities were determined by the ratios of the sample densities to that of fully dense Ti-6Al-4V (4.43 g/cm^3). The estimates of relative density are presented in Table II. This shows that the porous Ti-6Al-4V had relative density levels between 0.45 and 0.59.

Typical SEM images of the sintered compacts are presented in Figures 1a-d. These show clearly that the extent of sintering between powder particles increased with increasing sintering duration between 30 minutes and 24 hours. Also, the number of sintered powder particles was very low in the case of the compacts obtained after sintering for 30 minutes (Figure 1a). In contrast, most of the powder particles were sintered together after 24 hours of thermal exposure at 1000°C.

Table II. Relative Densities of Sintered Compacts Sintered at Different Durations, t, at 1000°C

Powder Size Range (microns)	Relative Density (ρ_s/ρ)			
	t = 30 minutes	t = 1 hour	t = 2 hours	t = 24 hours
125-150	0.49	0.51	0.50	0.59
150-180	0.48	0.49	0.47	0.51
180-212	0.47	0.47	0.50	0.49
212-250	0.47	0.46	0.45	0.49

The SEM images of the macrostructures also revealed the presence of a bimodal distribution of powder particles in each of the compacts (Figures 1a-d). In all cases, much smaller particles were mixed into the powder distributions. These particles were present on the Ti-6Al-4V powders, even before sieving and subsequent sintering (Figure 2). They are, therefore, artifacts of the atomization process that was used to produce the Ti-6Al-4V powders [9].

It is also interesting to note here that the smaller powder particles in Figures 1a-d occur in clusters. These clusters play a crucial role in the sintering process, as shown in bridge type bonds

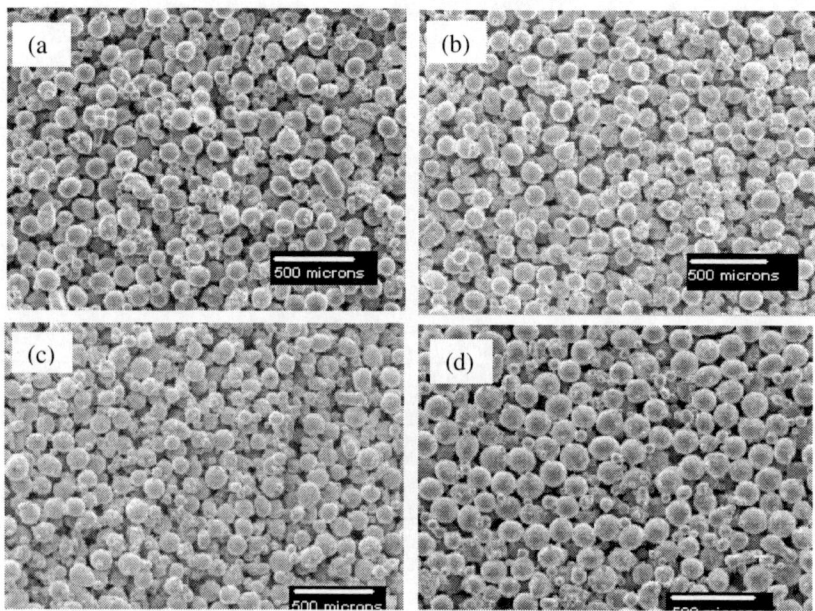

Figure 1: Sintered compacts for (a) 30 minutes; (b) 1 hour; (c) 2 hours; (d) 24 hours. All powder sizes are 125-150 μm, except d) 150-180 μm.

Figure 2: Unsintered powder (size 180-212 microns).

observed after 30 minutes (Figure 3a). These evolve into larger area bonds after durations of 1hour (Figure 3b), 2 hours (Figure 3c) and 24 hours (Figure 3d). It is important to note here that the number of bonds per unit area was much greater after increasing sintering duration to 24 hours (Figures 1a-d). Also, the size of the neck was significant in the regions in which the diffusion bonds were formed. Such necks are required for the elastic-plastic deformation of porous structures [10,11].

Furthermore, the curvatures that were observed in the regions of powder contact are consistent with those expected from surface diffusion-controlled processes [12,13]. Also, the smaller diameter titanium alloy spheres (bridging particles in Figure 3a) facilitated initial mass transport between adjacent larger particles. The images in Figures 3a-d suggest that they increase the amount of surface contact, leading ultimately to the prevalence of like-size large particle bonds (Figures 3b-d), in which diffusion bonds are formed between particles of the intended size within a particular powder size range.

The number of like-size particle bonds decreased with increasing powder particle size (Figures 4a-d). Also, only a few of the larger powder particles sintered successfully without the presence of the smaller powders. Furthermore, both bridge-type and like-size particle bonds were visibly larger and more developed, as sintering time increased (Figures 3a-d).

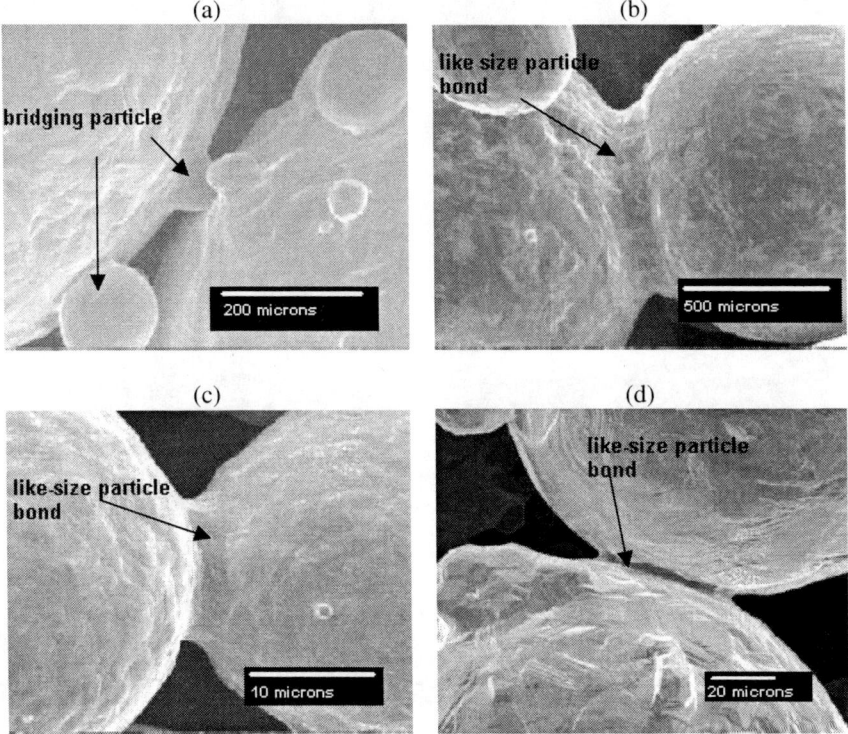

Figure 3: Bridge type bond at 30 minutes (a); large area bonds at 1hour (b); 2 hours (c); and 24 (d) hours. It should be noted that the most important aspect of increasing the sintering time was the increased number of bonds formed, not just the development of them.

Microstructure and Microchemistry
Higher magnification images of the sintered compacts revealed changes in the microstructure of the Ti-6Al-4V alloy. At room temperature, Ti-6Al-4V exhibits a two-phase structure that consists of an hexagonal-close-packed (HCP) crystal structure and a body-centered-cubic structure [10]. However, at temperatures above 980°C, this structure transforms to a body-centered cubic (BCC) structure. Upon cooling to room temperature (from above the β solvus), a transformed α/β structure is formed, consisting of α and β lamellae in a β matrix. The lamellar structure is evident in Ti-6Al-4V after sintering at 1000°C (Figure 5b). Analysis of both the

unsintered powder and sintered compacts by electron diffraction spectroscopy (EDS) also revealed that minor compositional changes occurred during the thermal exposure of the alloy. Increased amounts of nitrogen are apparent in the EDS analyses of the sintered specimens (Figure 6). The presence of nitrogen is not surprising, since nitrogen is the most abundant element in air. Also, titanium is highly reactive. Hence, these results are to be expected, given that nitrogen was not excluded from the dry inert argon atmosphere that was used in the sintering experiments.

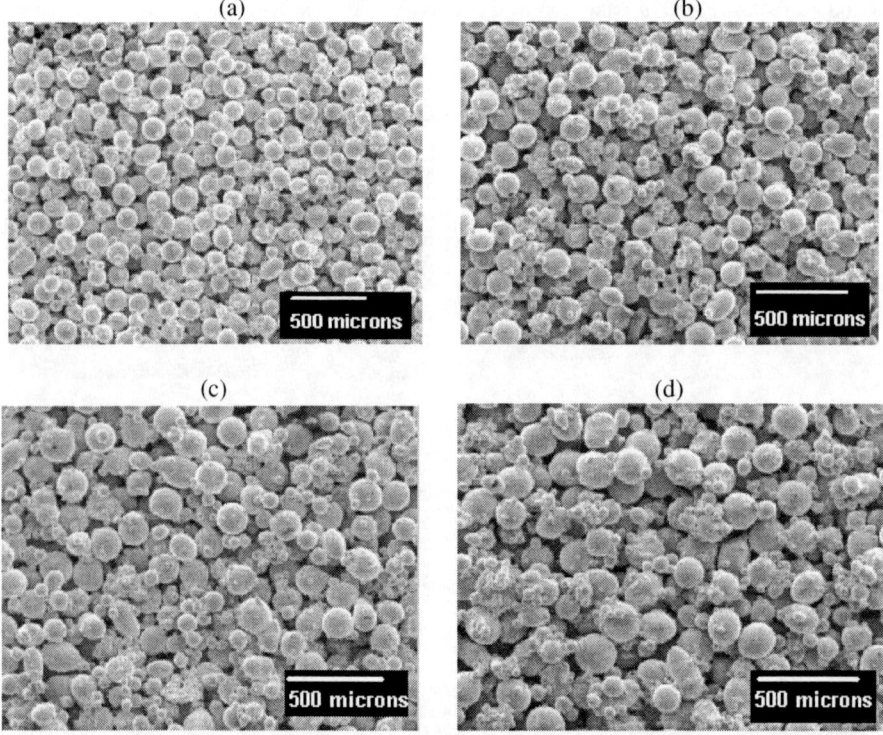

Figure 4: Four Ti-6Al-4V powder size ranges sintered at 1 hour for various particle size ranges: (a) 125-150 µm; (b) 150-180 µm; (c) 180-212 µm; (d) 212-250 µm.

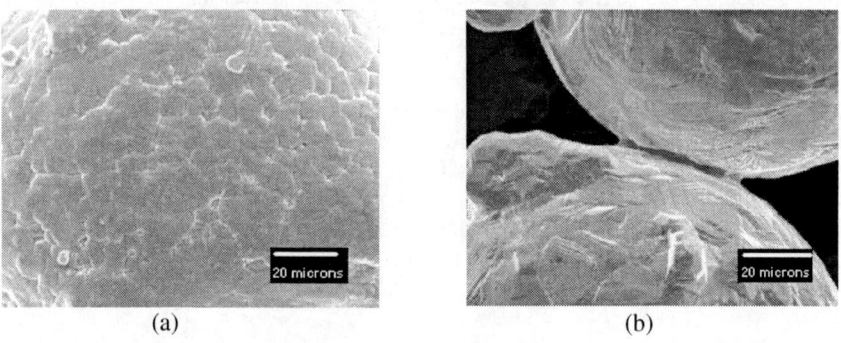

Figure 5: Microstructure of (a) unsintered powder, and (b) sintered powder (note lamellar structure of the sintered powder).

Experimental Procedures

Cylindrical specimens (diameter of ~20 mm, height of ~10 mm) were produced by sintering, as described in the prior section. After sintering, the specimens were tapped out of the crucibles. Before testing, the precise sample dimensions were measured with a micrometer. Compression testing was carried out in an Instron 8501 servo-hydraulic testing machine (Instron, Canton, MA). The specimens were tested under displacement control at a ramp rate of 3 mm/minute. The specimens were deformed continuously to failure (rupture or cracking) after grinding the surfaces flat to facilitate uniaxial loading with compression platens. Due to limited sample availability, only a portion of the processing conditions was tested. However, enough samples were tested to establish the effects of sintering duration and powder particle size on compression strength and deformation behavior.

Experimental Results

Compressive testing revealed significant strength improvement with increased sintering duration (Table III and Figure 7). Samples sintered for 30 minutes and 1 hour had relatively low strength levels compared to those sintered for 2 and 24 hours. Also, their stress-strain behavior under load was much more irregular (Curve I in Figure 7), when compared to that of the 2-hour and 24-hour samples. This irregular deformation behavior is most likely due to wide variability of different particle bond strengths. Moreover, the 2-hour samples had lower strength levels compared to those of the compacts produced after sintering for 24 hours (Table III).

The other types of deformation observed are shown as curve II and III in Figure 7. Curve II in Figure 7 corresponds to a sample in which crushing occurred after elastic-plastic deformation. No plateau stress was observed, although the unloading and loading curve indicates plastic deformation. Similarly, curve III in Figure 7 does not exhibit a significant plateau stress, following the onset of plasticity. However, the final failure point of curve III in Figure 7 corresponds to the onset of crack propagation. The above examples are not typical of the three-stage behavior that is commonly observed in metallic foams [10,11]. However, reasons for the observed differences are not yet fully understood.

Table III. Compressive Plastic Strengths for Porous Ti-6Al-4V (MPa)

Powder Size Range (μm)	Compressive Plastic Strengths for Various Sintering Duration (MPa)			
	30 minutes	1 hour	2 hours	24 hours
125-150	-	-	23.0	50.0
150-180	-	-	21.0	55.0
180-212	4.5	7.0	10.3	36.0
212-250	-	-	11.0	38.0

In view of the observed differences in stress-strain behavior, only the plastic/yield stress level can be used as a measure of strength that is applicable to the different sintered compacts. This was taken to correspond to the stress level at a strain of .002 in the current study. The resulting strength estimates are summarized in Table III. This shows that the plastic strength increases with increasing sintering duration. The plastic strength level also decreases with increasing particle strength. This is not surprising in view of the decreasing number of bridging bonds and like-sized particle bonds that occurs with increasing particle size (Figures 4a-d).

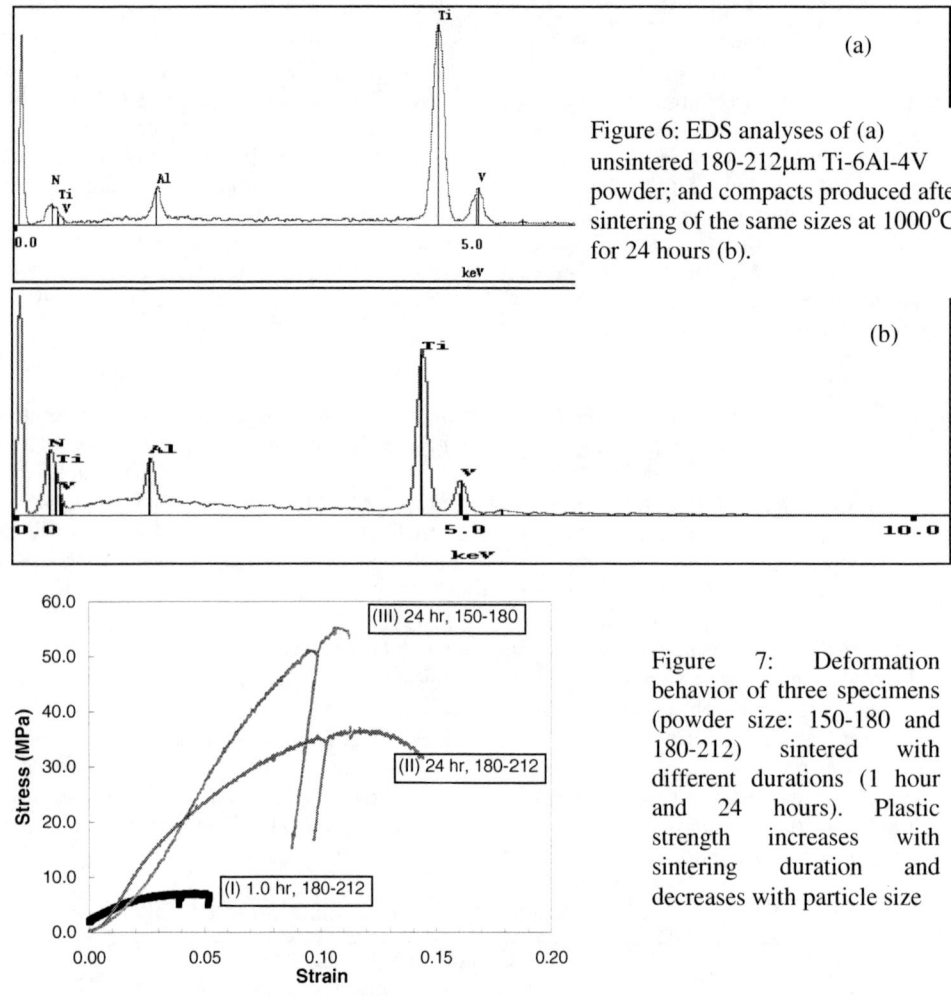

Figure 6: EDS analyses of (a) unsintered 180-212μm Ti-6Al-4V powder; and compacts produced after sintering of the same sizes at 1000°C for 24 hours (b).

Figure 7: Deformation behavior of three specimens (powder size: 150-180 and 180-212) sintered with different durations (1 hour and 24 hours). Plastic strength increases with sintering duration and decreases with particle size

Discussion

The current work shows that sintering can be used to produce porous Ti-6Al-4V compacts with moderate compressive strength levels (4.5 – 55 MPa). However, the measured strength levels are strongly dependent on particle size and sintering duration. Since the number of bonds increases with decreasing particle size, much higher strength levels are obtained from compacts produced with smaller particles (125-180 μm diameter). The increased bonding is also associated with the increased incidence of bridging bonds and like-sized particle bonds (Figures 4a-d).

It is of interest to compare the measured densities and strength levels with those reported by other researchers [1,5]. In the case of HIPing compacts [6,7], relative densities of ~ 0.5 to ~ 0.7

and strength levels up to 530 MPa have been obtained. However, it has been difficult to process ultra-lightweight Ti-6Al-4V structures with relative densities less than 0.5.

In the case of sintered compacts, a space-holder method has been used to produce sintered Ti-6Al-4V with porosities between 60 and 80% [1]. This method uses carbamide (urea) and ammonia hydrogen carbonate as space holders that can be used to produce angular or spherical pores. The space holders are removed during an initial thermal exposure at temperatures below 200°C. The compacts are then sintered at 1200 or 1400 °C to produce porous structures that exhibit elastic-plastic deformation that is similar to that of aluminum foams [10,11]. This suggests that sintering at 1200 and 1400 °C may give rise to improved elastic-plastic deformation characteristics.

Furthermore, recent work [7,8] suggests that hollow spheres may be sintered to produce ultra-lightweight Ti-6Al-4V structures with relative densities as low as 5-10% of the fully dense values. This is very exciting information, since it does suggest the possibility of fabrication of ultra-lightweight Ti-6Al-4V structures. However, the sintering conditions for the fabrication of hollow sphere foams are yet to be fully established. Also the elastic-plastic deformation behavior of such structures is yet to be studied in detail.

Summary and Concluding Remarks

Various sizes (125-250 μm) of Ti-6Al-4V powders were sintered in an argon atmosphere for varying durations. The samples were analyzed using SEM and tested to failure under compressive loads. Significant results obtained in this study are summarized at the following:

1. The porosity was not significantly affected by the sintering duration, at 1000°C, but the relative density and compressive strength of the porous compacts were found to vary with sintering time and powder sizes.

2. Increased sintering duration resulted in increased compressive plastic strength. Moreover, strength generally decreased with increasing powder size. This was due to the higher incidence of diffusion bonds that was facilitated by bridging particles and increased surface contact that occurred with decreasing particle size.

3. Sintering at 1000°C (for durations up to 24 hours) results in compacts with strength levels between 4.5 and 55.0 MPa. The highest strength level is obtained after sintering of 150-180 μm powders for 24 hours. This results in measured strength levels of ~55 MPa. The overall strength levels may be further improved by the optimization of sintering temperature and duration.

4. Relative densities of ~ 0.45 - 0.59 have been achieved by sintering at 1000°C without fillers. This suggests that the use of space fillers and the sintering of hollow spheres of Ti-6Al-4V may obtain much lower densities, and hence ultra-lightweight Ti-6Al-4V foams.

Acknowledgements

This work is supported by the Office of Naval Research, with Dr. George Yoder as Program Manager. Appreciation is extended to Mr. Nicholas Marcantonio for assistance with the initial stages of this work. The authors would like to thank Prof. Anthony Evans for useful technical discussions on metallic foams.

References

1. M. Bram, C. Stiller, H.P. Buchkremer, D. Stover and H. Baur, "High Porosity Titanium, Stainless Steel and Superalloy Parts," Advanced Engineering Materials, 2 (2000), 169-199.
2. C. Maniatopoulus, R.M. Pilliar and D.C. Smith, "Threaded versus Porous-Surfaced Designs for Implant Stabilization in Bone-Endodontic Implant Model," Journal of Biomedical Materials Research, 20 (1986), 1309-1333.
3. H.U. Cameron and H. Pilliar, "The Rate of Bone Ingrowth in Porous Metal," Journal of Biomedical Materials Research, 10 (1976), 295-302.
4. Y. Z. Yang, J.M. Tian, Z. Q. Chen, X. J. Deng, D.H. Zhang, "Preparation of Graded Porous Titanium Coatings on Titanium Implant Materials by Plasma Spraying," Journal of Biomedical Materials Research, 52 (2000), 333-337.
5. R. Martin, "Integral Porous Core Metal Bodies and In-Situ Method of Manufacture Therof," Patent 05564064, 1997.
6. N.G. Davis, D.J Teisen, C. Schuh and D.C. Dunand, "Solid-State Foaming of Titanium By Superplastic Expansion of Argon-Filled Pores," Journal of Materials Research, 16(2000), 1508-1520.
7. N. E. Baxter, T.H. Sanders, Jr., A.R. Nagel, C. Uslu, K.J. Lee (1997) *Metallic Foams From Hollow Spheres,* TMS Annual Meeting, p. 417-423.
8. O. Anderson, U. Waag, L. Schneider, G. Stephani, and B. Kieback, "Novel Metallic Hollow Sphere Structures," Advanced Engineering Materials, 2 (2000), 192-198.
10. L.J. Gibson and M.F. Ashby, Cellular Solids: Structure and Properties (Cambridge, United Kingdom, Cambridge University Press, 2nd Edition, 1997).
11. A.G. Evans, J.W. Hutchinson, N.A. Fleck, M.F. Ashby and H.N.G. Wadley, " The topological design of multifunctional cellular metals," Progress in Materials Science, 46 (2001), 309-327.
12. P.G. Shewmon, Diffusion in Solids (TMS, Warrendale, PA, 1989).
13. E.W. Collins, The Physical Metallurgy of Titanium Alloys (Materials Park, OH, 1984).

SECTION IV

MECHANICAL AND SERVICE PROPERTIES

AN INVESTIGATION OF COMPRESSIVE DEFORMATION MECHANISMS IN OPEN-CELL METALLIC FOAMS

J. Zhou, P. Shrotriya and W.O. Soboyejo

Princeton Materials Institute and Department of Mechanical and Aerospace Engineering
Princeton University, Olden Street, Princeton, NJ 08544

Abstract

This paper presents the results pf an investigation of compressive deformation mechanisms in Duocel open-cell aluminum foams. Dislocation slip bands were used to identify the sites in which plastic deformation was initiated and propagated in individual struts in foam samples with dimensions of 10 mm ×10 mm × 10 mm. The results from these studies are compared with those obtained from tests of larger samples with dimensions of 24.5 mm × 24.5 mm × 17.0 mm. It was found that plastic deformation occurred well before observable shape changes in the struts or cells. Plastic deformation in the foam structure was found to involve the plastic buckling of the struts parallel to the loading axis, and plastic bending of struts that have large angles with respect to the loading axis. However, plastic bending was observed to be the dominant deformation mechanism because very few struts were oriented parallel to the compressive loading axis.

Introduction

The initial studies of plastic deformation in closed-cell metallic foams were carried out by Thornton and Maggie [1,2] about 30 years ago. They had investigated plastic deformation in closed-cell metallic foams made from pure aluminum, an aluminum alloy containing 7% Mg, a 7075 aluminum alloy and zinc alloys. They found that the relationship between strength and relative density was non-linear, and that the measured strengths were scattered.

The most systematic studies of the mechanical behaviors of foams were carried out subsequently by Gibson and Ashby [3-5] in early 1980's. Following detailed studies of deformation mechanisms in various foams made from elastomeric polymer, elastic-plastic polymer and brittle materials, they proposed dimensional models that relate mechanical properties of foams with their relative densities. For foams made from elastic-plastic materials, their Young's modulus (E_f) and plastic collapse strength (σ_{pl}) are given by

$$\frac{E_f}{E_s} = C_1 \left(\frac{\rho_f}{\rho_s}\right)^2 \quad (1) \qquad \frac{\sigma_{pl}}{\sigma_{YS}} = C_2 \left(\frac{\rho_f}{\rho_s}\right)^{\frac{3}{2}} \quad (2)$$

where ρ_f is the density of a foam block, E_s, σ_{YE} and ρ_s are the Young's Modulus, yield strength and density, respectively, of the buck material from which the foam is fabricated; C_1 and C_2 are constants whose values should be determined by experiments.

In recent years, metallic foams have been developed for a wide range of potential commercial applications in ultra-lightweight-structures [6]. There have been a number of experimental studies of compressive deformation in metallic foams [7-11]. These include the use of surface mapping technology and computed x-ray tomography in the characterization of foam/strut/cell wall deformation [9,10].

However, in all prior work [1-10], the observations of deformation mechanisms in metallic foams were made only when the size-scale of the plastic deformation phenomena was large enough to be recorded by camera. Such observations were clearly limited by the resolution of cameras. Hence, since the onset of plastic deformation is likely to occur well before the observable shape changes in struts/walls or cells, a question arising from previous observations is how the plastic deformation (buckling or bending) is initiated and propagated. Furthermore, because plastic deformation occurs by the movement of dislocations, which in turn can leave dislocation slip bands as markers of plastic deformation in the struts of metallic foams, dislocation slip bands are excellent markers for the identification of the locations in which plastic deformation occurs.

In this study, scanning electron microscopy (SEM) was used to investigate the initiation and spread of plastic deformation in individual struts with open-cell 6101 aluminum (annealed) foams deformed under compressive loading. The initial and propagation of plastic deformation in foams were studied carefully along with shape changes in struts and cell morphologies that occurred with increasing compressive strain.

Materials

The ERG Duocel open-cell foams that were used in this study were fabricated from a 6101 aluminum alloy, in which Mg, Si and Fe are the main alloy elements. The chemical composition of alloy has been analyzed in previous study [11]. The foams were fabricated with a casting process that is similar to investment casting [6]. After formation, the foam blocks were subjected to annealing to stabilize their mechanical properties.

Foam blocks with 10 pores per inch were used. Here, the pores size is represented by the number of pores per inch (PPI). The relative density of studied foams is 0.82. Figures 1 shows the foam morphologies of a 10 PPI sample. It should be noted that the cells are significantly elongated in one direction, as indicated by double arrows in the image. Struts along the elongation directions were much longer than struts along the transverse direction. The elongation of cells and struts indicates that Duocel open-cell aluminum foams are not truly stochastic.

Experimental Procedure

In this study, small and large foam blocks of the 10 PPI foams were progressively loaded to different stages of compressive deformation. The loading process was controlled by displacement at a rate of 0.01 mm/sec. After each loading-unloading cycle, the tested sample was examined under a SEM to investigate the mechanisms by which the plastic deformation was initiation and propagation in the foam struts and blocks. Specimens with two dimensions were tested. The so-called small samples had dimensions of about 10.0 mm × 10.0 mm × 10.0 mm, while the larger samples had dimensions of about 24.5 mm × 24.5 mm×17.0 mm. All the specimens used in this study were obtained via electro-discharge machining (EDM).

Small samples were first tested to study the initiation of plastic deformation in individual struts. The limited number of struts in these small samples made it possible to check and trace the initiation and propagation of plastic deformation in each strut under SEM. Linear elastic deformation was studied at small load before yielding occurred in the tested small samples. The initiation of plastic deformation in each individual strut was identified once dislocation slip bands were observed on any portion of the strut. However, it is important to note that in plastic metals, the actual onset of plastic deformation occurs well before the onset of significant shape change. It is, therefore, possible for plastic deformation to occur in some struts, and cease after a certain mount of deformation that does not lead to observable shape changes. It is, therefore, incorrect to relate observable shape changes to the onset of plastic deformation in the struts. However, dislocation slip bands can always be observed when plastic deformation occurs by dislocation motion in grains within the struts. Since the grain sizes are on the order of millimeters (the same order as the strut thickness of studied foams) [11], the slip bands, therefore, can be observed easily in SEM. Hence, the identification of the dislocation slip bands on deformed struts provides an efficient and reliable method to determine the local conditions for the onset and spread of plastic deformation in individual struts. Following the study of compressive deformation mechanisms in small foam blocks, the plastic buckling sequence of a selected strut in a larger sample was recorded via a camera that was connected to a Questar QM1 Telescope (Questar is a trademark of the Questar Corporation of New Hope, PA).

Since the deformation mechanisms observed in small samples may not be typical of those in real foams, large samples with dimensions of about 24.5 mm × 24.5 mm×17.0 mm were examined to confirm the deformation mechanisms in observed small foam sample. Similar loading and unloading cycles (to those used in the testing of the small samples) were used to study the deformation mechanisms in large samples. Special efforts were made to identify the deformation bands that were observed in elastic-plastic polymeric foams and some closed-cell metallic foams. It should be noted that all loads were applied along the elongation direction, which is the strong direction, or close to the strong direction.

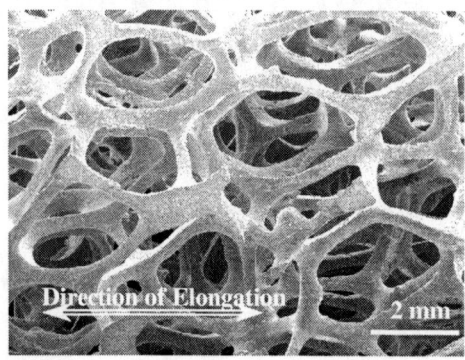

Figure 2: Macrostructure of open-cell 6101 aluminum foams along longitudinal section in a 10 PPI foam block. The foam cells are elongated in the direction indicated by arrows.

Results and Discussions

Plastic Collapse Deformation of a Small 10 PPI Sample
The complete stress-stain curve of a small sample is presented in Figure 2a. There are 31 loading/unloading curves in this diagram. After each unloading, the sample was examined via SEM. This was done to investigate possible changes of cell morphology and the macro/micro-deformation of foam struts. The stress-strain curve obtained for the first 10 cycles are presented in Figure 2b. It should be noted that the first three loading and unloading curves are within linear-elastic deformation regime. This is the region enclosed by the distorted elliptic curve in Figure 2b. Corresponding to the stress-strain curve shown in Figure 2a, the change of cell morphology and deformation of the foam struts are presented in Figure 3. The images $a, b, c...k$ in Figure 3 correspond to the sequence of letters in Figure 2. Also, for comparison, an SEM image of the same undeformed 10 PPI block is shown in Figure3l.

Stress-strain curves for the first three loading-unloading cycles are primarily identical. It should be noted that the loading curves and unloading curves overlay. Also, no dislocation slip bands were observed after the first three loading/unloading cycles in SEM. This is reasonable because the foam block is still in the linear elastic stage, and elastic deformation was completely recovered upon unloading.

Beyond the onset of local yielding, the measured stress-strain curve becomes non-linear (Figure 2), and the unloading curve deviates from the loading curve. This occurred when the sample was unloaded at a, as shown in Figure 2. The strain level at this point is about 0.009 prior to this unloading. There is no observable change of cell morphology, or macroscopic deformation of struts. However, dislocation slip bands were first observed on Strut I and strut II (in Figures 3a and 3b) under SEM. The evident details of dislocation slip bands are presented in Figure 4. Both strut I and II were close to the surfaces on which the compressive load was applied. They were oriented in a direction that was almost perpendicular to the loading axis. The sites where dislocation slip bands were observed were labaled in Figure 4.

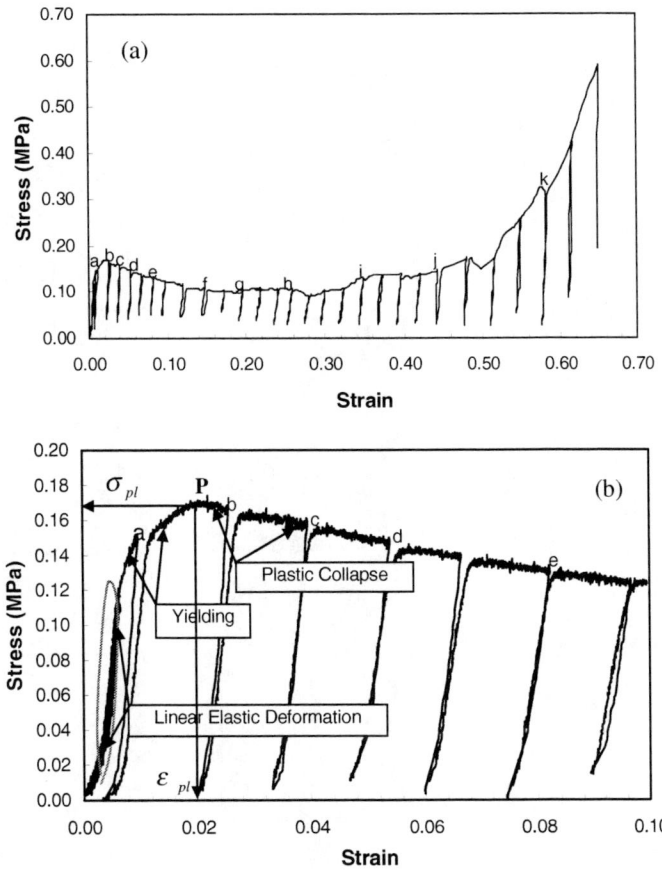

Figure 2: Stress-strain curve of small 10 PPI sample. (a) Complete stress-strain curve including 31 loading and unloading cycles; (b) details of stress-strain curve during early deformation stage. The linear elastic portion is enclosed by the distorted elliptic curve.

It is important to note that the slip bands were not homogeneously distributed, instead, the slip bands were only observed in two discrete regions in each deformed strut. The two regions were close to the two vertices to which the strut was connected. This suggests that plastic bending occurred on the deformed struts, resulting in the formation of plastic hinges. Subsequent observations revealed that the slip bands were limited within these two regions, even when the strain increased to about 0.50. The regions in which no slip bands were observed are also indicated by arrows in Figures 4a-c. Here, each figure consists of 4 sub-images at increasing magnifications. Each subsequent image presents an amplified picture obtained from the central part of the former one. The initial magnifications of the four parts are 15×, 50×, 200× and 1000×, respectively.

The highest stress was obtained at strain about $\varepsilon = 0.025$. This corresponds to point P in Figure 2b. Almost immediately after point P, the sample was unloaded at b, and examined in SEM. More struts were found to have participated in plastic deformation. Following this unloading

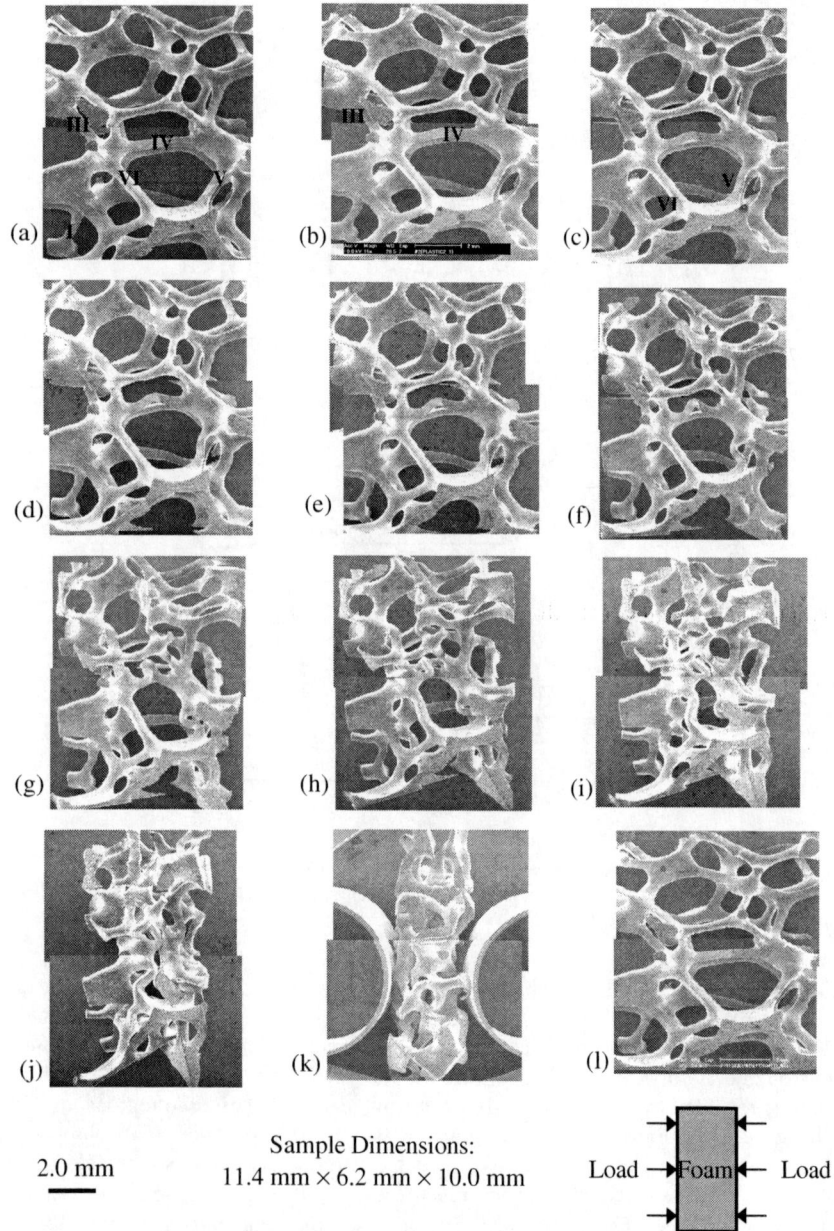

Figure 3. Change in cell morphology and deformation of individual struts at various strains for the small 10 PPI sample. Sequence letters correspond to the same sequence letter in Figure 2: ε = (a) 0.009, (b) 0.025, (c) 0.04, (d) 0.05, (e) 0.08, (f) 0.15, (g) 0.20, (h) 0.27, (i) 0.35, (j) 0.44, (k) ε = 0.58, (l) ε = 0

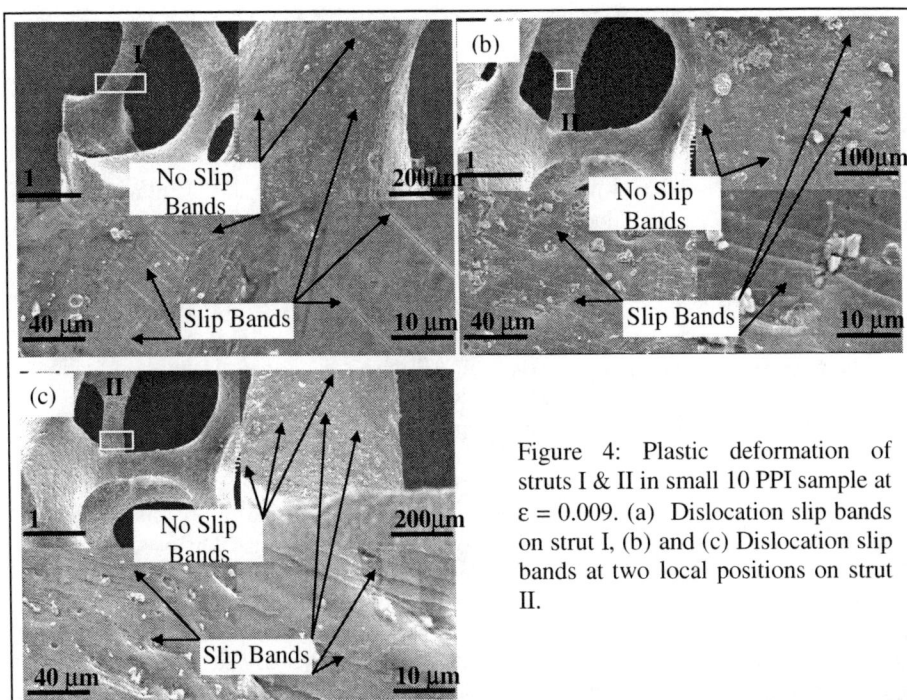

Figure 4: Plastic deformation of struts I & II in small 10 PPI sample at $\varepsilon = 0.009$. (a) Dislocation slip bands on strut I, (b) and (c) Dislocation slip bands at two local positions on strut II.

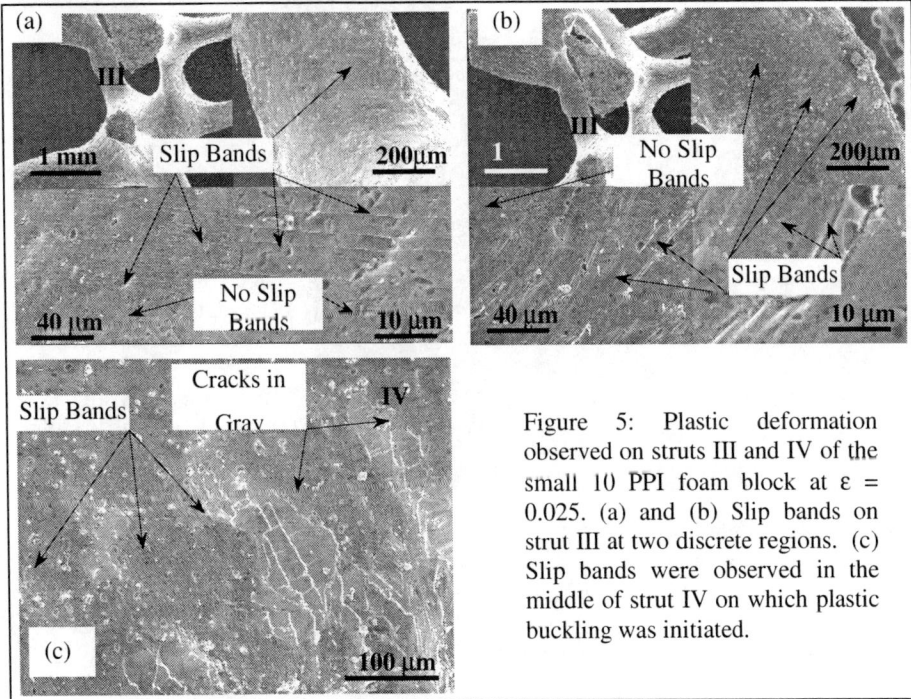

Figure 5: Plastic deformation observed on struts III and IV of the small 10 PPI foam block at $\varepsilon = 0.025$. (a) and (b) Slip bands on strut III at two discrete regions. (c) Slip bands were observed in the middle of strut IV on which plastic buckling was initiated.

sequence, dislocation slip bands were also observed on strut III, as shown in Figures 3a and 3b. Again, the slip bands were only observed at two discrete regions that are close to the two vertices to which Strut III was connected. It is interesting to note that dislocation slip bands were also presented in the middle of strut IV, as shown in Figure 5c. Comparing Figure 3b with Figure 3a, a slight bow can be observed to be developed in the middle of strut IV. This indicates the onset of plastic buckling on this strut. It should be noted that strut III was inclined at an angle of about 60 degrees with respect to the loading axis, while strut IV is almost parallel to the loading axis. When strain was increased to 0.039 (at c in Figures 2), plastic deformation was observed on strut V, as indicated in Figure 3c. This strut was deformed to reconsolidate the change in cell shape due to the post buckling of strut IV. It is interesting to observe that no deformation occurred on strut VI at this strain level. By a careful observation of the image before deformation, it can be shown that the angle between the load axis and strut V is larger than that between strut VI and the load axis. Strut VI is so strong that plastic deformation was only observed at strain larger than 0.152, as shown by f in Figure 3. Before the plastic deformation of the strong strut VI, the cell shape change was achieved by a combination of post-buckling of strut IV, and further plastic banding of strut V. When the strain was equal to 0.053 (d in Figures 2), slip bands were observed on all struts within this tested foam block (Figure 3d). Following the initiation of buckling in strut IV, the tested sample height decreased with further post-buckling and plastic bending of the deformed struts. Dislocation slip bands were also initiated and propagated along the remaining strut were either aligned perpendicular to the loading direction, or heavily twisted to bring the two vertices together, as shown by i in Figure 3. When strain reached about 0.600, the vertex and closed cell faces got to touch each other, and started to participate in the plastic deformation to densify the foam block. At the same time, struts twist so seriously that they could not be identified any more, as shown by k in Figure 3.

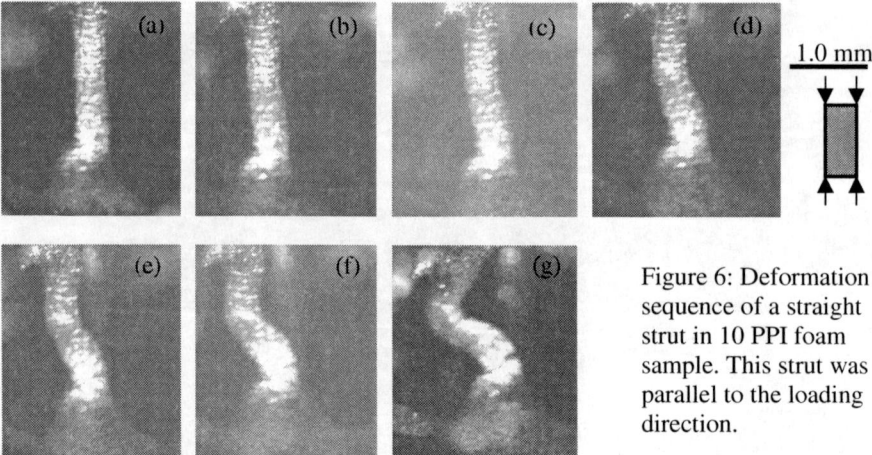

Figure 6: Deformation sequence of a straight strut in 10 PPI foam sample. This strut was parallel to the loading direction.

In summary, there were two types of plastic deformation modes in struts observed to initiate the plastic collapse of compressed open-cell foams, when the load was applied parallel to the cell elongation directions. The first deformation mode involved the bending of long struts that were not parallel to loading directions. Plastic bending was the major deformation mode because few struts were oriented parallel to the loading axis, strut I, II, III and V in Figure 3 belong to this deformation mode. Strut bending was by plastic failure mode that initiates plastic collapse in a compressed foam block [1-10]. The other significat failure mode was buclking of long struts oriented along the loading direction. This was evident with the half sinusoidal buckling wave form of strut IV in Figure 3. The image sequence in Figure 6 shows plastic buckling process of a selected straight strut that was oriented parallel to loading axis.

Plastic Collapse Deformation of Large 10 Sample
Figure 7a shows the cell morphology and foam structural of a large 10 PPI sample. Upon loading to a stain $\varepsilon = 0.100$ (Figure 7b), these aligned cells started to lose their initial alignment by plastic bending of long struts. In this tested sample, plastic bending of long struts dominated the deformation. In Figure 7b, circles were used to mark sites of local plastic deformation on the deformed struts. It should be noted that only those deformed struts that can be observed in SEM were marked. There were more deformed struts within the tested foam sample, and the distribution of the circles was homogenous. No deformation bands were observed even at strain rate larger than 0.100. This implies that the plastic deformation occurred homogenously through out the samples at this strain level. At a strain level of 0.200, the cell elongation was completely lost, as shown in Figure 7c.

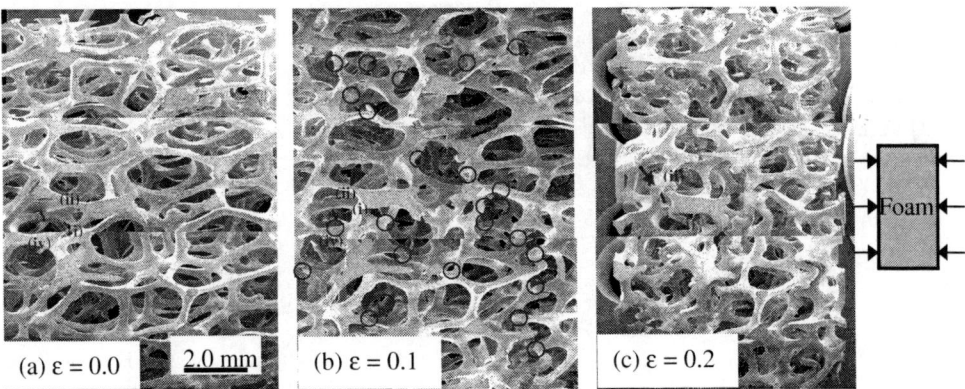

Figure 7: Deformation of large 10 PPI sample under compressive load in direction as indicated. (a) Cell morphology and foam structure before deformation. (b) Deformation at a stain $\varepsilon = 0.100$ and (c) deformation at a strain $\varepsilon = 0.200$. Cell A was deformed by consequent deformation of struts (i), (ii), (iii) and (iv); Sample Dimensions: 24.98 mm × 22.76 mm 17.03 mm

The shape changes in the selected cell through out the deformation are illustrated in Figure 7. Cell A consisted of struts (i), (ii), (iii) and (iv) (in Figure 7a). Strut (i) was first bended. Following the bending of strut (i) to certain extent, plastic buckling was initiated in strut (ii), as indicated by the arrow in Figures 7b. Finally, the buckling of strut (ii) was replaced by plastic bending on strut (iii). No change in shape of strut (iv) was observed at strain level of 0.200.

Summary and Concluding Remarks
Compressive deformation mechanisms were studied by using 10 PPI foam samples with two types of dimensions (10.0 mm × 10.0 mm × 10.0 mm and 24.5 mm × 24.5 mm×17.0 mm). Dislocation slip bands were used to identify initiation and propagation of compressive plastic

deformation on individual struts in foams. The obtained results are summarized as the following:
(1) Linear elastic deformation of a small foam block was observed and recorded.
(2) The onset of plastic deformation was well before the onset of shape changes in struts or cells.
(3) Both plastic bending and buckling were observed in the tested foam samples. Plastic bending was found to be the dominant deformation mechanism due to the strict requirements for the occurrence of plastic buckling.
(4) Plastic deformation in larger samples was homogenous, no deformation bands were observed in the large foam sample under compression.

Acknowledgements

This research is supported by The Office of Naval Research, with Dr. George Yoder as the Program Manager. The authors would like to thank Prof. Evans for stimulating discussion. We express appreciation to ERG for providing foam samples.

References

1. P.H. Thornton and C.L. Magee, "Deformation of Aluminum Foams," Metallurgical Transactions A, 6A (1975), 1253-1263.
2. P.H. Thornton and C.L. Magee, "Deformation Characteristics of Zinc Foam," Metallurgical Transactions A, 6A (1975), 1801-1807.
3. L. J. Gibson, M. F. Ashby, G. S. Schajer and C. I. Robertson, "The Mechanics of Two-Dimensional Cellular Materials," Proceedings of the Royal Society of London A, 382 (1982), 25-42.
4. L. J. Gibson, M. F. Ashby, "The Mechanics of Three-Dimensional Cellular Materials," Proceedings of the Royal Society of London A, 382 (1982), 43-59.
5. L.J. Gibson and M.F. Ashby, Cellular Solids: Structure and Properties (Cambridge, United Kingdom, Cambridge University Press, 2nd Edition, 1997)
6. M.F.Ashby, A.G. Evans, N.A. Fleck, L.J. Gibson, J.W. Hutchinson and H.N.G. Wadley, Metal Foams - A Design Guide (London, United Kingdom, Butterworths, 2000).
7. L.J. Gibson, "Mechanical behavior of metallic foams," Annual Review of Materials Science, 30 (2001), 191-227.
8. A.G. Evans, J.W. Hutchinson, N.A. Fleck, M.F. Ashby and H.N.G. Wadley, " The topological design of multifunctional cellular metals," Progress in Materials Science, 46 (2001), 309-327.
9. H. Bart-Smith, A-F. Bastwros, D.R.Mumm, A.G. Evans, D.J. Sypeck and H.N.G.Wadley:, "Compressive Deformation and Yielding Mechanisms in Cellular Al Alloys Determined Using X-ray Tomograpgy and Surface Strain Mapping," Acta Materialia 46 (1998), 3583-3592.
10. A-F. Bastawros, H.Bart-Smith, A.G. Evans, "Experimental Analysis of Deformation Mechanisms in A Closed-Cell Aluminum Alloy Foam," Journal of the Mechanics and Physics of Solids, 48 (2000), 311-322.
11. J. Zhou, C. Mercer and W. O. Soboyejo, "An Investigation of the Microstructure and Strength of Open-Cell 6101 Aluminum Foams," Submitted (2001).

FABRICATION AND CREEP STUDIES ON OPEN CELL NICKEL STRUCTURES

R.K. Oruganti and A.K. Ghosh

Department of Materials Science and Engineering
University of Michigan, Ann Arbor, MI-48109

Abstract

Cellular Ni-base alloy structures have potential for light to medium load bearing applications at elevated temperatures as these structures may be actively cooled by fluid flow. In this study open-cell Ni structures were fabricated using two different processes. A process to create honeycomb cells was developed that relied on diffusion bonding of thin walled tubes in which sintering forces were used to convert circular core geometry into hexagonal cores. Linear cellular structures with low density (~10% dense) were prepared by this method. The other process utilized woven wire mat to diffusion bond a layered stack to create a denser structure (56% dense). Room temperature deformation behavior of these structures and the compressive creep behavior of the honeycombs was characterized and the results compared with those of a solid material. FE analysis of creep deformation of ideal honeycombs was carried out to provide more insight into the nature of creep in such materials. It is shown that geometry of the honeycomb plays an important role in determining the creep response related to geometric 'weakening' and 'hardening' processes in the cellular structure.

Introduction

Cellular metals are of interest for light to medium load bearing applications. One such structure is a sandwich of honeycomb core with face sheets whose specific stiffness in bending is one of the highest known to mankind and it is therefore a structure suitable for efficient load bearing applications [1]. Fabrication of honeycomb sandwich for high temperature applications is of considerable interest. However, the necessary raw materials and processing are expensive, primarily because producing thin foils of high temperature materials is difficult and time consuming. Thus in this study we experimented with an easier method for fabrication of Ni honeycomb structures. For elevated temperature applications, often active cooling by fluid flow is desirable and thus open-cell structures of Ni-base alloys are also of interest. Woven wire mesh bonded structures under development can provide a unique combination of permeability and high temperature strength, which are also of interest here. Design for elevated temperature application of these materials requires a knowledge of the creep response and the mechanisms responsible for creep in such materials. The honeycomb geometry was chosen for this study since it is a simple linear cell structure amenable to analytical study.

The new fabrication technique for nickel honeycombs developed at the University of Michigan [2] involves diffusion bonding of stacked nickel tubes of thin wall section, in which an initially circular section is converted to a hexagonal geometry. The bonding process involves a transient liquid phase process utilizing electroless nickel coating (Ni- 8%P) [3]. A stack of tubes placed in a constrained die is shown schematically in Fig. 1a. When heated to the eutectic temperature of Ni-P coating, the latter melts and permits rapid diffusion of P into adjoining ligaments of Ni tube thus raising the melting temperature locally. Solidification of these regions leads to bonding of the tube walls. Surface tension forces from the liquid meniscus and creep load at the tube contact points (Fig. 2a) force the circular ligaments to lengthen and assume the hexagonal configuration as shown in Fig. 1c and 1d. The assumed geometry has similarity with the formation of triple points in metal grains and in soap films. In the latter case the transformation from spherical to hexagonal geometry is driven by surface tension and internal pressure within bubbles.

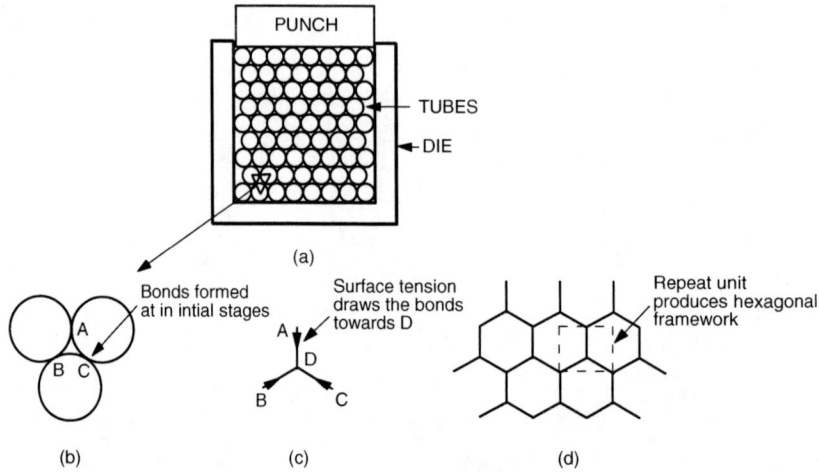

Fig. 1: a) Bonding set up for coated nickel tubes in a constrained die b), c) and d) Stages of sintering and development of hexagonal framework from circular tube geometry

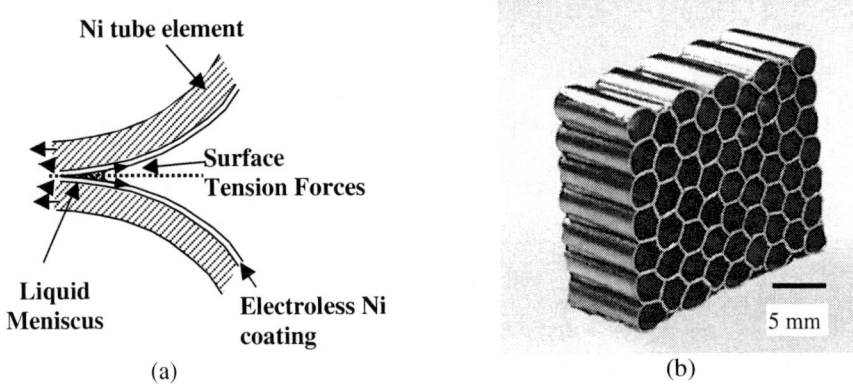

Fig. 2: a) Schematic of tube elements with liquid meniscus and developed forces b) Ni honeycomb fabricated by transient liquid phase bonding process

In elevated temperature use, the creep resistance of such honeycombs perpendicular to the cell axes may be inadequate and this would necessitate development of denser structures that have higher stiffness and strength. A relatively inexpensive method to produce such a structure is reported here which is to start with a stack of woven wire mesh and diffusion bond the layers to create a porous solid.

Elevated temperature mechanical properties - mainly creep - in such open cell structures has been dealt with in a limited way [4] and it is generally accepted that bending of the cell walls controls the overall creep response. Andrews et al. [4] have developed equations for steady state creep in honeycombs based on bending formulae. While the steady state is of fundamental importance, the primary and tertiary regions of the creep response need to be studied for a more complete picture. Finnie and Heller [5] point out that in pure creep bending of beams the time of approach to the steady state stress distribution might consume a large fraction of overall creep life of the beam. Moreover, a steady state analysis does not take into account the effect of macroscopic shape change on the moments induced in the structure by the applied forces. Closed form solutions for creep bending also ignore the deformation that occurs at the points of application of the load such as the regions around the nodes where cell walls meet.

This paper describes the results of creep studies carried out to address some of these issues. Constant load experiments were carried out on honeycombs fabricated using the new technique mentioned above. The experimental results are compared with FEM analyses of a representative section of the honeycomb and insights into the creep curve of a honeycomb thus derived are presented.

Experimental

Fabrication
Linear honeycombs were fabricated by bonding together tubes of commercially pure nickel (Ni-200) of outer diameter 3.17 mm and wall thickness 0.08 mm. The outer surfaces of the tubes were ground with 1000 grit SiC paper and subsequently coated with electroless nickel. The coated tubes were stacked in a graphite die with a light graphite punch on the top and the

bonding was carried out in vacuum at 900°C for 1/2 hour. Subsequent to the bonding run, the specimens were heat treated at 1000°C for 1 hour in order to homogenize the micro and macrostructures. The final structure (Fig. 2b) had a relative density of 9.8%.

The woven wire structure were fabricated by stacking nickel wire mesh of wire thickness 50 μm and pore opening 150 μm (shown in Fig 3a) in a graphite die and compressing under a uniaxial pressure of 0.05 MPa for 30 minutes at 900°C (see schematic - Fig. 3b). The structure so obtained, had a relative density of 56%. The microstructure is shown in Fig. 3c.

Testing

Constant load creep testing was carried out in air in an Instron frame equipped with a three zone furnace which maintained temperature constant to within 1 °C. Thermocouples fixed to the surfaces of the platens contacting the specimen ensured that the gradient over the specimen height was less that 1°C. LVDT's were used to measure displacement of the specimen. Tests were carried out at 600°C and 700°C. Creep data for the solid material were obtained from single tubes that were subjected to the same treatment as the cellular materials.

Unloading experiments were also carried out on the honeycombs subsequent to creeping in order to estimate the magnitude of creep recovery that occurs in such materials. In such experiments, once the specimen was seen to be creeping at a steady state, the load was reduced quickly to a value that was low enough to stop forward creep but high enough to still maintain contact with the platens and the corresponding strain time behavior was recorded.

Results and Discussion

Room temperature tests

Room temperature compression tests were carried out on honeycomb-type samples containing a small number of cells. These cells were not of fully formed hexagonal configuration but were closer to circular geometry. Nominal stress- engineering strain curves for these honeycombs are shown in Fig. 4a for both axial and transverse loading. The strength along the axial direction is substantially higher. This is due to load carrying tubular walls provide column support in the axial direction but deformation is mainly by bending for the transverse orientation. The peak stress for axial loading is quite high (27 MPa) and determined by a combination of yield and plastic buckling. This is followed by a stress drop and a subsequent stress rise when consolidation begins. Transverse loading showed a more gradual strain hardening as expected from bending and strain hardening of the material, without lateral constraint.

Compressive deformation behavior of the woven wire bonded (WWB) structure parallel and perpendicular to the original wire mesh planes are shown in Fig. 4b. Due to a denser structure, it is stronger and stiffer than the less dense honeycomb shown in Fig. 4a, with yield stresses rising to 75 MPa parallel to the layer-plane. This is a structural material of reasonable effectiveness. The higher strength for the layer-plane orientation is due to aligned continuous wires acting as columns, while transverse to layer-planes, wires are not continuous and develop resistance to local bending in regions of contact. More insight into the deformation behavior can be obtained from the work-hardening curves for the two directions (Fig. 5). It is seen that the work hardening curves can be divided into three regions - *a)* hardening rate decreases rapidly from a high value due to initiation of bucking in the wires, *b)* hardening rate reaches a minimum before *c)* the rate increases again. Region *b* represents a balance between the weakening processes such as column

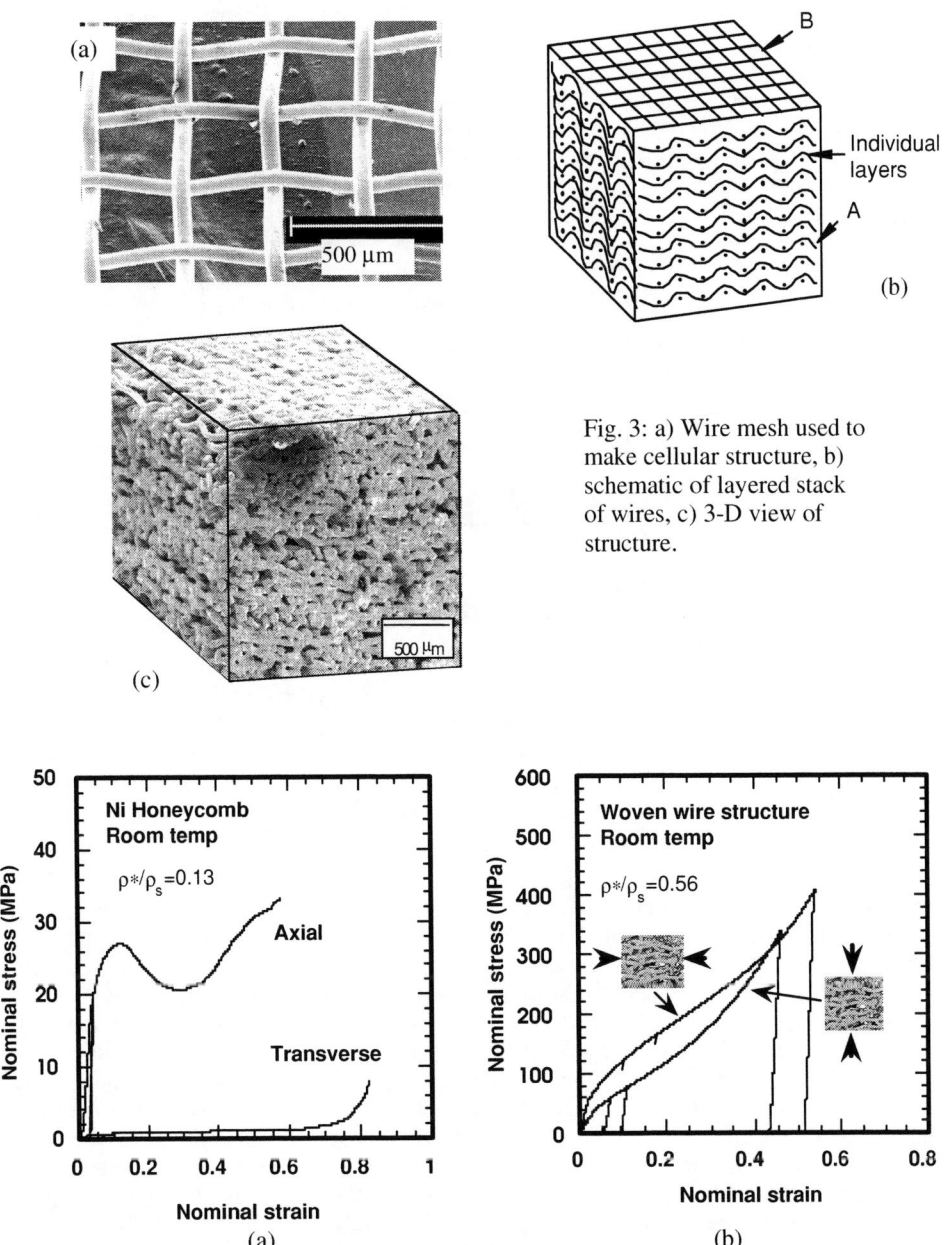

Fig. 3: a) Wire mesh used to make cellular structure, b) schematic of layered stack of wires, c) 3-D view of structure.

Fig. 4: Nominal stress strain curves for (a) nickel honeycomb deformed parallel and normal to the tube axes and (b) WWB structure parallel and normal to layer-planes.

buckling and fracture, and hardening process via ligament contact and plastic hardening. This is a gradual process since buckled columns suffer considerable displacement before starting to harden. The increase in hardening rate in region c indicates that consolidation and strain hardening of the material begin. For deformation normal to the layer-planes, the initial hardening rate is low since the primary deformation mechanism is shearing and local bending of wires. The extent of region b is also small since the unsupported bending ligaments are already small for this direction, and consolidation commences soon after this. The minimum value of hardening rate in both cases is about the same (~500 MPa) but reached after different amounts of strain, suggesting that the wires basic hardening behavior controls this level.

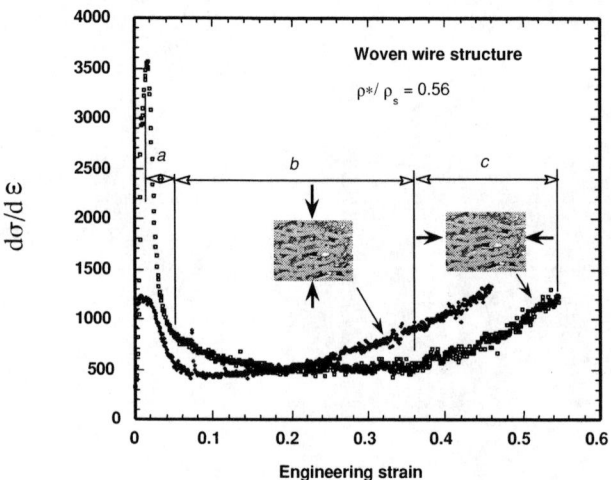

Fig. 5: Hardening rate curves for woven wire structure for deformation parallel and normal to layer-planes.

Creep

Experimental results Nominal strain (displacement divided by original specimen height) vs. time curves for the honeycomb are shown in Fig. 6a. All curves show an initial elastic part followed by a primary region and subsequently a steady state. The curve at the highest stress (0.27 MPa) shows an increase in strain rate as tertiary creep. The tertiary regime is interrupted by a second primary and steady regions which is followed by another repetion. This is shown in Fig. 6b with insets of photographs of the structure at various strains. It is seen that successive tertiaries in the curve correspond to collapse and shear instability in different layers of cells, indicated by arrows. The shape of the strain-time curve may thus be interpreted as follows. The phenomena that contribute to primary creep are work hardening and thickening of cell walls. But due to the change in geometry of the cells, bending moment on unsupported ligaments increase causing weakening of the structure. A balance in this competition results in the apparent steady state, which is then followed by overtaking of the hardening by the weakening processes resulting in the tertiary. Due to imperfections in the cell walls and nodes, additional shear localization of flow occurs in a band of cells until this layer of cells has collapsed. This process is then repeated in another layer of cells. Unlike this weakening effect, tertiary creep in solid metals involve necking and void formation. Papka and Kyriakides [6] have observed this kind of cell collapse in room temperature tests where the load displacement curve shows oscillations due to successive collapse of cells.

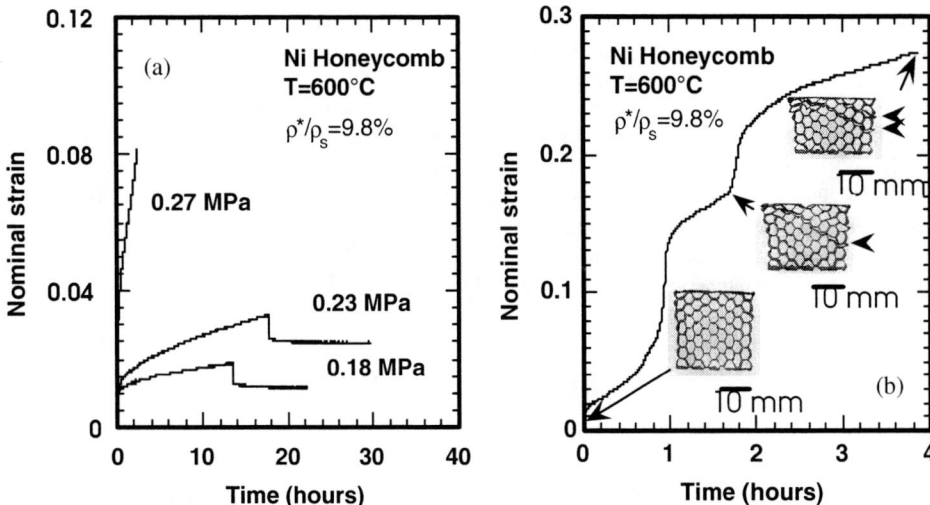

Fig. 6(a): Typical creep curves for nickel honeycomb showing elastic loading, primary, secondary, tertiary and unloading regimes, (b) Strain vs. time curve for nickel honeycomb showing collapse of cells contributing to the tertiary regime during creep deformation.

Minimum creep rate is plotted against nominal stress (load divided by overall specimen area) in Fig. 7. The creep behavior of the cellular structure in the axial direction follows a rule-of-mixtures behavior and the curve can be derived from that for the solid material by accounting for the missing material. Creep strength of the structures in the transverse direction is substantially lower since the creep mechanism is bending of inclined ligaments while in the axial case it is purely compression creep. In order to gain further understanding of the nature of creep in the transverse direction, finite element analysis of a representative section of the honeycomb was carried out.

Fig. 7: Minimum strain rate vs. stress data for nickel honeycomb and solid Ni material

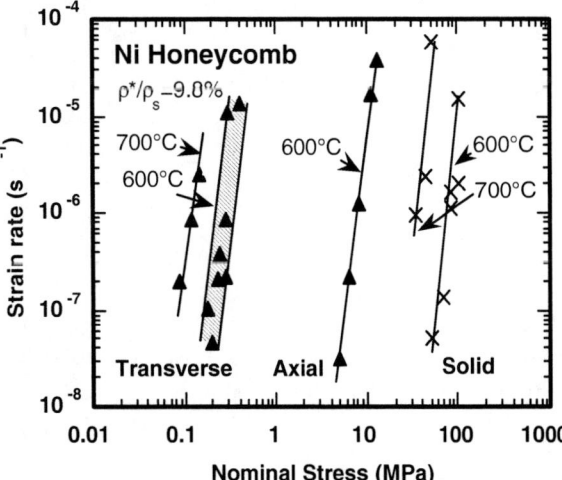

Finite Element Analysis In order to reduce computing time, a representative section of the honeycomb whose creep behavior would be reflective of creep of the entire structure was chosen. Fig. 8a shows a hexagonal honeycomb, with sets of horizontal and vertical dashed lines which were chosen such that they satisfy two conditions - 1) they are lines of symmetry and 2) the region between them can be repeated to obtain the overall structure. The shaded regions with unidirectional lines (identified by numbers 1,2,3,4) are repeating units and one such unit is shown enlarged in Fig. 8b with the appropriate boundary conditions. Surfaces A and B are allowed to move only in the X_1 and X_2 directions respectively. Surface C was constrained to remain horizontal and a concentrated force F was applied as shown. ABAQUS finite element software was utilized for the analysis. Creep simulation was carried out using the VISCO procedure which required input of parameters for the power law equation, $\dot{\varepsilon} = K\sigma^n$, where $\dot{\varepsilon}$ is the strain rate, σ is the stress, K and n are material parameters. The parameters used in the model were as follows. Material parameters: $K = 2.4 \times 10^{-20}$ MPa^{-1}s^{-1}, $n = 7$: Geometrical parameters: $t = 0.15$ mm, $h = l = 1.7$ mm, $\theta = 30°$ (these values are the average of many measurements on the actual honeycombs.). These values of t, h and l give a theoretical honeycomb relative density of 10.1%.

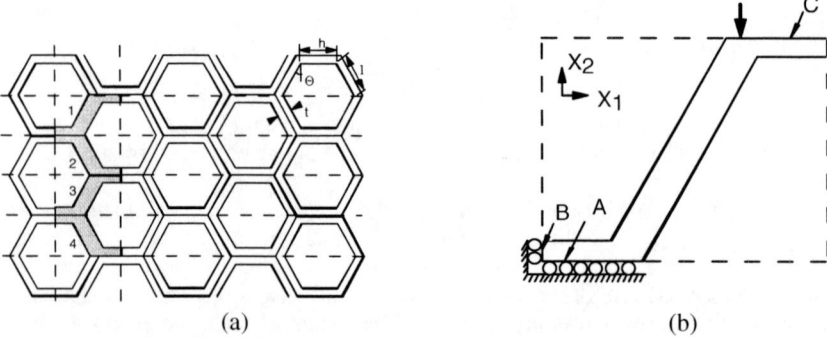

(a)　　　　　　　　　　　　　　　(b)

Fig. 8: a) Schematic of hexagonal honeycomb showing representative unit cell that is used for FE modeling b) isolated representative unit cell showing boundary conditions.

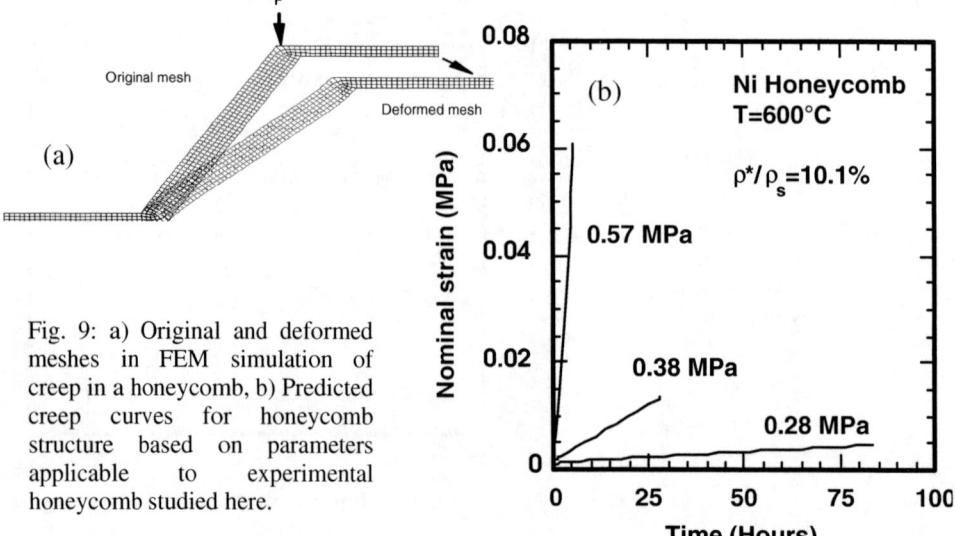

Fig. 9: a) Original and deformed meshes in FEM simulation of creep in a honeycomb, b) Predicted creep curves for honeycomb structure based on parameters applicable to experimental honeycomb studied here.

The FE analysis was carried out under constant load simulating a creep experiment and recording displacement vs. time behavior for a selected node on the surface of the structure. A comparison of the original and deformed meshes is shown in Fig. 9a. Bending near the nodes occurs in opposite senses and the deformation is symmetrical about the midpoint of the inclined ligament. Slight thickening also occurs in the inclined ligament adjacent to the nodes (this is not very apparent in this low magnification picture). The displacement -time data are plotted in Fig. 9b as nominal strain-time curves. The nominal stresses are indicated alongside the curves. The unique features of the obtained curves are as follows:

1) The elastic loading part is followed by a short primary creep region having decreasing slope.
2) The primary creep is followed by a region where the creep rate is relatively steady. The extent of this steady state regime reduces as applied load increases.
3) A third region follows this steady behavior where the creep rate increases, this being more evident at higher loads within short time scales considered.

Since the power law model used here assumes no work hardening behavior, the primary that is observed in the FEM simulations must arise from other effects. One is the gradual decrease in driving force for creep as the elastic strains are replaced by creep strains. This would cause the creep rate to decrease continuously. Another possible reason is the role of axial deformation of the ligaments in creep. As is seen from Fig. 8(b), the applied load would have a component parallel to the ligament which would cause compression. At the beginning of deformation, this component would be large and would progressively decrease as deformation proceeds. The thickening effect due to this axial deformation in the initial stages would cause the creep rate to decrease. Unless a 'weakening process initiates, this would cause the creep rate to decrease asymptotically to zero value. Evidently, the appearance of a steady state and the subsequent increase in creep rate observed in the simulations points to the occurrence of a weakening mechanism which in the present case is the increase in moment arm for the given load. As the ligament deforms and the angle that it makes with the vertical increases, the moment arm of the applied force increases thereby offsetting the aforementioned 'hardening' processes and eventually causing an increase in creep rate. This is similar to the collapse of cells seen in the experimental studies.

Minimum creep rate data from the FEM simulations are plotted in Fig. 10a. The shortfall in the experimental strengths from the FEM simulations could be due to the defects in the cell structures such as cell wall curvature, and deviation of the included angles between ligaments from those of a regular hexagon (see Fig. 10b for a schematic comparison of an ideal regular honeycomb and one fabricated using the technique developed here). It has been shown [7] that the elastic modulus of honeycombs can be reduced substantially by the cumulative action of such defects and it is expected that they would have a similar effect on the creep strength as well. Predictions from the following equation from Andrews et al. [4] are also included in Fig. 10a for comparison.

$$\dot{\varepsilon}_1 = K \frac{1}{(n+2)} \left(\frac{l}{t}\right)^{n+1} \left(\frac{2n+1}{n} \sigma_1^* \sin\theta \left(\frac{h}{l} + \sin\theta\right)\right)^n \frac{\sin\theta}{\cos\theta} \quad (1)$$

where $\dot{\varepsilon}_1$ is the strain rate of the honeycomb, σ_1^* is the remote stress on the honeycomb and all the other symbols have the meanings assigned earlier in this paper. The predictions from this equation show a weaker structure probably due to neglect of the compressive component of the applied force on the deformation of the inclined ligaments at least near the nodes.

Summary and Conclusions

A new technique for fabrication of low-density nickel honeycombs and a denser structure from nickel wire mesh are outlined. The fabrication of honeycombs rely on a surface tension-driven liquid phase bonding process. Similar technique was applied to the bonding of wire mesh structures, using electroless Ni (Ni-8%P) coating. Strengths at room temperature for the dense wire structure are found to be reasonably high, and thus it may be considered a viable structural material. Creep studies on the honeycombs show that normal three stages of creep deformation occurs, however during tertiary creep collapse of cells begin to occur. It is a form of shear instability which disappears after a small increment of strain, and again it leads to another primary creep region. Thus a series of creep cycles occur at regular strain intervals.

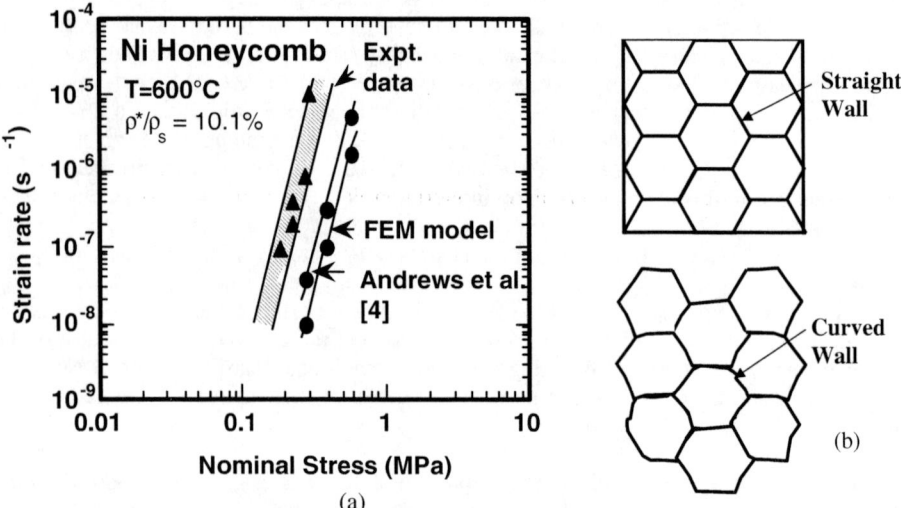

Fig. 10: a) Comparison of theoretical and experimental creep curves for transverse loading of nickel honeycombs b) A schematic illustration of ideal straight wall (top) and experimental hexagonal honeycombs with imperfect walls (bottom).

Because of the change in moment arm with wall bending deformation and simultaneous wall thickening, creep in transversely loaded honeycomb can lead to a balance between geometric 'weakening' and 'hardening' processes. FEM analysis of ideal honeycomb structures verifies these mechanisms of deformation and explains the experimental observations. The measured creep strengths are however lower than the theoretically predicted values. This is attributed to the defects in the structure of the honeycombs. Further work on the role and extent of the primary and tertiary regimes will be carried out for a complete understanding of creep in honeycombs.

Acknowledgements

The authors gratefully acknowledge the support of this research through ONR grant DOD-G-N00014-97-1-0510, program manager, Dr. S.G. Fishman; and through a contract from DARPA/NIST, (contract no: F000763), program monitors: W. Coblenz, K. Lyons and R. Sriram.

References

[1] L. J. Gibson and M. F. Ashby, Cellular Solids, Pergamon press, Oxford, 1988.
[2] R. K. Oruganti and A. K. Ghosh, " A Technique to Fabricate Hexagonal Honeycombs", (University of Michigan, Unpublished research)
[3] G. A. Di Bari, ASM Handbook, Vol. 5, Surface Engineering, ASM International (Materials Park, OH, 1994), 201-212.
[4] E. W. Andrews, L. J. Gibson and M. F. Ashby, " The Creep of Cellular Solids, Acta Materialia, 47 (10) (2000), 2853-2863.
[5] Iain Finnie and William R. Heller, Creep of Engineering Materials, McGraw Hill, New York, 1959.
[6] S.D. Papka and S. Kyriakides, " Experiments and Full-Scale Numerical Simulations of In-Plane Crushing of a Honeycomb", Acta Materialia, 46 (1998), 2765-2776
[7] A. E. Simone and L. J. Gibson, " The Effects of Cell Face Curvature and Corrugations on the Stiffness and Strength of Metallic Foams", Acta Materialia, 46(11) (1998), 3929-3935.

PROPERTIES OF MARAGING STEEL HONEYCOMB UNDER COMPRESSIVE QUASISTATIC AND DYNAMIC LOADING

Alethea M. Hayes, David L. McDowell, and Joe K. Cochran, Jr.

School of Materials Science and Engineering, Georgia Institute of Technology
771 Ferst Dr. N.W., Atlanta, GA 30332-0245 USA

Abstract

Two-dimensional linear cellular solids or honeycombs are known for their high-strength to weight ratios compared to other low-density architectures. Georgia Tech has developed a process that produces honeycomb from maraging steel compositions, including the 18% Ni series, which have high yield and ultimate strength after heat treatment. This steel incorporated into a honeycomb structure with a relative density of 20-25%, produced very high compressive strength under both quasi-static and dynamic loading conditions. Both quasi-static and dynamic experiments were conducted for the out-of-plane direction of the maraging steel honeycombs. The dynamic data for out-of-plane loading were generated by the split Hopkinson pressure bar at strain-rates on the order of 10^3 s^{-1}. Peak compressive strengths, collapse modes, and strain-rate sensitivity are examined in this study and discussed in light of simple models for brittle and ductile honeycombs. Minmal strain-rate sensitivity was found for the maraging steel honeycomb material.

Introduction

This paper focuses on the mechanical behavior of ordered metallic cellular materials with extended prismatic cells, otherwise known as linear cellular alloys (LCAs). As distinguished from traditional metal honeycombs, the manufacturing process for LCAs enables more complex in-plane cell morphologies to achieve desired functionality. The mechanical behavior of honeycombs has been widely studied, mainly with regard to serving as core material for lightweight structural sandwich beams. Conventional metal honeycombs (most often aluminum) have been fabricated using multistep processes involving sheet machining, layup and bonding, followed by stretching. In this work we consider LCAs formed in two primary steps by extrusion of powders, followed by a chemical reaction/sintering process to form near fully dense walls; this process was developed by the Lightweight Structures Group at Georgia Tech [1]. In this method a ceramic paste is extruded through a die and is then reduced in a hydrogen atmosphere to obtain a metallic honeycomb. The ceramic paste primarily consists of metal oxides and water, with a small percentage of lubricants and binders. This precursor offers flexibility to process a wide range of metallic compositions, as well as a variety of cell geometries. For example, metal oxides of Co, Ni, Cu, Zn, Mo, W, Mn, and Nb are commercially available and can readily be reduced by hydrogen.

An advantage of this processing method is that even high strength, refractory alloys can be extruded with small web thickness and fine detail shapes, in contrast to direct hot metal extrusion. Several maraging steel compositions have been processed to date, including analogs to commercial grade alloys 350, 200, and a low Mo grade [2]. A subsequent aging process is applied to the maraging steel in order to form the desired martensitic structure. Figure 1 shows a cross section of a square cell maraging steel LCA with an 8x8 array of cells, with a relative density of about 25%.

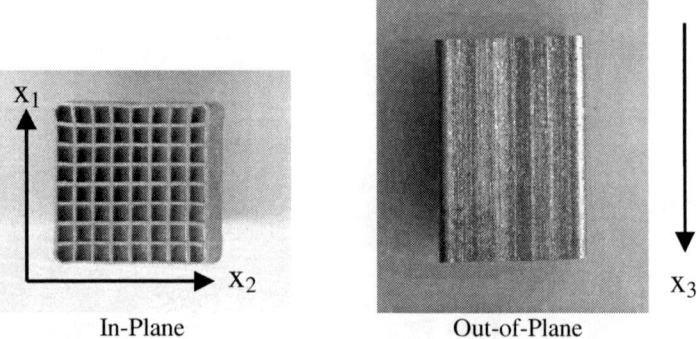

Figure 1: Extruded maraging steel LCA reduced from oxide powders and reduced to achieve an 8x8 square cell array (left); side view of extruded section (right).

A subsequent aging process is applied to the maraging steel in order to form the desired martensitic structure. Figure 1 shows a cross section of a square cell maraging steel LCA with an 8x8 array of cells, with a relative density $r = \rho^*/\rho_s$ of about 25%, where ρ^* and ρ_s are the densities of the LCA and solid cell wall materials, respectively. The relative density of square cell LCAs in this study typically lies between 21% and 25%.

Behavior in Compression

Common to many ductile honeycombs or foam materials is a response characterized by several regimes for in-plane compression: initial elastic deformation, initial yielding corresponding to plastic cell wall buckling, a relatively flat extended stress plateau regime, and finally densification after substantial crushing [3]. Generally, honeycombs exhibit the highest stiffness and strength under out-of-plane loading (i.e., along the cell axes) because the cell walls are initially subject to extensional or membrane stresses only; in-plane loading promotes bending of cell walls, with the governing mechanism of failure depending on cell shape [3-4]. The out-of-plane effective elastic stiffness is given by $E_3^* = E_{33}^* = r E_s$, where E_s is the Young's modulus of the solid cell wall material.

Properties or LCAs processed by the direct reduction method (DRM) are listed and compared to commercial grades of maraging steel in Table I. The alloying elements impart high strength [5] combined with limited ductility. Two of the nominal compositions produced by the direct reduction process involved the type 18 Ni-Co-Mo maraging steel, termed 18Ni(350) and 18Ni(200). Their compositions are, respectively, 18Ni-12Co-4.8Mo-1.4Ti-63.8Fe and 18Ni-8.5Co-3.3Mo-0.2Ti-70Fe. The 350 grade was selected as the highest strength material and the 200 grade was selected because of the improvement in ductility. These compositions were made from Fe_2O_3, NiO, Co_3O_4, TiH_2, and powder Mo. For the case of the Low-Mo composition, 18.4Ni-12.2Co-2.9Mo-1.4Ti-65.1Fe, the molybdenum was added by using the oxide form MoO_3, instead of pure metal powder.

The direct reduction-processed maraging steel LCAs were aged at 480°C for 5 hours in an inert atmosphere for the 18Ni(350) and 18Ni(200) grades. The Low-Mo material was also subjected to this aging thermal process, hereby termed HT1, and a second heat treatment HT2 was applied for some of the specimens. The HT2 process included a homogenizing step at 900°C for an hour to drive all the elements into solution before aging at 480°C for five hours. The two different heat treatments produced different properties for the Low-Mo composition are listed in Table I. Table II lists the basic mechanical properties for the nominal commercial 18Ni(350) and 18Ni(200) compositions and the direct reduction method (DRM) compositions. The relatively lower cell wall material densities, failure strains, and ultimate tensile strengths of as-reduced LCA cell wall materials in comparison to analog commercial alloys are due to residual porosity in the cell walls for the LCA material. The residual porosity for as-reduced maraging steel LCA cell wall materials ranged from 3-7%. Porosity is known to reduce strength and ductility significantly, with a drop of as much as 30%-40% in ductility and 20% in strength relative to the fully dense case for porosity levels on the order of 5% [6].

Out-of-Plane Quasistatic Compression

Honeycombs exhibit the highest directional strength in the out-of-plane (x_3) direction. Standard screw drive test machines were employed in displacement-control at room temperature. Hardened steel platens were used to compress the maraging steel specimens. The square cross section specimens had a length to diameter aspect ratio of 3:2 (longer along cell axes). Both 5x5 and 8x8 arrays of square cells were used in specimens to explore size effects. The relative density of these specimens was nominally 25%, with cell size of approximately 1.84 mm and cell wall thickness of 0.23 mm.

Due to limited cell wall ductility, only small plastic rotation of cell walls (a few percent strain) preceded cell wall fracture near joints. Figure 2 shows a quasistatic engineering stress-strain curve for a Low Mo (HT1) maraging steel specimen in compression, along with a sequence of images corresponding to various points along the stress-strain curve. An extended shear/kink band was observed to form following the initial instability load, as shown in Frame 4. The collapse plateau stress following the initial instability was approximately half the value of the initial peak strength. Densification occurred at around 60% engineering strain. McFarland [7] first noted this shear collapse mode for low density metal honeycombs compressed in the out-of-plane direction.

Table II presents the mean peak collapse stress levels in quasistatic compression, along with information on cell wall porosity and types of LCA defects, for maraging steels DRM 18Ni(200), DRM 18Ni(350), DRM Low-Mo(HT1), and DRM Low-Mo(HT2). Note that the cell wall porosity levels for LCAs are considerably higher than those of extruded strip specimens presented in Table I.

Table I: Mechanical properties of commercial maraging steels and Direct Reduction Method (DRM) maraging steel strip specimens processed at Georgia Tech.

Material	σ_{UTS} (MPa)	σ_{ys} (MPa)	E (GPa)	Grain Size (μm)	Vickers (HV)	ε_f (%)	Porosity (%)
18Ni(200)[a]	1345–1585	1310–1550	181	-	435–940[b]	6 – 12	nil
18Ni(350)[a]	2468	2427	194	-	660[b]	8	nil
DRM 18Ni(200)	1200	1150	224	10-30	402	2	0.72
DRM 18Ni(350)	1675	1600	160	5-30	534	3	0.30
DRM Low-Mo (HT1)	1200	1150	N/A	15-30	302	2	1.03
DRM Low-Mo (HT2)	1225	1175	N/A	10-20	451	1	nil

[a] Nominal Commercial with Heat Treatment: 815° C for 1 hr and 480° C for 3 hours
[b] Converted values from Hardness Testing [8]

Table II: Experimentally measured mean peak collapse strength for quasistatic out-of-plane compression of maraging steel LCAs.

Material	Cell Array	$r = \dfrac{\rho^*}{\rho_s}$	Cell Wall Porosity (%)	$(\sigma^*_{pl})_3$ (MPa)	Known Defects
DRM 18NI(200)	8x8	0.25	5.0	397	Multiple Seam Cracks Larger Grains
	4x4	0.22	7.5	406	Slightly Skewed Cells
DRM 18Ni(350)	8x8	0.25	5.3	515	Skewed Cells Cell Wall Ruptures
	5x5	0.25	3.7	551	Residual Wall Joints
DRM Low-Mo(HT1)	4x4	0.21	5.2	375	Skewed Cells Large Cell Wall Corrugation
DRM Low-Mo(HT2)	4x4	0.23	4.5	424	One Large Continuous Pore Slight Wall Corrugation

Specimen size effects were examined by comparing out-of-plane compression results for 5x5 and 8x8 square cell DRM 18Ni(350) maraging steel LCAs with the same cell size. Indeed, our previous studies comparing 3x3, 5x5, 8x8 and 10x10 LCAs subjected to both in-plane and out-of-plane loading have shown that size effects appear to diminish for 8x8 arrays and above; they become pronounced for specimens with fewer than 8x8 arrays of cells. However, it is noted that the number of specimens in the present study is perhaps too limited to be too conclusive regarding size effects.

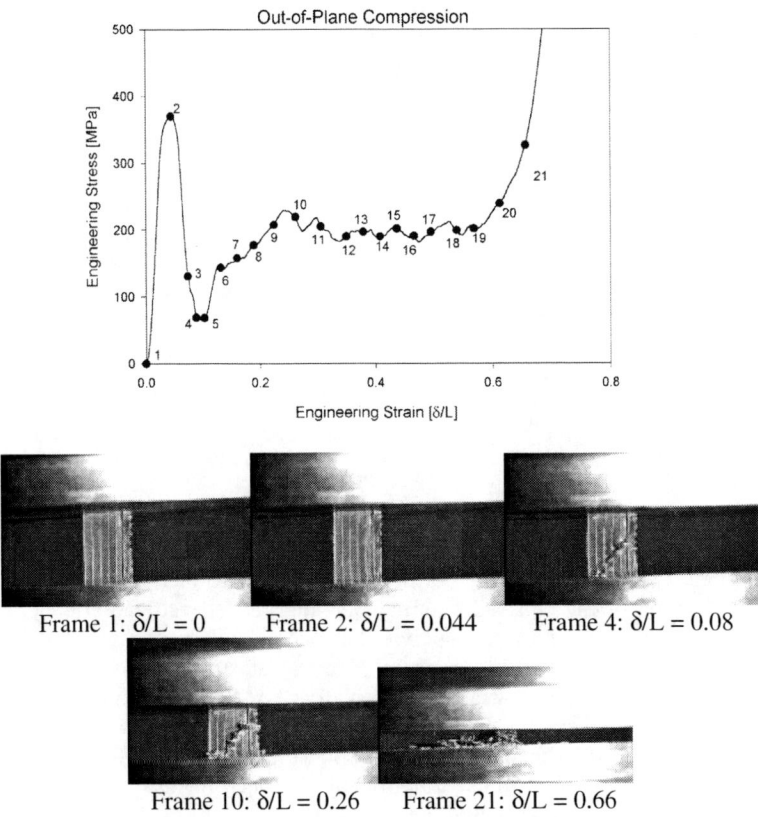

Figure 2: Out-of-plane nominal stress-strain behavior (top) at an imposed strain rate of 8.35×10^{-3} s^{-1}, and (bottom) corresponding images at various stages (frames) of compression of a 6x6 square cell DRM Low-Mo(HT1) maraging steel LCA.

Figure 3 compares the out-of-plane specific energy absorption and plateau stress of maraging steel specimens subjected to quasistatic out-of-plane compression with that of other cellular materials, including stochastic metal foams and randomly packed hollow sphere metal foams [3]. Note that the specific energy absorption approaches the upper bound solution for a bank of axially loaded tubes [3] (also an ordered honeycomb structure, in essence), and the plateau stress is one to two orders of magnitude higher than that of stochastic metal foams. In Fig. 3 it is seen that the specific energy absorption for in-plane quasistatic compression is significantly less than that of out-of-plane compression and is well below the upper bound case for an axially loaded bank of tubes, but is still comparable to some of the higher energy absorbing metal foams reported in the literature.

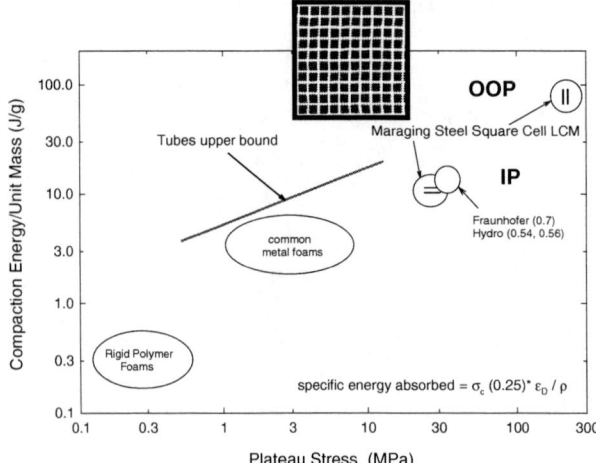

Figure 3: Specific energy absorption in compression as a function of plateau stress. Note that out-of-plane (OOP) compression of a 10x10 array square cell DRM 18Ni(350) maraging steel honeycomb with 25% relative density approaches the extrapolated upper bound solution for a bank of tubes [3], in contrast to the in-plane (IP) case (see Fig. 1).

Dynamic Compression

Three structural deformation mechanisms have been identified under high strain rate conditions for honeycomb structures: localization of deformation into bands where local strain rates are much higher than the average strain rate of the specimen, micro-inertia [9] related to the rotation and lateral movement of the cell walls during buckling, and densification coupled with shock wave dynamics [4]. Since strain rate sensitivity has been identified at rates of 10^2 s^{-1} for cellular solids, evaluating maraging steel linear cellular alloys at strain rates up to 10^3 s^{-1} helps to determine if this honeycomb is strain rate sensitive. A Split Hopkinson pressure bar was employed to generate strain rates ranging from 900 to 2500 s^{-1} at room temperature in order to assess strain sensitivity and dynamic energy absorption for LCAs. Although the stresses and strains are not axially uniform as in the simplification of one-dimensional wave propagation, the average stress-strain relationship is still quite useful [10].

Tungsten platens were used on each side of the specimen to protect the ends of the pressure bars for the high strength maraging steel specimens, similar to modifications for testing ceramics with little impedance mismatch. Polymer sleeves were used to hold the platens parallel to the system, and the interfaces were well lubricated to minimize frictional effects. Various pressure settings for the striker bar gun produced the velocities need to achieve different strain rates. Using this technique, only the initial compression response was captured.

The strain could not be measured up to densification because of the limitations of the Split Hopkinson apparatus; consequently the strain values only reach 0.35 – 0.40. Post-crush examination of specimens compressed at different strain rates reveal a plastic collapse mode very similar to that of the quasistatic compression experiments. Comparing to the quasistatic experiments, the initial plastic buckling appears to be more uniformly distributed through the LCA specimen under dynamic loading, but shear localization is still observed, leading to stress-strain curves and energy absorption characteristics very similar to the quasistatic case for this class of materials. The failure mode appears to follow that of a brittle honeycomb material [11], i.e., formation of a kink band that propagates through the specimen after a pronounced peak collapse stress.

Dynamic compression stress-strain curves for DRM 18Ni(200) maraging steel are shown in Figure 4, and compared to quasistatic results. There is a modest increase of peak stress level with increase of the strain rate. In this regard, the maximum difference between the highest strain rate of 2500 s^{-1} and the quasistatic data was about 15%. DRM 18Ni(350) behaved similarly. The dynamic peak compressive stress was 50-60 % higher than the quasistatic peak stress, and the energy absorption for the dynamic case is substantially greater than the quasistatic case.

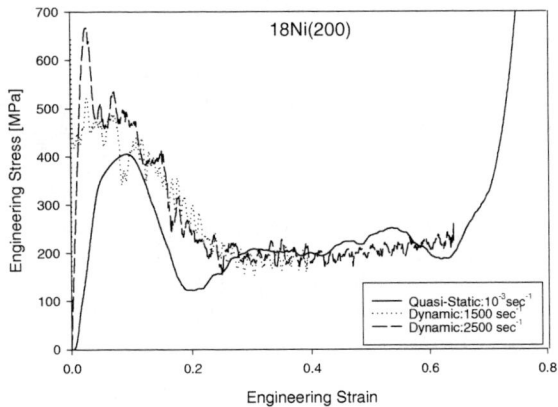

Figure 4: Comparison of out-of-plane compression behavior of DRM 18Ni(200) maraging steel square cell LCAs at various strain rates.

Maraging steels are not considered to be strain rate sensitive and many fcc nickel alloys are strain rate sensitive. This study indicates that, to first order, the cellular structure of maraging steel LCAs does not give rise to strain rate sensitivity (at least for the range of strain rates considered) in the range of strain rates of 10^3 s^{-1}. This implies that the strain rate sensitivity of the cell wall material manifests rate dependent LCA behavior. The dynamic compression was very similar to that of the quasistatic case. During compressive impact, shear bands formed and fracture occurred. The elastic portions of the dynamic curves are invalid due to the transient nature of the impact problem; however, the overall initial peak response was broader compared to the quasistatic case. This may be due to more uniform cell wall collapse behavior.

Models for Initial Collapse Strength Under Out-of-Plane Quasistatic Compression

Wierzbicki and Abramowicz [12] extended collapse models based on compatible plastic hinge modes to several cell geometries, including square cellular solids. A compatible collapse mode was identified that incorporates supplementary plastic hinges at the cell corners and a limited amount of cell wall extension at the corners. Through minimization of the collapse load with respect to the wavelength of buckling, Wierzbicki and Abramowicz [12] formulated a relationship for the plastic buckling strength of square cell honeycombs as

$$\frac{(\sigma_{pl}^*)_3}{\sigma_{ys}} = 13.97 \left(\frac{t}{l}\right)^{5/3} \tag{1}$$

Because it assumes a kinematical plastic collapse mode, this result pertains more closely to the post buckling plateau strength than the initial collapse/yield strength. Reckoning that initial yielding for out-of-plane loading of a square cell LCA occurs when

$$\frac{(\sigma_{pl}^*)_3}{\sigma_{ys}} = r = 2\frac{t}{l}, \tag{2}$$

this offers a lower bound on the stress level for plastic collapse.

Table III lists the theoretical prediction for square cell honeycombs given in Equations (1) and (2), and compares them to measured values of the peak collapse stress. The accuracy of the prediction is considerably better for the lower strength alloys (Low-Mo and 18Ni(200)) than for the highest strength material, 18Ni(350). This can perhaps be understood as being related to lower defect sensitivity for the lower strength, lower hardness alloys; moreover, the yield strengths used in Equation (1) are based on extruded strip specimens (Table 1) with considerably lower porosity than the LCA cell walls, which would lower the yield strength and predictions of plastic collapse strength by 10-20%.

It is reckoned that although the cell wall ductility of DRM maraging steels is relatively low, as listed in Table I limited plastic yielding and bending of the cell walls indeed does occur as a precursor to cell wall fracture, and any tension-compression asymmetry is not pronounced. Hence, the ductile collapse models in Equations (1) and (2) appear to offer reasonable correlation with maraging steel LCA experiments. The measured values of peak collapse stress lie within the bounds of Equation (1) from above and Equation (2) from below. It is noted that the plateau strength level observed for strain levels greater than 20% or so approximately follow Equation (2).

Table III: Theoretical and measured initial plastic collapse strength for quasistatic out-of-plane compression of maraging steel LCAs

Material	Cell Matrix	Theoretical $(\sigma_{pl}^*)_3$ Eq. (1) [MPa]	Theoretical $(\sigma_{pl}^*)_3$ Eq. (2) [MPa]	Measured Peak Stress $(\sigma_{pl}^*)_3$ [MPa]
18Ni(200)	8x8	577	288	397
	4x4	462	253	406
18Ni(350)	8x8	803	400	515
	5x5	795	400	551
Low-Mo(HT1)	4x4	394	242	375
Low-Mo(HT2)	4x4	499	270	424

Acknowledgements

This work is sponsored by DSO of DARPA (N00014-99-1-1016) under Dr. Leo Christodoulou and by ONR (N0014-99-1-0852) under Dr. Steven Fishman. The mechanics effort within this program is indebted to J. Cochran, T.H.B. Sanders, Jr. and K.J. Lee, Georgia Tech colleagues engaged in the processing effort. Justin Clark, a graduate student in the School of Materials Science and Engineering, is acknowledged for providing properties listed in Table I. The assistance of Dr. M. Zhou and his graduate students in helping to set up the Hopkinson bar experiments and dealing with special issues related to high strength maraging steel alloys is very much appreciated. Undergraduate students J. Tidwell (LCA processing) and and T. Akiskalos (mechanical testing) are gratefully acknowledged for their assistance.

References

1. Cochran, J., Lee, K.J., McDowell, D., Sanders, T.H., et al., 2000, "Low Density Monolithic Metal Honeycombs by Thermal Chemical Processing" (Paper presented at the 4th Conference on Aerospace Materials, Processes, and Environmental Technology, September, Huntsville, AL).

2. Hayes, A., Compression Behavior of Linear Cellular Steel, M.S. Thesis, School of Materials Science and Engineering, Georgia Institute of Technology, Atlanta, GA.

3. Ashby, M.F., Evans, A., Fleck, N.A., Gibson, L.J., Hutchinson, J.W., Wadley, H. N.G., Metal Foams – A Design Guide (Butterworth-Heinemann, Boston, 2000).

4. Gibson, L.J., Ashby, M.F., Cellular Solids: Structures & Properties (2nd Edition, Cambridge University Press, Cambridge, UK, 1997).

5. Magnee, A., Drapier, J.M., Dumont, J., Courtouradis, D., Habraken, L., Cobalt Containing High –Strength Steels, Brussels, Belgium, 1974.

6. Bocchini, G.F., "The Influence of Porosity on the Characteristics of Sintered Materials," The International Journal of Powder Metallurgy, 22 (3) (1986) 185-202.

7. McFarland, R.K., "Hexagonal Cell Structures under Post-Buckling Axial Load," A.I.A.A. Journal, 1 (1963) 1380-1385.

8. Boyer, H.E., editor, Hardness Testing (ASM International, Metals Park, OH, 1987).

9. Wierzbicki, T. and Ambramowicz, W., "On the Crushing Mechanics of Thin-Walled Structures," International Journal of Solids and Structures, 29 (1983) 3269-3288.

10. Wu, E., Jiang, W., "Axial Crush of Metallic Honeycombs," International Journal of Impact Engineering, Vol. 19, Nos. 5-6, 1997, pp. 439-456.

11. Zhang, J. Ashby, M.F., "The Out-of-Plane Properties of Honeycombs," International Journal of Mechanical Sciences, 34 (6) (1992) 475-489.

12. Wierzbicki, T. and Abramowicz, W., <u>Manual of Crashworthiness Engineering</u> (Report of MIT Center for Transportation Studies, Cambridge, MA, 1987).

MECHANICAL TESTING OF IN718 LATTICE BLOCK STRUCTURES

David L. Krause*, John D. Whittenberger*, Pete T. Kantzos**, and Mohan G. Hebsur**

*National Aeronautics and Space Administration, Glenn Research Center
Cleveland, Ohio 44135
**Ohio Aerospace Institute, Glenn Research Center, Cleveland, Ohio 44135

Abstract

Lattice block construction produces a flat, structurally rigid panel composed of thin ligaments of material arranged in a three-dimensional triangulated truss-like structure. Low-cost methods of producing cast metallic lattice block panels are now available that greatly expand opportunities for using this unique material system in today's high-performance structures. Additional advances are being made in NASA's Ultra Efficient Engine Technology (UEET) program to extend the lattice block concept to superalloy materials. Advantages offered by this combination include high strength, light weight, high stiffness and elevated temperature capabilities. Recently under UEET, the nickel-based superalloy Inconel 718 (IN718) was investment cast into lattice block panels with great success. To evaluate casting quality and lattice block architecture merit, individual ligaments and structural subelement specimens were extracted from the panels. Tensile tests and structural compression and bending strength tests were performed on these specimens. This paper first presents metallurgical and optical microscopy analysis of the castings. This is followed by mechanical test results for the tensile ligament tests and the subelement compression and bending strength tests. These tests generally showed comparable properties to base IN718 with the same heat treatment, and they underscored the benefits offered by lattice block materials. These benefits might be extended with improved architecture such as face sheets.

Introduction

In 1999 the National Aeronautics and Space Administration created the Ultra Efficient Engine Technology research program at the John H. Glenn Research Center to develop revolutionary turbine engine propulsion technologies. The program seeks innovation and improvements in high temperature engine materials and structures. Superalloy lattice block materials have been proposed in the program for static turbine engine components of a supersonic exhaust nozzle, including possible use in actuated panels, exhaust nozzle flaps, and side panel structures.

Lattice block materials have been developed and patented by JAMCORP (Jonathan Aerospace Materials Corporation) of Wilmington, Massachusetts. JAMCORP's method of fabrication produces three-dimensional structures of various configurations with internal matrices of triangles acting as space frames. The structures look similar to bridge trusses, bar joists, or geodesic domes, except that they are on a meso- or micro-scale. This architecture gives high specific strength and stiffness for most materials. JAMCORP has produced lattice block in a variety of parent materials including steels, aluminums, plastics, rubber, ceramics, and specialty alloys. In NASA's research program, the strength and lightweight benefits of nickel-based superalloy lattice block are being evaluated for high temperature applications.

IN718 Lattice Block Panels

After considering several metals with good general casting properties, Inconel 718 (IN718) was chosen to demonstrate feasibility in the first superalloy casting trials. Hebsur describes casting and processing details in a paper of this symposium [1]. A rectangular flat panel lattice block configuration was selected that resulted in trimmed integral panels 131 mm wide by 290 mm long and 11 mm in height. Layout of the three-dimensional triangulated space frame produced 14 transverse and 31 longitudinal repeating cells, with one cell through the thickness. Material ligaments averaged 1.6 mm in diameter and were connected at lightly reinforced nodes. A typical panel is illustrated in Figure 1.

This geometry was chosen primarily to facilitate the casting process and not to optimize the structure for carrying loads. The arduous flow path from the peripheral gating through the tiny ligaments was considered to be the first challenge in demonstrating the technology, with mechanical efficiency to be improved upon later. Simple linear analysis of the panel as configured (Figure 1) showed that specific longitudinal bending stiffness would be 36 % higher than a solid plate; specific bending strength would be 14 % higher.

The superalloy lattice block panels were fabricated by JAMCORP using investment casting in

Figure 1: IN718 lattice block panel 5 after removal of peripheral casting gates; (a) size is 131 mm by 290 mm by 11 mm, (b) enlarged top view shows nodal reinforcement.

shell molds made from wax patterns. Initially there were numerous problems in making and assembling the wax patterns. During the six-month development program, ten IN718 castings were successfully produced. After careful inspection, investment casting risers and gates were cut off, and the panels were subjected to a heat treatment as follows: initial hot isostatic pressing (HIP) at 1120 °C for four hours at 103 MPa in inert argon atmosphere to close internal porosity, followed by one hour at 1010 °C and an argon gas quench; then eight hours at 718 °C and a furnace cool to a hold at 620 °C for at total time at 718/620 °C of 18 hours.

Panel Physical Characteristics

Overall physical inspection of the ten panels revealed only minor visual defects. Three of the panels were missing one or two ligaments (out of over 2250 per panel) due to lack of fill in the shell molds. Maximum out-of-plane warping seen on the panels was 0.2 to 1.0 mm. Measuring and weighing eight panels obtained an estimate of the density of the IN718 lattice block. The values ranged from 1.17 to 1.26 Mg/m^3 with an average of 1.21 Mg/m^3 and a standard deviation of 0.028 Mg/m^3; this average density corresponds to about 15 % of the density of IN718.

Metallurgical examination of unetched, polished, random sections taken from two lattice block panels revealed that the majority of the ligaments were defect free as illustrated in Figure 2a; however a few ligament sections did possess significant amounts of surface connected porosity (Figure 2b) which was not closed by the HIP'ing. Out of a total of 110 ligament cross sections, 85 were whole, 16 had one small casting pore and 9 had multiple (two or more) pores. Figure 2 also illustrates that the ligaments were not cylindrical in cross section, as all possessed "ears" which derived from the split mold used to make the wax pattern for investment casting.

Average diameter, standard deviation and maximum/minimum values were determined from photomicrographs of ligament cross-sections. Two determinations of diameter were made on each 90° section (Figure 2a: perpendicular to the ears and just below the ears) and one measurement of diameter was taken from the diagonal sections (Figure 2b: minor ellipse axis). These diameter values are presented in Table I along with diameters obtained from 166 micrometer measurements of ligaments cut for tensile test specimens. Clearly there are no major discrepancies among the values listed in Table I, which confirms the reproducibility of the investment casting process used for lattice block.

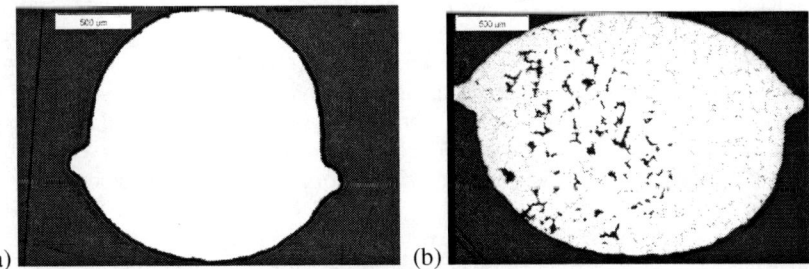

Figure 2: Photomicrographs of unetched, polished IN718 lattice block ligaments from panel 2; (a) 90° cross-section, (b) body diagonal.

Table I Diameter of Ligaments in IN718 Lattice Block

Panel No.	Method	Avg. Diam. (mm)	Std. Dev. (mm)	Max./Min. Diam. (mm)
1	photomicrographs	1.62	0.071	1.76 / 1.52
2	photomicrographs	1.64	0.071	1.80 / 1.54
3	micrometer	1.63	0.045	1.74 / 1.51

Mechanical Testing Description

Mechanical testing of portions (ligaments and subelements) of three IN718 lattice block panels has been performed. Samples for testing were electrodischarged machined (EDM'ed) from the panels to provide suitable specimens for ligament tensile and subelement structural testing.

Ligament Tensile Testing

Several longitudinal ligaments, approximately 0.3 m in length, were EDM'ed from both the top and bottom faces of one heat-treated IN 718 panel for tensile testing of individual ligaments. Each of these lengths was then cut into ~0.15 m sections, where one half was reserved for room temperature tensile testing and the other half set aside for 650 °C testing. Each 0.15 m piece of ligament contained 8 nodes, spaced about 19 mm apart, where the face/body lattice block diagonals intersect the length. Maximum and minimum diameters were measured for each section between nodes, and all these values were then averaged to give an average diameter for calculation of cross sectional area of the ligament specimen assuming an uniform cylindrical geometry.

Tensile testing both at room temperature and at 650 °C was conducted using a screw driven universal test machine at an engineering strain rate of 0.005 per minute with a probe type extensometer measuring deformation between a central pair of nodes. Offset (0.2%) yield strength and Ultimate Tensile Strength (UTS) were calculated on the basis of the average diameter for each sample. The experimental setup and testing details are further described in Reference [2].

Subelement Structural Testing

Two lattice block panels were machined into smaller subelements for structural compression and bend testing, as indicated in Figure 3. Bend test specimens were oriented to represent the major (longitudinal) and lesser (transverse and diagonal) axes of the panels, with compression test specimens extracted from the remaining areas. Specimen nomenclature is as follows: the first character represents the lattice block panel from which it was cut, the second character defines the specimen's orientation relative to panel direction, and the third character is a sequential number as defined in Figure 3. All subelement testing was performed at room temperature.

Because of the difficulty in measuring experimental ligament strains in the subelements, finite element analysis of each configuration was performed to relate experimental loads to ligament stresses, and experimental deflections to ligament strains. The linear elastic analyses were performed using beam elements for ligaments, with no additional reinforcement modeled at nodes; all ligament connections were assumed to be fully moment-carrying. Boundary conditions reflected the actual test conditions of rigid rollers for load application and reaction. Ligament

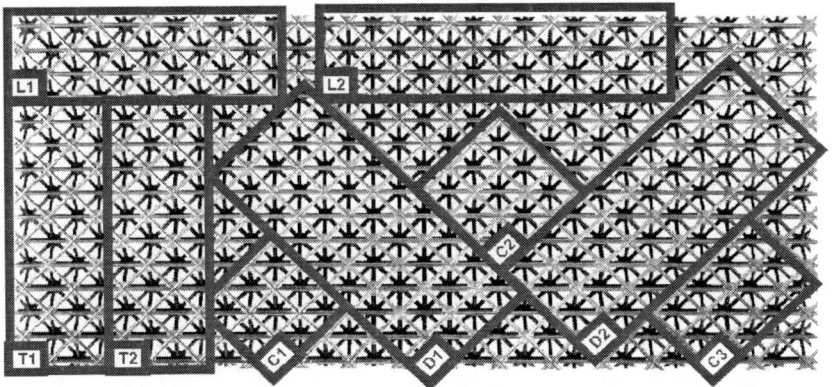

Figure 3: Lattice block panel machining layout and numbering for subelement specimens.

diameters were defined as the average 1.6 mm diameter, and node spacing was set at the average measured panel values: 18.72 mm longitudinally, 9.38 mm transversely, and 9.39 mm through the thickness. The material modulus was set at 191 GPa, and Poisson's ratio was defined as 0.29, based on reported data in the Aerospace Structural Metals Handbook [3].

The subelement structural testing was performed using a hydraulic test machine with all-digital controls. A hydraulically actuated wedge grip was outfitted with a universal joint fixture that permitted axial vertical load to be transmitted, but the joint released rotational degrees of freedom in the horizontal plane (see Figure 4). Total applied force was measured with a calibrated load cell on the hydraulic cylinder; deflections were measured with both the system's linear variable differential transformer and a dial indicator with 0.0254 mm resolution. Rigid top and bottom platens were used to sandwich specimen 1C1 for the one subelement compression strength test performed. For subelement bend tests, a combination of 8 mm and 32 mm solid stainless steel rollers were used with fixtures to apply loads and support the specimens. Four-point bending strength tests were completed for specimens 1L1 and 1T1; a three-point bending strength test was done for specimen 1D1 due to the odd number of engaged nodes on the top face.

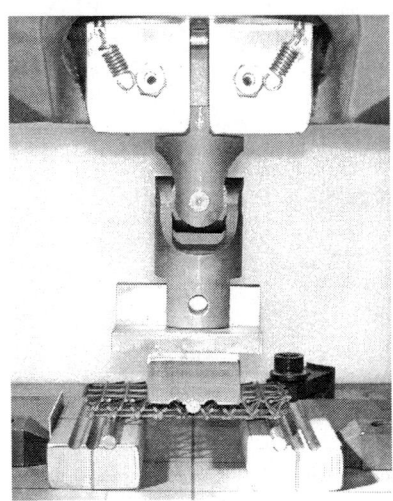

Figure 4: Experimental setup for subelement bend strength testing (specimen 1D1 shown).

All strength testing was performed in displacement control, with data being recorded at 0.025 mm increments in the linear behavior range, and generally five to ten times that step size in the plastic region. Tests were terminated when little significant strength remained after fracture of multiple ligaments, or when the limits of the test fixture were reached. Additionally, bending fatigue tests were performed on subelements; these tests are described in Reference [2].

Experimental Results

Ligament tensile testing and subelement structural testing produced somewhat different results. In general, the ligament tests revealed lower yield and ultimate strengths and little elongation, while the subelement tests showed high strengths and elongations.

Ligament Tensile Test Results

The average tensile properties measured at room temperature and 650 °C for individual IN718 lattice block ligaments are given in Table II. Five tests were initially conducted at room temperature with 100 mm long ligament test sections; four of these tests failed prematurely during the elastic deformation, while the fifth test only reached 0.08 % plastic strain before failure. Seven of the broken pieces were retested, because they had undergone only room temperature elastic deformation and were of sufficient length for gripping and the use of the extensometer between a pair of nodes. These retested samples had test section lengths between 29 to 70 mm, and all specimens deformed plastically at room temperature with the total strain at failure ranging from 1.3 to 8.9 %. Therefore a size effect appears to exist, where longer lengths of ligament will contain a serious casting defect (Figure 2b), but shorter lengths do not.

Table II Tensile Properties of IN718 Ligaments

Test Temp. (°C)	Gage Length (mm)	0.2% Yield Strength (MPa) Avg./Std. Dev.	Ult. Tensile Strength (MPa) Avg./Std. Dev.	Elongation (%) Avg./Std. Dev.
RT	100	- / -	627[1]/ 150	- / -
RT	29-70	757 / 24	827 / 32	4.3 / 2.9
650	100	630[2]/ -	582[1]/ 111	- / -
RT[3]		760	860	5
RT[4]		830	970	16
650[4]		650	720	16

[1]Fracture stress [3]See reference [3]
[2]Single value   [4]See reference [4]

Based on the room temperature experience with 100 mm long test sections, it was expected that few, if any, of the six 650 °C tests would demonstrate yielding before failing. However four specimens did plastically deform beyond the proportional limit, but only one sample experienced sufficient deformation for calculation of the 0.2% yield strength (Table II).

Examination of the literature values for the tensile strength of investment cast IN718 indicates that these properties are strongly dependent on processing and the specific heat treatment, where, for example, the Aerospace Structural Metals Handbook [3] reports room temperature yield strengths from 585 to 1000 MPa. Minimum room temperature strength and ductility properties (Table II) are given in [3] for investment cast IN718 which was annealed at 1093 °C, air-cooled and aged at 720 °C for 8 hours, then furnace-cooled to 620 °C and held for a total of age of 18 hours. Comparison of these minimums to the values obtained for the ductile ligaments indicates that investment cast lattice block can have tensile properties near the acceptable range.

Bouse and Behrendt [4] have reported room temperature and 650 °C tensile results for conventionally cast and HIP'ed (@ 1120 °C) IN718, and these values are also reported in Table II. While the 0.2% yield strength properties for the IN718 lattice block ligaments compare favorably with the conventionally cast/HIP IN718 at both temperatures, neither the ultimate tensile strengths nor tensile elongations do. Presumably the low tensile ductilities found in the IN718 ligaments are due to casting defects like that seen in Figure 2b, which localize deformation to a smaller volume with a reduced load bearing area. This, in turn, lowers the calculated ligament ultimate tensile strength because of the lesser load required to fracture a diminished cross section.

Subelement Structural Test Results

As previously described, beam-element finite element analysis of each specimen configuration was performed to relate loads, ligament stresses, and ligament strains (Figure 5, for example). Maximum and minimum ligament stresses reported here were determined by multiplying the measured experimental loads by an appropriate analytically determined factor; likewise, ligament strains were derived by multiplying experimental deflections by the appropriate factor. Because the analyses were linear elastic, they did not predict the inelastic buckling phenomena of compression ligaments nor second order effects that occurred in most of

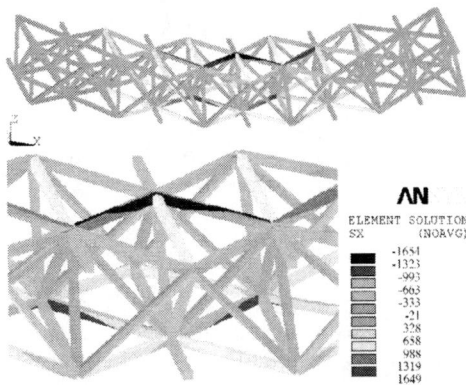

Figure 5: Representative finite element analysis; ligament longitudinal stresses (MPa) for specimen 1T1 with 2.60 kN load applied.

the strength tests. The consequence of using the linear analyses' factors on the experimental data beyond the yield point is many-fold; two effects are noted here. First, calculated maximum stresses for some compressive ligaments are under-reported, as buckled members released load that was redistributed to other compression members to maintain equilibrium. And secondly, local strains for buckled ligaments are grossly under-reported (buckled member curvature radii as small as three times the ligament diameter were common), while tension ligament strains are over-reported due to the decreased specimen stiffness.

The plots and data that follow are not modified to account for these plastic deformation effects. However, the general conclusions reached are little affected, since structural failure of the lattice block subelements for bend tests was defined by the tensile fracture of tension ligaments, not by loss of stiffness due to compressive ligament buckling. The compression strength test results are somewhat affected, but since this type of loading is not an efficient use of the lattice block architecture, and since this loading is not applicable to any of the proposed uses for lattice block, it has not been analyzed further.

Compression Strength Test Results. Specimen 1C1 was a four-cell lattice block subelement. The top face grid of ligaments was located one-half cell longitudinally and one-half cell transversely from the bottom face grid. This arrangement produced a net lateral force on the structure when subjected to a vertical, through-thickness pressure load. The result was that some of the ligaments were loaded in tension, even though the global loading was compressive. Upon loading, specimen 1C1 exhibited initial stiffening thought to be fixture-related. This was followed by nearly linear response up to a maximum load of 24.0 kN. The specimen deformed plastically after this point to a thickness compression of 1.87 mm, roughly 20 % of its total thickness, where the test was terminated with no ligament fractures apparent. Approximately one-half of the diagonal ligaments in the thickness direction were buckled. The calculated peak compressive stress of 1060 MPa occurred at 0.59 % strain; the peak tensile stress of 570 MPa occurred at 0.32 % strain. Ligament strains at test termination were 1.92 % in compression and 1.03 % in tension. Figure 6 shows the stress-strain diagram for the compression strength test.

The analysis predicted much higher stiffness than that observed. For the linear portion of the test (up to 20 kN load), experimental deflections were 2.44 times the analytical prediction. This may be related to early buckling of compression ligaments or not developing full ligament end moments at nodes.

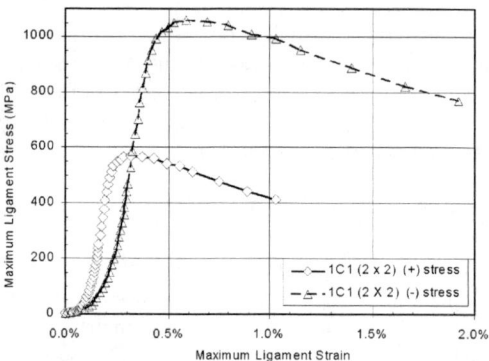

Figure 6: Stress-strain diagram for compression strength test of specimen 1C1.

<u>Bending Strength Test Results.</u> One bending strength test was performed for subelements in each of the three lattice block panel strength axes. The calculated stress-strain diagrams for the three tests are shown in Figure 7. As expected, the four-point loaded longitudinally oriented specimen 1L1 carried the largest maximum load due to its advantageous arrangement of ligaments. Linear behavior was evident until approximately 5.00 kN and 0.73 mm; at this load the analysis predicted 0.66 mm total deflection, about 9 % stiffer than experimentally measured. The smaller deflections predicted by the finite element analyses might be due to the boundary conditions imposed. The analysis assumed vertical supports at the end nodes of all four longitudinal ligament rows. But because of the deflected shape, two of the outer-most nodes on each end of the specimen actually lifted off of the support roller, an effect not modeled that would allow greater deflections.

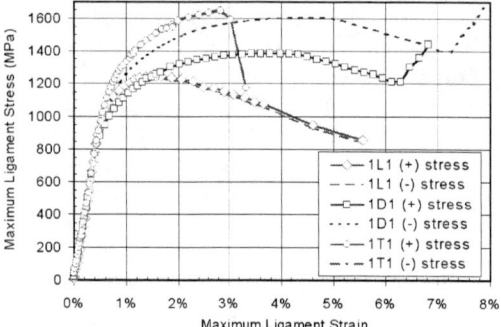

The peak load carried by specimen 1L1 was 6.37 kN, and the test ended at the maximum deflection of 7.53 mm, when the fixture limits

Figure 7: Stress-strain diagrams for three lattice block subelement strength tests.

were reached. The maximum deflection-to-span ratio was 1/9.94. One ligament was believed to have fractured based on audible detection soon after peak load was reached, although the fracture could not be visually observed. The center eighteen cells of top longitudinal and diagonal ligaments were highly buckled, along with most compression vertical diagonals in this area. The buckled ligaments are readily visible in Figure 8. The calculated peak ligament stresses of 1250 MPa in tension and 1230 MPa in compression occurred at 1.56 % strain. Strains at test termination were 5.55 % in tension and 5.48 % in compression.

Specimen 1D1 was the diagonally oriented specimen and was a three-point bend test because of node alignment. It carried a moderate load due to its longitudinal arrangement of face diagonal ligaments. Linear behavior was exhibited until approximately 2.50 kN and 0.65 mm; at this load the analysis predicted 0.72 mm total deflection, about 12 % less stiff than experimentally measured. The smaller experimental deflections might be due to the extra material reinforcing the ligaments at the connecting nodes. The analysis did not model this extra stiffness.

Beyond this linear range, a broad deflection curve for specimen 1D1 gradually peaked at a maximum load of 4.20 kN, and the test ended at the fixture-limited maximum deflection of 9.73 mm. The deflection-to-span ratio at this point was 1/9.54. No ligaments were believed to have fractured based on experimental observation. The specimen was oriented with two cells across the width on the top face and three across the bottom face. This asymmetrical cross-section about the horizontal neutral axis led to higher stresses and strains in the top face ligaments than in the bottom face ligaments. The center four cells of top longitudinal and diagonal ligaments were highly buckled at the end of the test, with only minor buckling of the vertical diagonals in this area. The calculated peak tensile ligament stress was 1390 MPa at 3.80 % strain; the calculated peak compressive stress was 1600 MPa at 4.39 % strain. Strains at test termination were 6.83 % in tension and 7.89 % in compression.

Figure 8: Post-test photographs of specimen 1L1 showing buckled compression ligaments; (a) top view, (b) side view.

The final bending strength test was the four-point loaded transversely oriented specimen 1T1, which carried the smallest maximum load due to its lack of longitudinally oriented face ligaments. The subelement behaved linearly until approximately 1.70 kN and 0.84 mm; at this load the analysis predicted 1.07 mm total deflection, which was 28 % less stiff than experimentally measured. The extra specimen stiffness might be due to the unmodeled nodal reinforcement. This effect would be particularly large for specimen 1T1, as all extreme fiber stress was transmitted in a zigzag fashion in the top and bottom face ligaments through the effectively reinforced nodes. This would be somewhat offset by the increased flexibility due to lift-off from the outer roller supports as described earlier for specimen 1L1.

The peak load carried by specimen 1T1 was 2.60 kN, and this was quickly followed by test termination at the maximum deflection of 4.65 mm, when multiple tension ligaments fractured. The deflection-to-span ratio at failure was 1/16.1. Ligament breaks were not detected prior to the final stroke step, and ligament buckling also was not observed. The calculated peak ligament stress of 1650 MPa in tension and compression occurred at 2.84 % strain. The actual stresses were likely lower, again due to the presence of the nodal reinforcement material. A maximum strain 3.34 % was reached just prior to failure.

Concluding Remarks

Superalloy lattice block materials are being investigated for aero-propulsion use. Successful lattice block panels have been investment cast in Inconel 718 superalloy. The promising preliminary results hint at beneficial use for a variety of applications, and have led to plans for additional testing with IN718 as well as higher temperature aerospace alloys.

Porosity or other casting defects caused size effect, premature failure and low ductilities in ligament tensile testing. Poor ligament properties in areas of highest stress would reduce the quality of components fabricated from lattice block material. For this reason the early IN718 lattice block casting practice has to be improved.

The first subelement test under compression gave surprising results. Early inelastic buckling of compression ligaments caused large structural deflection. This phenomenon could be advantageous for some applications but not for others. Revising lattice block geometry or ligament cross-section could promote or lessen their occurrence. Casting defects have little effect on compressive strength.

For structural elements carrying mainly bending loads, strength testing of bend specimens showed that lattice block architecture has several important benefits besides inherent light-weight and high stiffness. These include: damage-tolerant design, evident from reasonable strength even with fractured and buckled ligaments; large, readily-observable deflections prior to breaking; and, load spreading ability due to many load path redundancies. For structures in bending, only a fraction of the loaded material is at the highest stress level; for this reason, reduced strength related to the size effect was not observed. The bending strength tests were generally limited by base material ultimate tensile strength, rather than the structure of the lattice.

Acknowledgements

The authors wish to acknowledge the dedication provided by the NASA UEET program that funded this research, especially the guidance and support provided by Mr. Robert D. Draper, Ms. Catherine L. Peddie, and Dr. Robert J. Shaw of the Glenn Research Center. We also would like to recognize the enthusiasm and commitment provided by Mr. Jonathan Priluck of JAMCORP in advocating and improving lattice block materials, and producing all materials for the test program. Finally, the valuable technical assistance and mechanical skills provided during test development and execution by Mr. Stephen J. Smith of Akima Corporation are gratefully acknowledged.

References

1. M. Hebsur, "Processing of IN-718 Lattice Block Castings," (Paper presented at the TMS Third Global Symposium, Processing and Properties of Lightweight Cellular Metals and Structures, Seattle, Washington, 17-21 February 2002).

2. D. L. Krause, J. D. Whittenberger, P. T. Kantzos, and M. G. Hebsur, "Mechanical Testing of IN718 Lattice Block Structures," NASA TM-2001-211325, 2001.

3. W. F. Brown Jr., H. Mindlin, C. Y. Ho, eds., <u>Aerospace Structural Metals Handbook</u>, vol. 4 (West Lafayette, IN: CINDAS/USAF CRDA Handbooks Operation, Purdue University, 1996), Code 4103.

4. G.K Bouse and M.R. Behrendt, "Superalloy 718 - Metallurgy and Applications," ed. E.A. Loria (Warrendale, PA: The Minerals, Metals & Materials Society, 1989), 319-328.

THE COLLAPSE AND ENERGY ABSORPTION OF EGG-BOX PANELS

Marc Zupan, N. A. Fleck, M. F. Ashby

Department of Engineering, University of Cambridge, Trumpington St., Cambridge, CB2 1PZ, United Kingdom

Abstract

The collapse modes of aluminum egg-box cores with aluminum face sheets are investigated in uniaxial compression. The effects of the geometry and relative density of the core upon the through-thickness stiffness and strength are explored. The energy absorbing capabilities of the egg-box panels are measured and compared with competing energy absorbing structures.

Processing and Properties of Lightweight Cellular Metals and Structures
Edited by Amit Ghosh, Tom Sanders and Dennis Claar
TMS, 2002

Introduction

Energy absorbing structures are widely used in transportation for both human protection and to protect the transportation system itself. One class of absorber utilizes the dissipation of energy by plastic work; it includes honeycombs, metal foams and thin walled crush tubes [1-4]. The requirements of increased fuel efficiency and economy dictate the development of an energy absorbing structure that is lightweight, cheap and easy to manufacture.

Energy absorbing material structures are designed to convert kinetic energy into plastic dissipation [3,4] while maintaining the force induced on the structure below a critical value. The area under the stress versus strain curve up to the densification strain (ε_D) for a material structure measures the absorbed work per unit volume (W_v) [3, 4]. Thus, in selecting a material to absorb a given amount of energy and occupy the minimum volume, the best choice is that with the largest value of W_v. For a minimum weight design we seek instead the largest energy absorption per unit weight (W_w), where $W_w = W_v/\rho$, in terms of the density ρ.

Cold-pressed panels of egg-box geometry have been developed by Cellbond Composites Huntingdon U.K., for energy absorbing applications, Fig 1. The work presented in this paper describes the study of two different geometric designs of egg-box and three different relative densities. Experimental measurements of the compressive response were made for three choices of constraint on the top and bottom faces of the egg-box. Additionally, stacks of several egg-box panels were tested to determine the effect of layering upon the collapse response. The quasi-static energy absorbing capabilities of the egg-boxes was measured and compared with those of metal foams.

Specimen Preparation and Test Method

Egg-Box Geometries

Egg-box panels were prepared from commercially pure and fully-annealed Al 1050 H11 alloy for the two geometries listed in Fig. 1. The panels were composed of truncated cones having four-fold in-plane symmetry. Specimens were manufactured from different thickness Al sheets to achieve a variety of relative densities ($\bar{\rho}$). By cold pressing Al 1050 H11 sheets of thickness 0.500 mm and 0.800 mm, relative densities of 0.044 and 0.028 were achieved for geometry 1 and 0.092 and 0.054 for geometry 2. A third relative density was achieved for both geometries by etching egg-box panels with concentrated phosphoric acid. Specifically, specimens of geometry 1 and of initial relative density $\bar{\rho} = 0.044$ were etched to $\bar{\rho} = 0.035$ with final sheet thickness of 0.650 mm. Samples of geometry 2 with an initial relative density $\bar{\rho} = 0.054$ were etched to $\bar{\rho} = 0.039$, with a sheet thickness of 0.340 mm.

Test method

Samples were tested in uniaxial compression at a nominal strain rate of 10^{-4} s^{-1} strain rate under three different constraint conditions. First, egg-box sandwich panels were constructed by bonding 2 mm thick Dural Al Alloy face sheets to the crowns of the egg-box panels with two component polyurethane adhesive CIBA XB5090-1/XB5304. The face sheets were chosen to be substantially thicker and stronger than the egg-box panels so that the face sheets remained elastic during the transverse compression. Second, to simulate the response of an infinite sheet undergoing zero in-plane strain, specimens were tested with lateral constraint. This was achieved by testing egg-box panels with no face sheets in a confining container with internal dimension equal to the external dimensions of the specimen. Third, egg-box panels were tested free from lateral constraint except the contact friction between the platens of the load frame and the crowns of the egg-box panels. In all cases specimens 5 cells by 5 cells were tested and have been shown to be sufficiently large to eliminate size affects during testing.

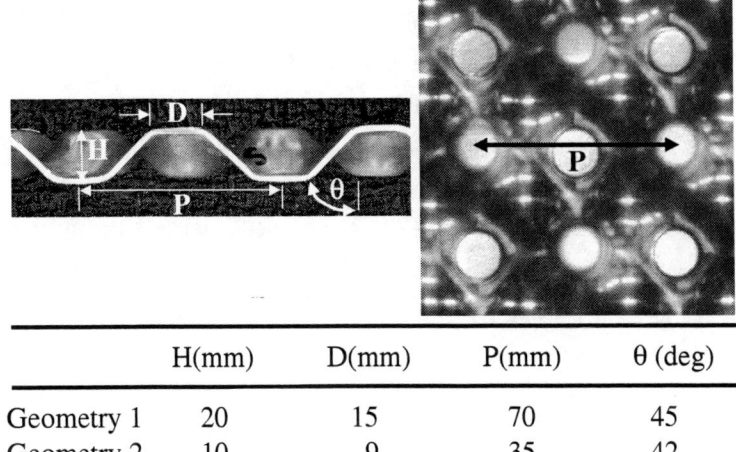

	H(mm)	D(mm)	P(mm)	θ (deg)
Geometry 1	20	15	70	45
Geometry 2	10	9	35	42

Figure 1: The two egg-box panel geometries studied. The egg-box panels were cold pressed from commercially pure Al 1050 H111.

In order to construct thick layers of egg-box core, layers of egg-box can be stacked on top of each other. To explore the effect of stacking on the compression response, two and three layers of panel were stacked and tested in each of the three constraint states. For the sandwich configuration and the unconstrained configuration the panels were stacked peak to peak and bonded by the crowns between each layer. No bonding between layers was used in the constrained multiple layer experiments; the panels were stacked peak to peak and held in place only with the lateral constraining box.

The compression tests were performed using a screw driven load frame with a 25 kN load cell. The axial displacement was measured with a linear potentiometer and a laser extensometer; combination of the two measurements ensures uni-axial loading of the specimen. The densification strain (ε_D) was defined as the strain where the stress level reaches a value 10% higher than the initial peak stress. The energy absorbed per unit volume (W_v) and the energy absorption per unit weight (W_w) were determined by integrating the area under the nominal stress versus strain curve up to the densification strain.

Experimental Results and Discussion

Deformation Response

Monotonic compression tests on Al egg-box panels have been conducted for two geometries at three relative densities. Typical stress versus strain responses for geometry 2 under the different constraint conditions are shown in Figure 2. We deduce that the response of the egg-box panels is sensitive to the lateral constraint. Both the bonded sandwich panels and the unconstrained panels display an initial peak in the stress level, with a subsequent softening, only rising again at the densification strain (about 0.55). The magnitude of the peak stress for the bonded sandwich panels is significantly higher than that for the unconstrained samples: bonding of the core to the face sheet inhibits the initiation of plastic buckling which is the subsequent mode of collapse. Conversely, the constrained specimens do not show a pronounced peak and the deformation is characterized by a long flat plastic deformation response after the elastic loading: the collapse mode is by a travelling circular hinge from each crown of the panel.

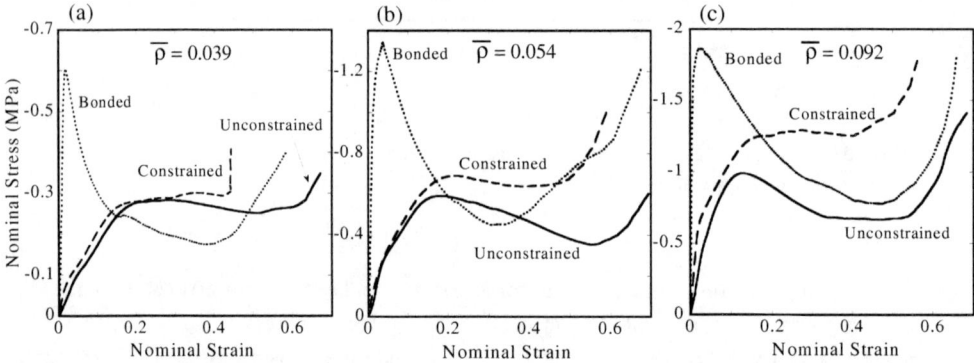

Figure 2: The stress vs. strain response for egg-box panels of geometry 2 as a function of lateral constraint and relative density.

The dependence of the core modulus and peak stress upon relative density is shown in Figure 3. In all cases the bonded panels display a higher stiffness and peak stress than the unconstrained and constrained cases. Figure 3 shows that the stiffness and peak stress of the egg-box panels scale with relative density to the power of 2 and 1.5, respectively.

Multiple Layer Deformation Response

Egg-box panels of geometry 2 with $\bar{\rho} = 0.092$ were used to explore the effect of multiple panels upon the compression response. Figure 4 shows the measured response of one, two, and three layers of panels: for all stackings the unconstrained and bonded panels exhibit a peak in the stress while the constrained shows a smooth transition from the elastic loading to a flat plastic deformation plateau. Figure 4(c) shows that the multiple unconstrained layers are stiffer; this is attributed to the fact that the stack has been bonded together. But, as seen in Figure 4(a) and (c), the presence of the multiple layers reduces the yield strength; it is thought that this is a statistical effect, with the weakest panel of the stack dictating the macroscopic yield strength. The stacked constrained specimens, Fig 4(b), collapse by two simultaneous modes. The crowns of the egg-box in contact with the loading platen deform by a traveling hinge mechanism, whilst the remaining undulations collapse by an un-symmetric buckling mode.

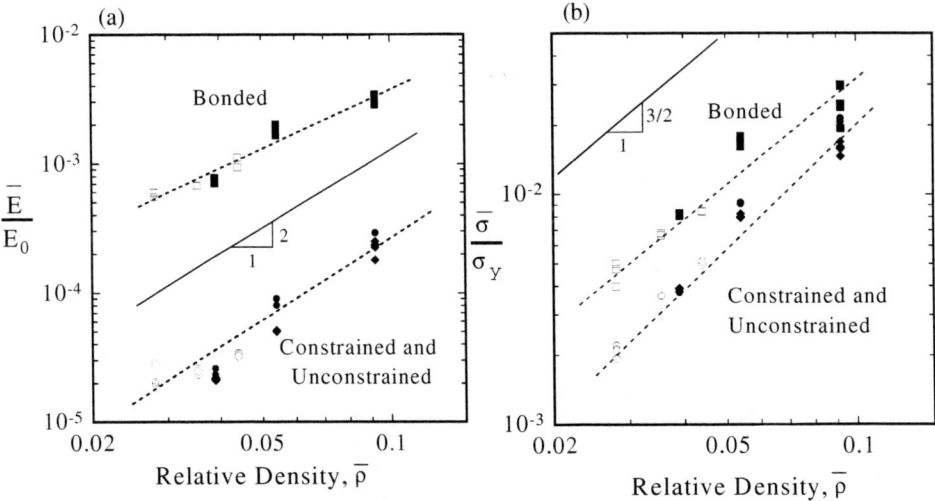

Figure 3: The dependence of effective modulus and peak stress as a function of relative density. Unfilled data points are geometry 1 measurements and filled are geometry 2.

Energy Absorption

The energy per unit mass and per unit volume were measured for each of the loading configurations. These measured values of energy absorption and peak stress are compared with those for metallic foams in Fig. 5. We conclude that the egg-boxes and metallic foams have

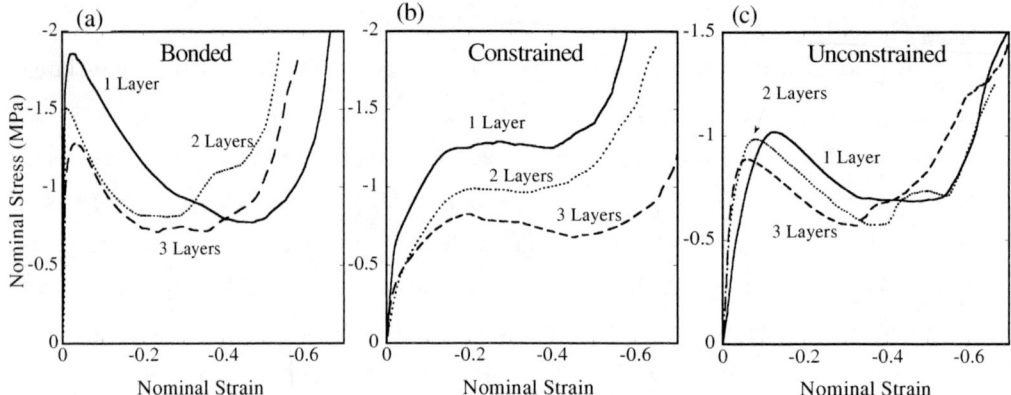

Figure 4: The result of multiple egg-box panel layers on deformation. Several layers reduce the yield level exhibited by the egg-box panels, and smoothes out the peak exhibited by the bonded and unconstrained specimens.

comparable energy absorptions per unit volume, but the egg-boxes are much lighter than the metallic foams for the same level of energy absorption.

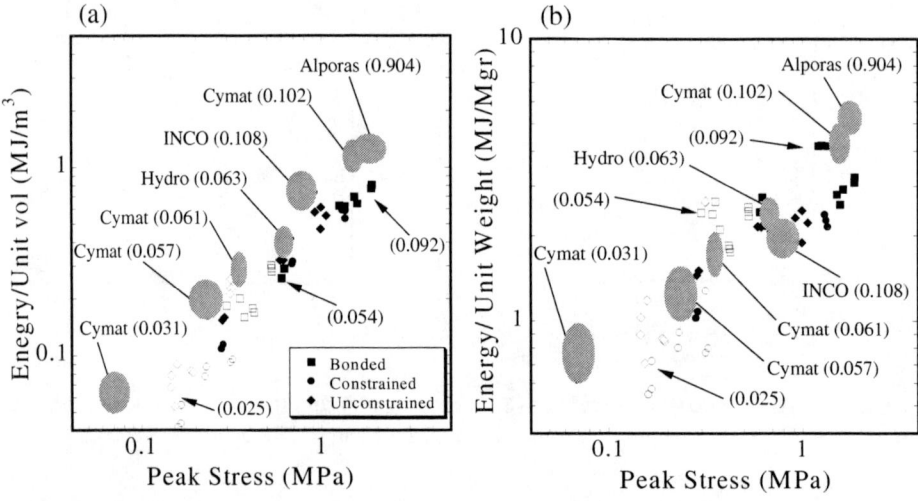

Figure 5: The energy absorbed per unit volume and weight for egg-box panels and currently available metal foams. The relative density of each of the metal foams and some of the egg-box panels are shown in parentheses. Unfilled data points are geometry 1 measurements and filled are geometry 2.

Concluding Remarks

Two egg-box geometries have been tested at three different relative densities in bonded sandwich panels, and in constrained and unconstrained loadings. The effective elastic modulus, peak stress and energy absorption have been measured for the monotonically loaded panels. The deformation character of the egg-box panels is sensitive to the nature of imposed lateral constraint. The dependence of peak stress and effective modulus scale with relative density to the power of 1.5 and 2, respectively. The egg-box has superior energy absorption per unit mass to that of metallic foams.

Acknowledgments

The authors would like to thank Mike Ashmed and Olivier Bedus of CellBond Composites Ltd in Huntingdon, U.K. for their input and for generously supplying the material used in this work. Financial support from DARPA, ONR are gratefully acknowledged.

References

1. J.G. Olivera and T. Wierzbicki, "Crushing Analysis of Rotationally Symmetric Plastic Shells", Journal of Strain Analysis 17 (1982), 229-236
2. A.G. Mamalis, et al. "Extensible Plastic Collapse of Thin-Wall Frustra as Energy Absorbers", Int. J. Mech. Sci. 28 (1986), 219-226
3. Gibson, L.J. and Ashby, M.F., Cellular Solids Structures and Properties, 2nd ed. (Cambridge, United Kingdom: Cambridge University Press, 1997)
4. Ashby, M.F., Evans A.G., Fleck N.A., Gibson L.J., Hutchinson J.W., and Wadley H.G., Metal Foams: A design Guide (Boston, MA: Butterworth-Heinemann, 2000)

MECHANICAL PROPERTIES AND IN-SITU DIFFRACTION STRAIN MEASUREMENTS IN ALUMINUM – MULLITE MICROSPHERE SYNTACTIC FOAMS PRODUCED BY LIQUID METAL INFILTRATION

Dorian K. Balch, David C. Dunand

Department of Materials Science and Engineering
Northwestern University
Evanston, IL 60208-3108 USA

Abstract

High-strength aluminum syntactic foams are fabricated using a liquid metal infiltration technique. Unalloyed and 7075 aluminum are infiltrated into a packed bed of hollow mullite-ceramic spheres. The resulting syntactic foams have relative densities between 0.45 and 0.60 (1.25 – 1.65 g/cm^3). High compressive strengths are observed (100 – 250 MPa), with exceptional energy absorption exceeding 50 MJ/m^3 during densification. In-situ measurements of lattice strains within the mullite phase during compressive loading are performed using high-energy synchrotron x-rays, permitting investigation of sphere stresses and foam damage evolution during loading.

Introduction

Metallic foams are an attractive class of structural materials due to their low densities, high specific stiffnesses, high energy absorbing capabilities, and good mechanical and acoustic damping capacities, among other attributes.[1-3] This combination of properties makes metallic foams an attractive choice for such structural applications as foam sandwich cores, fireproof and sound-damping panels, energy-absorbing packaging, and underwater buoyant structures. A particular class of foam structure, syntactic foams, consists of hollow spheres embedded in a continuous matrix. Such foams are primarily made with polymeric matrices and spheres,[4-6] but metallic syntactic foams containing hollow ceramic spheres can also be fabricated using traditional composite fabrication techniques.[7-9] Such aluminum/alumina sphere, aluminum/silica–alumina sphere, and magnesium/carbon sphere foams have higher densities than conventional metallic foams, but have the advantages of higher strengths, isotropic mechanical properties, and excellent energy-absorbing capabilities due to extensive strain accumulation at relatively high stresses.[7]

The present work illustrates one route for production of aluminum syntactic foams via liquid metal infiltration of packed beds of ceramic microspheres. This method produces syntactic foams with relative densities between 0.45 and 0.60, having minimal unintentional porosity between spheres. By changing the matrix alloy we demonstrate that considerable increases in the strength of these foams are readily achievable. Finally, using synchrotron x-ray diffraction experiments measuring the strains present in the ceramic spheres, we show that considerable stresses are borne by the spheres during compressive loading, indicating that careful tailoring of sphere material and geometry may lead to further improvements in the mechanical properties of metallic syntactic foams.

Experimental Methods

Fabrication and Characterization
Commercial-purity aluminum and 7075 aluminum were infiltrated into packed beds of ceramic microspheres using a custom-built vacuum/pressure infiltrator, shown in Figure 1. The microspheres (Envirospheres PTY Ltd., Lindfield NSW, Australia) were composed of a mixture of 45 vol% crystalline mullite (3 Al_2O_3–2 SiO_2) and 55 vol% amorphous silica (SiO_2), and had diameters of 15 to 75 μm, wall thicknesses of 5 to 10 μm, and densities of 0.4 to 0.6 g/cm^3. The method of infiltration is described in detail elsewhere,[10] and is summarized here. A 50 mm diameter graphite crucible was tap-packed with microspheres to a height of 80 – 120 mm. An ingot of aluminum was placed above the microspheres, separated by a 3 mm thick layer of alumina felt (Zircar Ceramics Inc., Florida, NY). The felt prevented contact between the aluminum and the spheres before pressure was applied, and also partially filtered the aluminum oxide layer during infiltration. After flushing with argon, the infiltrator was heated under vacuum until the aluminum melted and formed a seal across the width of the crucible. Argon gas was then rapidly introduced into the chamber, reaching a pressure of 3.5 MPa within 30 seconds, thereby forcing the molten aluminum into the spaces between the hollow microspheres. Solidification occurred under pressure at ca. 10°C / min, with the resulting foam having minimal unintentional porosity.

The 7075 aluminum foam was heat-treated after sectioning to a standard peak-aged (T6) temper.[11] Homogenization was performed at 470°C for 24 hours, followed by a water quench. Peak-aging occurred at 120°C for 36 hours. During the time between homogenization and aging, approximately 3 hours, the samples were stored at −75°C in a dry ice/ethanol bath to minimize room temperature aging, which can have a deleterious effect on mechanical properties.[11]

Samples for metallography, density measurement, and mechanical testing were machined from the as-infiltrated cylinders using a low-speed diamond saw. Metallographic samples were polished using SiC paper, followed by 6 and 1μm water-based diamond suspensions. Density measurements were performed using a helium pycnometer.

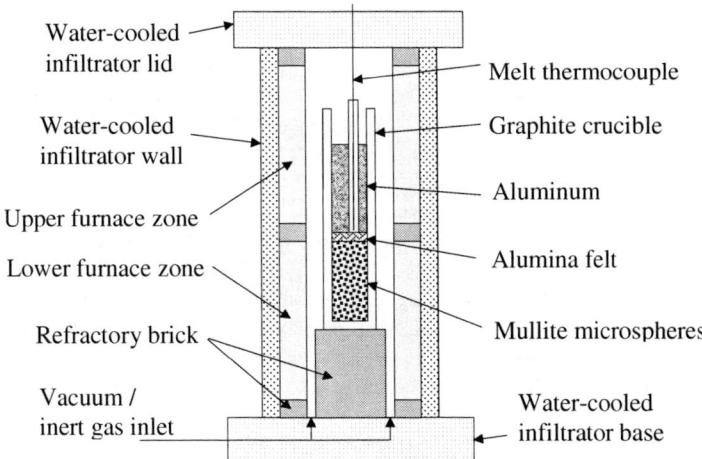

Figure 1: Schematic of the vacuum / pressure infiltrating unit

Mechanical Testing and Diffraction Measurements

Compression testing was performed on parallelepiped samples with dimensions 9.39 x 4.47 x 3.94 mm (pure Al) and 11.06 x 5.22 x 4.91 mm (7075-T6 Al) tested to engineering strains in excess of 60%. Testing was carried out at constant crosshead speed with an initial strain rate of 10^{-3} sec^{-1}. Strains were calculated from the crosshead displacement, adjusted for deflection of the load frame. A compression cage with oil-lubricated tungsten carbide platens was used to ensure proper alignment and minimize sample barreling.

In-situ x-ray diffraction measurements during compression testing were performed at the DuPont-Northwestern-Dow Collaborative Access Team (DND-CAT) at Sector 5 of the Advanced Photon Source (Argonne National Laboratory). Using a table-top load frame (Interactive Instruments, Inc., Scotia, New York), the compressive stress on a pure aluminum foam sample was varied from 0 to –100 to 0 MPa in steps of 20 MPa, during which diffraction measurements took place at constant applied stress. The general set-up for this experiment has been described in detail elsewhere.[12] The 14.98 x 6.77 x 6.63 mm parallelepiped sample, while subjected to a constant uniaxial load, was irradiated for 900 seconds with a monochromatic 65 keV (λ = 0.019 nm) x-ray beam with a square cross-section of 0.5 x 0.5 mm aligned perpendicular to the sample compression axis. The 1.66 mm^3 diffracting volume contained ca. 23,000 microspheres. The transmitted Debye-Scherrer diffraction cones from the mullite grains present in this volume were recorded as rings using a CCD camera, as was one Debye-Scherrer cone from a molybdenum powder standard attached to the sample for calibration purposes. The CCD camera (MAR Inc., Evanston IL) was positioned at a distance of 708 mm from the sample, and had a 132 mm diameter detector with 16 bit intensity readings over an orthogonal array of 64.4 x 64.4 µm pixels. Changes in the diameters of the diffracted mullite Debye-Scherrer rings during compressive loading are used to calculate the strains present in the mullite grains; the analytical procedure used is discussed in detail in Reference 12, as is the use of the calibration standard to remove errors caused by sample movement or x-ray energy fluctuations.

Results

Micrographs of the pure aluminum and 7075-T6 syntactic foams are shown in Figure 2. The microspheres are uniformly distributed, with an average diameter of 45 µm. Some infiltrated spheres are seen, as well as some sphere fragments formed during packing or infiltration. The densities measured by pycnometry for these foams were 1.41 and 1.64 g/cm^3 respectively, corresponding to relative densities of 0.52 and 0.59 as compared to fully dense matrix material. The engineering stress – strain curves for both foams are shown in Figure 3. Typical foam behavior is observed,[1,2] namely an initial elastic region continuing to a peak stress, a densification plateau at roughly constant stress, and an upward turn of the stress-strain curve at the completion of densification. For the pure aluminum foam, a typical diffraction pattern showing multiple mullite rings from the {101}, {231}, {240}, {022}, and {060} lattice planes, four aluminum {111} diffraction spots, and a strong {110} ring from the molybdenum powder standard is shown in Figure 4. The applied compressive stress vs. mullite lattice strain curves (as measured using the overlapping {231} / {240} lattice plane reflections in the diffraction experiment) are shown for both the axial and transverse directions in Figures 5 (a) and (b).

Discussion

Microstructure and Mechanical Properties

As seen in Figure 2, infiltration of the spaces between microspheres was very nearly complete, with no visible porosity remaining between spheres. There are some infiltrated spheres,

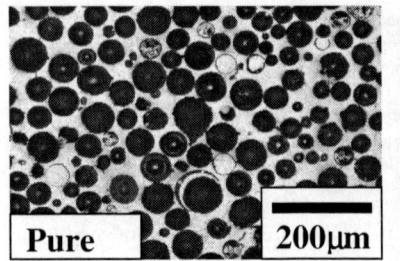
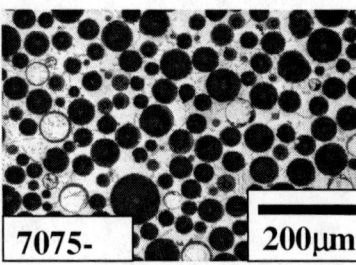

Figure 2: Micrographs of pure Al and 7075-T6 Al syntactic foams

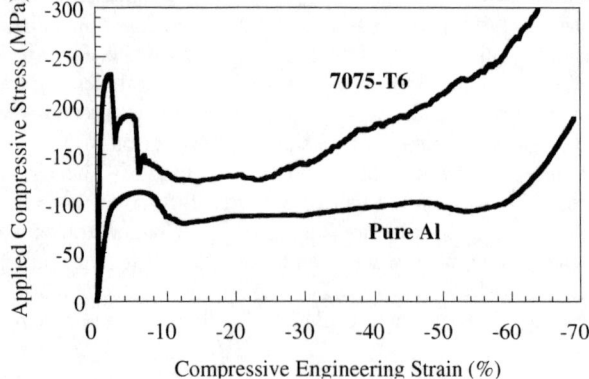

Figure 3: Syntactic foam engineering stress – strain curves

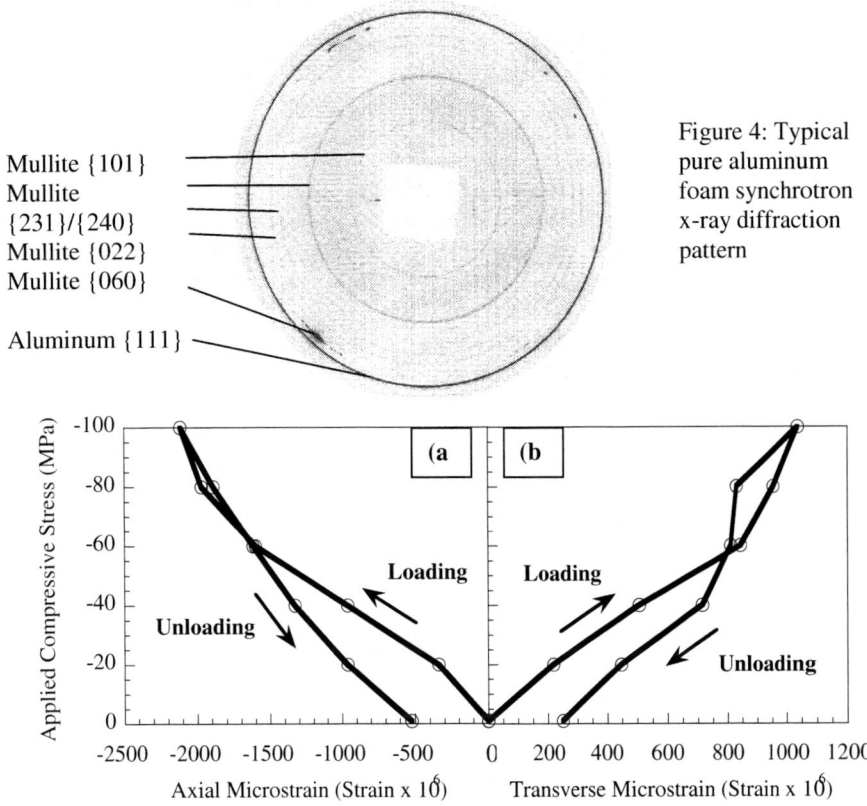

Figure 4: Typical pure aluminum foam synchrotron x-ray diffraction pattern

Figure 5: Applied stress vs. mullite lattice strain curves: (a) axial, (b) transverse

approximately 13% and 11% of the total sphere population in the pure aluminum and 7075-T6 foams, respectively. These spheres were either cracked prior to or during infiltration, or had thin regions of amorphous silica that dissolved in the molten aluminum.[13] Evidence for the dissolution of some silica is seen in the formation of blocky gray inclusions, which are most likely silicon in the case of the pure Al foam, and Mg_2Si in the 7075-T6 foam.[14] In both foams, some merged or nested spheres are observed, but these occur infrequently and should have a negligible effect on bulk mechanical properties.

The stress-strain curves shown in Figure 3 show the behavior typically seen in cellular and porous materials.[2] The pure aluminum foam reached a peak stress of −111 MPa before the onset of densification. The densification plateau stress was between −80 and −100 MPa until an engineering strain of ~60% was reached; beyond this point, densification was essentially complete and stress rose continuously with increasing strain. The 7075-T6 foam reached a considerably higher stress, −232 MPa, prior to the onset of densification. Following two sharp drops in stress, densification proceeded at −120 to −140 MPa until an engineering strain of ~25% was reached. The pure Al and 7075-T6 foams exhibited different modes of failure at the peak stress and during densification. The pure Al foam began to barrel at the center of the sample prior to reaching the peak stress, leading to the rounded shape of the peak in the stress –

strain curve. This was in contrast to the sharp primary peak and slightly more rounded secondary peak in the 7075-T6 curve, which were the result of two 45° shear-type bands of collapsed spheres forming in the specimen due to the reduced ductility of the peak-aged matrix. The lower strain at which densification appeared complete and the region of slowly increasing stress during compression from 25 to 60% engineering strain in the 7075-T6 foam were also likely a result of the reduced matrix ductility. After a brief initial period of microsphere collapse and limited matrix plastic flow, higher stresses were required to fully deform the less ductile matrix into the voids created by microsphere cracking.

Energy absorption during densification was exceptionally high for both foams. The energy absorbed during compression to the densification strain was 55 MJ/m^3 (at ε = 60%) and 36 MJ/m^3 (at ε = 25%) for the pure Al and 7075-T6 Al foams, respectively. These values are one order of magnitude higher than those observed for conventional aluminum foams,[15-17] and are comparable to energy absorption values for steel foams produced by powder metallurgy.[18] These steel foams have similar relative densities (0.38 – 0.64), but far greater absolute densities (3.2 – 5.0 g/cm^3), than the aluminum foams discussed here.

Mullite Lattice Strain Measurements

Figures 5(a) and (b) show the average axial and transverse lattice strains measured in the mullite grains within the microsphere walls as a function of applied stress on the foams, using the {231} / {240} lattice plane reflections. The axial strains (Fig. 5a) are measured in those grains with lattice planes oriented perpendicular to the applied load; i.e. the normal to the analyzed lattice planes is parallel to the applied load. The transverse strains (Fig. 5b) are measured on planes whose normals are perpendicular to the applied load; thus, the axial strains are due to direct compressive loading, while the transverse strains are due to Poisson expansion under applied load. In the following discussion, average strains and stresses will be considered, without accounting for spatial variations of stress within the microspheres. The {231} / {240} rings overlap in the diffraction pattern due both to the similar d-spacings of the planes (0.341 and 0.338 nm, respectively), and to geometrical broadening caused by the sample thickness. They were therefore analyzed as a single ring. A separate analysis on the less-intense {101} ring, not presented here for brevity, yielded results in good agreement with the analysis of the {231} / {240} rings.

As seen in Figures 5(a) and (b), both axial and transverse strains are significant in magnitude, reaching values of –2110 $\mu\varepsilon$ in the axial and 1040 $\mu\varepsilon$ in the transverse directions for an applied stress of –100 MPa. In both the axial and transverse directions, the lattice strains increase roughly linearly up to –60 MPa applied compressive stress. From –60 to –100 MPa compressive stress both strain curves deflect upwards, indicative of a change in loading of the mullite grains. During unloading the axial strains decrease smoothly with a steeper slope than was seen during loading. The steeper slope, as well as the deviations in the transverse strain unloading curve, may be due to sphere damage and matrix plasticity altering the partitioning of load within the foam. After unloading the residual strains present are –520 and 250 $\mu\varepsilon$ in the axial and transverse directions, respectively.

The stresses present in the mullite grains can be calculated from the measured strains using the following equations:[19]

$$\sigma_1 = \frac{E}{1+v}\varepsilon_1 + \frac{vE}{(1+v)(1-2v)}(\varepsilon_1 + \varepsilon_2 + \varepsilon_3) \tag{1a}$$

$$\sigma_2 = \sigma_3 = \frac{E}{1+v}\varepsilon_2 + \frac{vE}{(1+v)(1-2v)}(\varepsilon_1 + \varepsilon_2 + \varepsilon_3) \tag{1b}$$

where σ_1 is the axial principle stress, σ_2 and σ_3 are the transverse principle stresses, ε_1 and ε_2 (= ε_3) are the measured strains, E = 220 GPa is the Young's modulus,[20] and ν = 0.27 is the Poisson's ratio.[20] The calculated stresses during loading are shown in Figure 6 in the axial and transverse directions, assuming that the strains measured using the {231} / {240} overlapping rings can be considered as the average strains in the mullite, i.e. the {231} and {240} planes within the mullite lattice are neither particularly stiff nor compliant. It is apparent in Figure 6 that the stresses behave in a roughly linear fashion up to ca. −60 MPa, but deflect at −80 and −100 MPa to values of lower magnitude than would be expected from extrapolation of the linear region. This is likely indicative of cracking in the sphere walls at the higher applied stresses, which decreases the average stress borne by the spheres. Furthermore, Figure 6 shows that the mullite stresses are considerably higher than the applied stress: axial stresses in the mullite grains reach −370 MPa, with transverse stresses of 180 MPa, at an applied stress of −100 MPa. Loading of the sphere walls is therefore considerable, and is likely due both to load transfer from the more compliant matrix and to direct loading through sphere-to-sphere contact points.

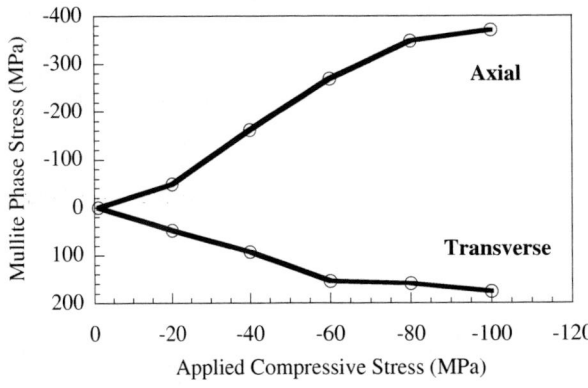

Figure 6: Mullite phase stress vs. applied stress in the pure Al foam

Aluminum Diffraction Spot Evolution

Due to the large size of the matrix grains relative to the diffracting volume, diffraction from the Al {111} planes (d = 0.234 nm) results in a small number of elongated spots rather than complete rings, as for the mullite. Due to the lack of a complete ring, strains cannot be easily calculated for the aluminum phase. It is possible, however, to gain insight into the behavior of the matrix by examination of the shapes and intensities of the Al {111} diffraction spots at various applied stresses. Figures 7(a) and (b) show two images taken from the diffraction patterns obtained at −60 MPa and −100 MPa during loading. Each image is of the most intense Al {111} diffraction spot (located in the lower-left quadrant of Figure 4), and both images are at the same magnification and intensity scaling. The diffraction spots prior to loading and at −60 MPa (Fig. 7a) are qualitatively the same, as are the spots at −100 MPa (Fig. 7b) and after unloading. Figure 8 shows the variation in peak intensity (taken as the maximum pixel intensity) for this diffraction spot as a function of the applied stress. Not shown for brevity are the variations in subtended angle (taken as the angle around the ring of the spot area with intensity above an arbitrary threshold, chosen here as 5 times the background intensity), and full width at half-maximum (taken in the radial direction on the diffraction pattern), of the same diffraction spot; these quantities evolve in a very similar fashion with applied stress. As can clearly be seen in Figures 7 and 8, the angle subtended by the diffracted spot, its maximum intensity, and its FWHM begin to increase at an applied load of −80 MPa, reach

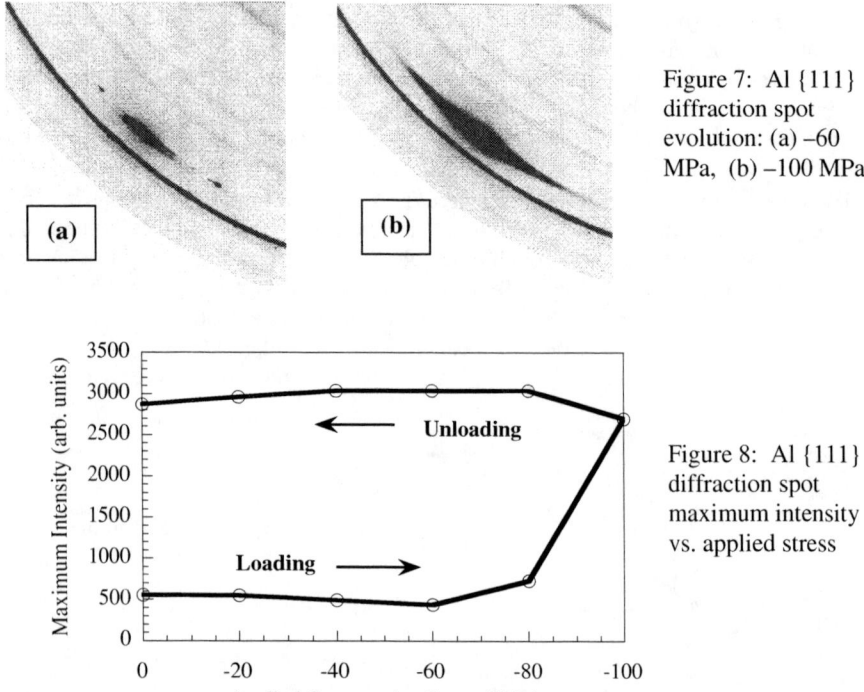

Figure 7: Al {111} diffraction spot evolution: (a) –60 MPa, (b) –100 MPa

Figure 8: Al {111} diffraction spot maximum intensity vs. applied stress

values near their maximums at –100 MPa, and remain at their elevated values during unloading to 0 MPa. This asterism can be attributed to a combination of plastic deformation of the aluminum matrix (leading to dislocation movement and subgrain development, where the initial aluminum grain diffracting to that spot has developed a mosaic structure), and bending or rotation of the aluminum grain. The slight degree of misorientation in the newly formed subgrains leads to a smearing out of the diffraction spot, while the increase in intensity can be explained by the increase in subgrain misalignment reducing the amount of extinction present.[21]

The fact that the aluminum matrix undergoes plastic deformation would generally indicate that the spheres are bearing more of the applied load, through the many sphere-to-sphere contact points present as a result of the fabrication method. Increased load transfer to the spheres would manifest itself, however, as a change in slope towards increased lattice strains in the mullite with increases in applied stress for the foam, and a corresponding increase in the magnitude of the calculated mullite strains and stresses shown in Figures 5 and 6. Instead, the opposite trend is seen- the changes in the slopes of lattice strain and phase stress curves are indicative of reduced loading in the mullite at –80 and –100 MPa. The implication is that the increased load transfer due to plastic flow in the matrix is more than compensated by the load transfer decrease due to cracking in the microsphere walls, and the aluminum matrix therefore bears relatively more of the applied load after the initiation of damage in the foam.

Conclusions

A liquid metal pressure infiltration technique was used to fabricate aluminum-matrix syntactic foams with minimal porosity between mullite microspheres and excellent compressive properties. The energy-absorbing capability of these foams is exceptionally high, and can be easily tailored to specific applications through proper choice of matrix, microsphere materials, and microsphere geometry. The microsphere reinforcement experiences significant loading during foam compression, reaching stresses nearly four times greater than the applied stress, as measured by synchrotron x-ray diffraction experiments. The synchrotron experiments also show that both aluminum matrix plasticity and microsphere cracking are present at the highest applied stresses, with a resulting decrease in overall load transfer to the microspheres.

Acknowledgements

DKB acknowledges the Department of Defense for support in the form of an NDSEG Fellowship. We thank Mr. Paul Seshold (Envirospheres PTY Ltd.) for providing microspheres. Diffraction experiments were performed at the DuPont-Northwestern-Dow Collaborative Access Team (DND-CAT) Synchrotron Research Center located at Sector 5 of the Advanced Photon Source, and the help of the DND-CAT staff is greatly appreciated. DND-CAT is supported by the E.I. DuPont de Nemours & Co., The Dow Chemical Company, the U.S. National Science Foundation through Grant DMR-9304725 and the State of Illinois through the Department of Commerce and the Board of Higher Education Grant IBHE HECA NWU 96. Use of the Advanced Photon Source was supported by the U.S. Department of Energy, Basic Energy Sciences, Office of Energy Research under Contract No. W-31-102-Eng-38.

References

1. M.F. Ashby, A. Evans, N.A. Fleck, L.J. Gibson, J.W. Hutchinson, and H.N.G. Wadley, Metal Foams: A Design Guide (Boston, MA: Butterworth Heinemann, 2000)
2. L.J. Gibson and M.F. Ashby, Cellular Solids, Structure and Properties, 2nd Ed. (Cambridge, UK: Cambridge University Press, 1997)
3. J. Banhart, "Manufacture, Characterization, and Application of Cellular Metals and Metal Foams," Prog. Mater. Sci. 46 (2001), 559-632
4. F.A. Shutov, "Syntactic Polymer Foams," Adv. Polymer Sci. 73 / 74 (1986), 63-123
5. N. Gupta, Kishore, E. Woldesenbet, and S. Sankaran, "Studies on Compressive Failure Features in Syntactic Foam Material," J. Mater. Sci. 36 (2001), 4485-4491
6. P. Bunn and J.T. Mottram, "Manufacture and Compression Properties of Syntactic Foams," Composites 24 (7) (1993), 565-571
7. M. Kiser, M.Y. He, and F.W. Zok, "The Mechanical Response of Ceramic Microballoon Reinforced Aluminum Matrix Composites Under Compressive Loading," Acta Mater. 47 (9) (1999), 2685-2694
8. P.K. Rohatgi, R.Q. Guo, H. Iksan, E.J. Borchelt, and R. Asthana, "Pressure Infiltration Technique for Synthesis of Aluminum – Fly Ash Particulate Composite," Mater. Sci. Eng. A244 (1998), 22-30
9. I.W. Hall and C. Poteet, "High Strain Rate and Other Properties of a Carbon Microsphere Reinforced Magnesium Alloy," J. Mater. Sci. Lett. 15 (1996), 1015-1017
10. A. Mortensen and I. Jin, "Solidification Processing of Metal Matrix Composites," Intern. Mater. Rev. 37 (3) (1992), 101-128

11. J.E. Hatch, ed., Aluminum: Properties and Physical Metallurgy, (Metals Park, OH: ASM International, 1984), 134-199

12. A. Wanner and D.C. Dunand, "Synchrotron X-Ray Study of Bulk Lattice Strains in Externally Loaded Cu-Mo Composites," Metall. Mater. Trans. 31A (2000), 2949-2962

13. R.Q. Guo and P.K. Rohatgi, "Chemical Reactions Between Aluminum and Fly Ash During Synthesis and Reheating of Al - Fly Ash Composite," Metall. Mater. Trans. 29B (1998), 519-525

14. Metals Handbook, Ninth Ed., Volume 9: Metallography and Microstructures, (Metals Park, OH: ASM International, 1985), 351-388

15. A.E. Markaki and T.W. Clyne, "The Effect of Cell Wall Microstructure on the Deformation and Fracture of Aluminum-Based Foams," Acta Mater. 49 (2001), 1677-1686

16. F. Han, Z. Zhu, and J. Gao, "Compressive Deformation and Energy Absorbing Characteristic of Foamed Aluminum," Metall. Mater. Trans. 29A (1998), 2497-2502

17. T. Miyoshi, M. Itoh, T. Mukai, H. Kanahashi, H. Kohzu, S. Tanabe, and K. Higahi, "Enhancement of Energy Absorption in a Closed-Cell Aluminum by the Modification of Cellular Structures," Scripta Mater. 41 (10) (1999), 1055-1060

18. C. Park and S.R. Nutt, "PM Synthesis and Properties of Steel Foams," Mater. Sci Eng. A288 (2000), 111-118

19. G.E. Dieter, Mechanical Metallurgy, 3rd Ed. (New York, NY: McGraw-Hill Inc., 1986), 51

20. Engineered Materials Handbook, Volume 4: Ceramics and Glasses, (Materials Park, OH: ASM International, 1991), 761-765

21. B.D. Cullity, Elements of X-Ray Diffraction, 2nd Ed. (Reading, MA: Addison-Wesley, 1978), 260-295

VIBRATION PROPERTIES OF CLOSED-CELL ALUMINUM FOAM

H.F. von Bremen, V.S. Sokolinsky, S. Rajaram, H. Shen,
J.A. Lavoie, Y. Zhou, and S.R. Nutt

Center for Composite Materials, University of Southern California,
Los Angeles, CA 90089-0241

Abstract

Many lightweight structural applications experience dynamic excitations. Therefore, it is important to assess the vibration properties of aluminum foam. A simple experimental layout consisting of a shaker and an accelerometer is used to investigate the free vibration response of a cantilever aluminum foam beam. The natural vibration frequencies of the beam were measured over 20–5000 Hz. The Young's modulus of the aluminum foam is calculated with the aid of the classical beam theory using measured natural frequencies of cantilever specimens. A comparison between the foam modulus values obtained using the vibration approach and experimental values of the modulus measured from the unloading static slope are in close agreement (within 4 %). It was found that the closed-cell aluminum foam under consideration possesses almost isotropic structure. The vibration study shows that the presence of randomly located internal and external voids in the material structure has minor effect on the vibration response of the aluminum foam specimens. The Young's modulus of the metal foam can be accurately and effectively measured using the natural frequencies of the cantilever sample. Supported by Honda R&D Americas.

Introduction

Many industrial and defense-related applications have a need for lightweight materials and structures that combine high load-bearing capacity with efficient energy absorption under a wide range of vibration and impact conditions. Ultralight, closed-cell aluminum (Alporas-type) foams used as a core material in sandwich structures have the potential to meet this challenging demand. Olurin [1] found Alporas foam to be almost isotropic, with modulus and yield strength varying with specimen orientation by ± 15%. Motz and Pippan [2] reported that the deformation mechanism of Alporas foam loaded in tension differed from that for compression. The dependence of Young's modulus and Poisson's ratio on the microstructure and density of solid materials with a closed-cell cellular structure was studied by Roberts and Garboczi [3]. Their finite element results for several isotropic random models were in good agreement with the experimental data, confirming analytically the power law dependence of modulus on density of foamed solids.

This paper reports on the vibration response of closed-cell aluminum foam produced by direct metal foaming [4, 5], and offers a simple vibration approach for measuring the Young's modulus of a metal foam. The setup used for experiments is described, and compression properties of the foam are given. Next, the experimental study of the frequency response of three foam cantilever specimens is given. Finally, the results are discussed and conclusions drawn.

Experimental Setup

The free vibration response of a cantilever aluminum foam beam was measured using a simple setup consisting of a signal generator, amplifier, accelerometer, shaker table, and computer data acquisition system. The signal generator produced a sine sweep signal of constant amplitude and slowly varying frequency. The signal was then amplified and delivered to the shaker table that generated the excitation force. A miniature Dytran accelerometer (2.5 grams, model 3025AG) with a sensitivity of 100 mV/g was attached to the tip of the cantilever (Figure 1). The signal from the accelerometer was recorded using the computer data acquisition system.

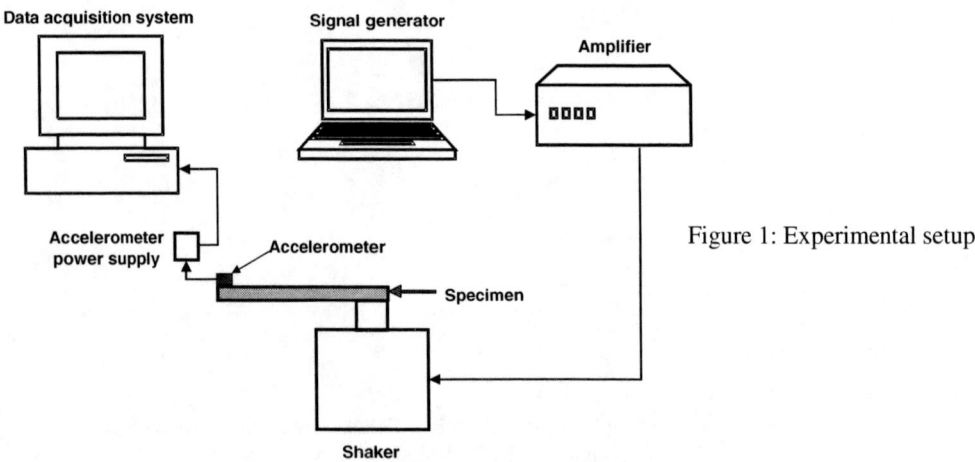

Figure 1: Experimental setup.

Compression Response of Aluminum Foam

Compression tests of aluminum foam at room temperature were conducted in a universal test machine in displacement control at a rate of 1.5 mm/min. The Young's modulus was measured by loading the foam into the plastic range, then unloading and taking the modulus from the unloading slope [6]. The average values of measured mechanical properties of the foam are given in Table I. The stress–strain curves for loading in the direction parallel to the foam expansion (longitudinal) and perpendicular to the foam expansion (transverse) appear in Figure 2. Five specimens of size 25x25x20 mm were tested for each direction.

Table I Mechanical Properties of Closed-cell Aluminum Foam

Direction	Relative Density	Density, Mg/m^3	Young's Modulus, GPa	Compression Strength, MPa
Longitudinal	0.134	0.362	1.35	4.404
Transverse	0.133	0.359	1.292	3.872

The compression curves in Figure 2 and the data in Table I show that the closed-cell aluminum foam under consideration exhibits almost isotropic properties in the linear range of material response. These data are consistent with that for closed-cell Alporas foam of Ref. [1].

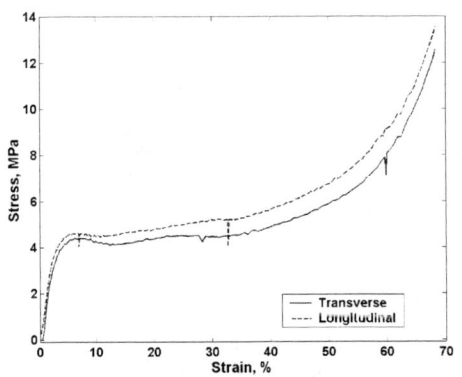

Figure 2: Foam compression curves.

Vibration Analysis and Discussion

A schematic of the three aluminum foam specimens used in the experiments is shown in Figure 3. The specimens were taken from a 300x300 mm (12"x12") board of aluminum foam produced by direct metal foaming [4, 5]. The three specimens were cut sequentially starting from the edge of the aluminum foam board towards its center. Specimen #1 was taken near the edge of the board, and had the most condensed structure and the smallest cell size among the three specimens. The reason for this is that the perimeter of the foaming aluminum loses heat due to contact with the crucible walls, resulting in smaller bubbles. Specimen #2 was taken from the part of the foam board that did not have any defects on its surface. However, after the specimen was cut from the board, two internal defects (voids) were found at one of the specimen surfaces.

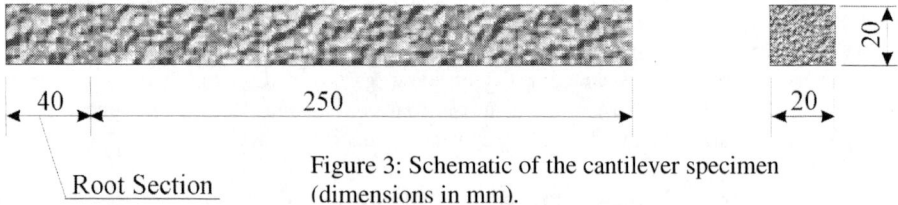

Root Section

Figure 3: Schematic of the cantilever specimen (dimensions in mm).

The voids formed in the direction of the foam expansion, and the longest void was nearly 20 mm. Specimen #3 was cut from the part of the foam board that contained sizable voids (12x7 mm) on its external surface. One object of the study was to determine whether these structural defects influenced the vibration response of the foam specimens. A photograph of a foam specimen clamped in the shaker rig is given in Figure 4.

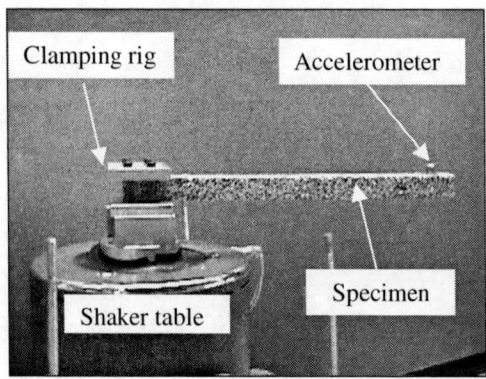

Figure 4: Cantilever foam specimen in testing rig.

The natural frequencies of the cantilever beams were measured over the range 20–5000 Hz, with each specimen tested in two orientations. First, the specimen was clamped so that the foam expansion direction was perpendicular to the direction of oscillation of the shaker table, then the axis of foam expansion was aligned parallel to the direction of oscillation. Experimental acceleration versus frequency response of the three cantilever beam specimens is shown in Figures 5–7.

From Figures 5(a) and 6(a), and Figures 5(b) and 6(b) observe that the frequency responses of Specimens #2 and #3 are almost identical for both, the perpendicular and parallel experimental configurations, respectively. Moreover, in the case of the vibration plane parallel to the expansion axis (Figures 5(b) and 6(b)), the locations of the natural frequencies are clearly defined. This is not the case for Specimen #1 (Figure 7(b)), where the frequency response for the parallel configuration exhibits split peaks. As a result, the exact locations of the higher frequencies can not be clearly established, and can only be estimated as an average value within a certain frequency range. Comparison between Figures 5, 6 and 7 shows that in the case of

Specimen #1 the natural frequencies are shifted to the right. Furthermore, the maximum acceleration amplitudes for Specimens #2 and #3 are attained at the second natural frequency in both experimental configurations (Figures 5, 6). In the case of Specimen #1, vibrations in the plane perpendicular to the foam expansion axis also produce the maximum amplitude at the second natural frequency, just as for the other two specimens (Figure 7(a)). However, when the axis of foam expansion and the vibration plane are parallel, the character of the frequency response of Specimen #1 changes significantly (Figure 7(b)). Here, the maximum acceleration amplitude is attained at the first natural frequency, and the acceleration amplitudes of higher vibration modes are comparable in magnitude with the lower modes.

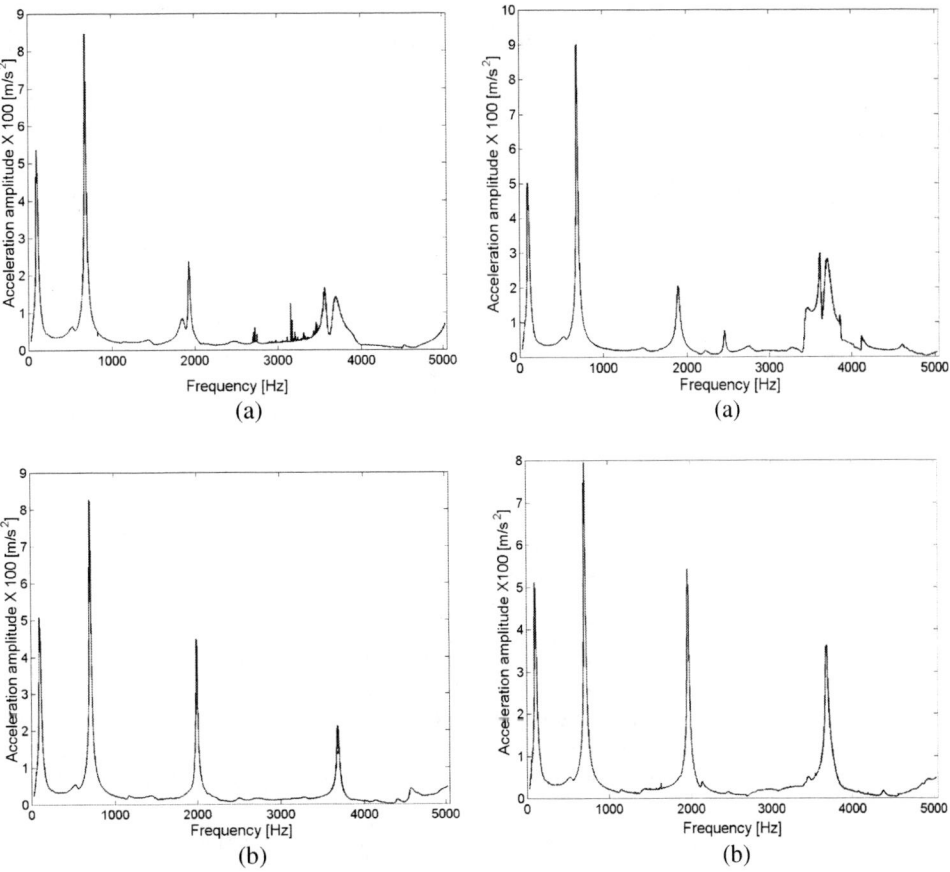

Figure 5: Experimental acceleration vs. frequency response of Sample #2 with the axis of foam expansion:
(a) perpendicular to the bending plane;
(b) parallel to the bending plane.

Figure 6: Experimental acceleration vs. frequency response of Sample #3 with the axis of foam expansion:
(a) perpendicular to the bending plane;
(b) parallel to the bending plane.

Tables II and III show the measured natural frequencies for the three specimens, where natural frequencies correspond to the acceleration amplitude peaks in Figures 5–7. It is seen that the experimental results for Specimens #2 and #3 are close to each other, with the maximum difference less than 2 % in both configurations of the specimen. Thus, taking into consideration the presence of randomly located internal (Specimen #2) and external (Specimen #3) voids in the structure of the two specimens, it appears that the influence of the structural defects has minor effect on the vibration response of the aluminum foam specimens.

It can be observed from the tables that in general, the natural frequencies of Specimen #1 are higher than the frequencies for the two other specimens in both experimental configurations. The greatest difference (up to 13.5 %) is observed for the fundamental (lowest) frequency when Specimen #1 vibrates in perpendicular configuration and for the highest frequency (up to 15.6 %) in the parallel configuration. This is attributed to the fact that Specimen #1 had the most condensed cell structure and the smallest cell size which resulted in a higher bending stiffness than that of the other specimens.

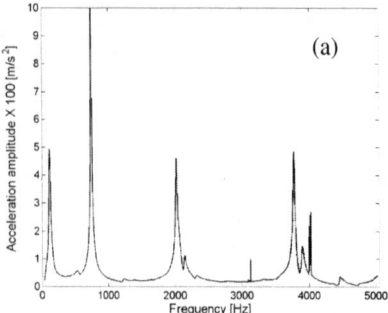 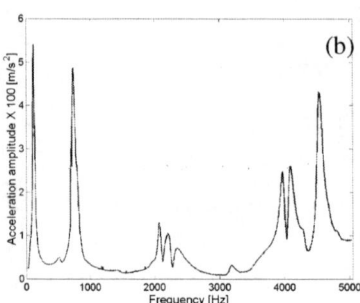

Figure 7: Experimental acceleration frequency response of Sample #1, with the axis of foam expansion: (a) perpendicular to the bending plane; (b) parallel to the bending plane.

By comparing the data listed in Tables II and III note the similar response for vibrations in either direction (the discrepancy does not exceed 4 % for Specimens #2 and #3, and 13 % for Specimen #1). This supports the conclusion, made earlier on the basis of the compression tests, that the aluminum foam under consideration behaves isotropically. Also, the vibration tests demonstrate that the foam sample cut from the edge of the board is less isotropic than those nearer the middle.

Table II Natural frequencies with vibration plane perpendicular to expansion direction

Mode	Experimental Frequencies, [Hz]		
	Specimen #1	Specimen #2	Specimen #3
1	110	98	97
2	726	682	689
3	2018	1929	1894
4	3773	3559–3694	3613–3700

Table III Natural frequencies with vibration plane
parallel to expansion direction

Mode	Experimental Frequencies, [Hz]		
	Specimen #1	Specimen #2	Specimen #3
1	113	100	99
2	744	709	706
3	2066–2355	1992	1963
4	3971–4540	3685	3682

The experimental values of natural frequencies in Tables II and III can be used to calculate the Young's modulus. The Euler-Bernoulli classical beam theory [6,7] yields the following equation for the cantilever (fixed-free) beam

$$E_f = \left(\frac{\omega_n}{k_n^2}\right)^2 \frac{\rho_f A}{I} \qquad (1)$$

where E_f is the Young's modulus of the foam; ρ_f is density; A is the cross section area; I is the second moment of inertia of the beam cross section; and k_n is determined by the following transcendental equation:

$$\cos k_n L \cosh k_n L = -1 \qquad (2)$$

where L denotes the beam span. The Young's modulus of the foam was calculated for each of the measured natural frequencies in both configurations using Equations (1) and (2). The results are presented in Tables IV and V.

Table IV Calculated Young's modulus of the foam with
vibration plane perpendicular to expansion direction

Mode	Young's Modulus, [GPa]		
	Specimen #1	Specimen #2	Specimen #3
1	1.4490	1.1501	1.1268
2	1.6072	1.4183	1.4475
3	1.5838	1.4472	1.3951
4	1.4418	1.3320	1.3541
Mean value	1.5204	1.3369	1.3309
Mean value of the Young's modulus from the three specimens:			**1.3961**

Tables IV and V show that according to the vibration measurements and the classical beam theory, the variation in Young's modulus with the cantilever specimen orientation is less than 9%. The main cause of the variation is due to Specimen #1, for which the values of Young's modulus differ by about 14%. If Young's modulus is determined from the results of Specimens #2 and #3 only, the variation of modulus with the specimen orientation is about 5%.

Table V Calculated Young's modulus of the foam with vibration plane parallel to expansion direction

Mode	Young's Modulus, [GPa]		
	Specimen #1	Specimen #2	Specimen #3
1	1.5291	1.1975	1.1737
2	1.6878	1.5328	1.5198
3	1.9004	1.5433	1.4987
4	1.8341	1.3753	1.3731
Mean value	1.7379	1.4122	1.3913
Mean value of the Young's modulus from the three specimens:			1.5138

Young's modulus for the foam can be estimated following Ref. [6] as

$$E_f \approx \alpha_2 E_s \left(\frac{\rho}{\rho_s} \right)^n \qquad (3)$$

where E_s is the Young's modulus of the aluminum alloy used to make the foam board ($E_s \approx 70$ GPa); ρ/ρ_s is the foam relative density (see Table I); and, according to the compression tests and the vibration results presented here, if n is taken to be 2, then α_2 has a value between 1.06 and 1.40.

A comparison between the experimental values of the foam modulus in Table I with the values calculated from the measured natural frequencies in Tables IV and V reveals that agreement is within ~ 8 % perpendicular to the foam expansion axis and ~ 12 % parallel to it. It may be noted that the influence of the condensed edge material on the compression test results was essentially less than on the vibration measurements due to the random choice of the compression specimens, their small size and larger number with respect to the vibration specimens. If according to the last statement, the values of the Young's modulus are calculated without taking into consideration the vibration results of Specimen #1 (see Tables IV and V), the discrepancy between the compression and vibration tests for both directions does not exceed 4 %. Therefore, the Young's modulus of the metal foam can be accurately and effectively measured using the natural frequencies of the cantilever sample.

Concluding Remarks

Vibration response of closed-cell aluminum foam produced by direct metal foaming was investigated. The study leads to the following observations.
1. The foam material near the edge of the board has the most condensed cell structure and the smallest cell size. This leads to higher natural frequencies of the cantilever beam specimen made of the material near the board edge, and as a result, higher values of the Young's modulus.

2. The vibration study shows that the presence of randomly located internal and external voids in the material structure has minor effect on the vibration response of the aluminum foam specimens.
3. The Young's modulus of the aluminum foam calculated from the classical beam theory using the measured natural frequencies of a cantilever beam vary with the orientation of the specimen by 8.5 %. This confirms the conclusion based on the conventional compression tests, that the closed-cell aluminum foam under consideration is almost isotropic.
4. The vibration experiments demonstrate that the foam at the edge of the board is less isotropic than that in nearer the middle.
5. A comparison between the experimental values of the foam modulus obtained from the unloading static slope and the values calculated from the classical beam theory using the measured natural frequencies of a cantilever beam are in close agreement (within 4 %). Therefore, the Young's modulus of the metal foam can be accurately and effectively measured using the natural frequencies of the cantilever sample.

Acknowledgments

The authors gratefully acknowledge the financial support from the M.C. Gill Corporation. Special thanks are given to Dr. Z. L. Song for his help in preparation of the test specimens. The authors are also thankful to Dr. L. Vaykhansky for his valuable notes that helped to improve the paper.

References

1. O.B. Olurin, N.A. Fleck, and M.F. Ashby, "Deformation and Fracture of Aluminium Foams," Material Science and Engineering, A291 (2000), 136–146.

2. C. Motz, and R. Pippan, "Deformation Behaviour of Closed-Cell Aluminium Foams in Tension," Acta Materialia, 49(2001), 2463–2470.

3. A.P. Roberts, and E.J. Garboczi, "Elastic Moduli of Model Random Three-Dimensional Closed-Cell Cellular Solids," Acta Materialia, 49(2001), 189–197.

4. Z.L. Song, L.Q. Ma, J.W. Zhao, and D.P. He, "Effects of Viscosity on Cellular Structure of Foamed Aluminum in Foaming Process," Journal of Materials Science, 35(2000), 15-20.

5. Z.L. Song, J.S. Zhu, L.Q. Ma, and D.P. He, "Evolution of Foamed Aluminum Structure in Foaming Process", Material Science and Engineering, A298(2001), 137-143.

6. M.F. Ashby et al., Metal Foams: A Design Guide (Oxford, UK: Butterworth-Heinemann, 2000), 40-77.

7. I.H. Shames, and C.L. Dym, Energy and Finite Element Methods in Structural Mechanics (New York, NY: McGraw-Hill, 1985), 323-335.

3D MICRO TOMOGRAPHY (µCT) OF CELLULAR METALS USING AN UP TO 320KV X-RAY TUBE

B. Illerhaus[1], E. Jasiūnienė[1], A. Kottar[2], J. Goebels[1]

[1] Federal Institute for Material Testing and –Research (BAM),
Unter den Eichen 87, 12205 Berlin, Germany
[2] Institute of Material Science and Testing, Vienna University of Technology, Austria

Abstract

Cellular metals were investigated by 3D micro tomography (3DµCT) using an up to 320kV micro focus X-ray tube. The tomograms were investigated with 3D image processing methods. Aluminium foams as well as samples of hollow iron spheres were investigated. The aim was to enhance the quality of the foam by non destructive testing and to generate software tools which will be able to link the measured properties to mechanical parameters of foams. A way to characterise the density and the homogeneity at one glance is proposed. Internal deformation of samples after subsequent pressing is shown. A software tool calculates the partial shift in 3D and allows to combine it with other foam properties. The pores are detected and the pore size distribution is calculated. Also the centre of gravity and the volume of each pore is stored and thus allows us to generate a FEM model of a real foam.

Introduction

Cellular metals, which are produced as foams or as regular structures, will be used more and more in the aircraft and auto industry [1]. Due to the very complex structure of cellular metals, micro tomography (µCT) is a superior method to determine their inner integrity. For characterization of metallic foams a fully three dimensional investigation of the objects is required with high spatial resolution in all directions. A common method to achieve high spatial resolution is using X-ray tubes with a very small focus and magnification technique. For larger components the difference in absorption between the surface metal sheet and the foam can be reduced by using higher X-ray energies. X-ray tubes with higher energies allows us to investigate foams of higher absorbing materials.

The definition of underlying principles of foaming and of material strength, the generation of simple models and the calculation of complex structures needs the geometrical description of foams. The discussion on the underlying features of mechanical stability of metallic foams is still ongoing. For comparison with the theoretical calculations, the objects have to be described in different ways. Relevant microstructural features are: localised mean density, the pore size and pore size distribution, mean thickness of the walls, etc. Starting from a high resolution 3D-CT image all these features can be calculated. The aim of this study was a detailed structural analysis of the foam using 3D computed tomography, interpretation of different features of the metallic foams starting from highly resolved 3D tomograms. For these purposes several software programs were developed.

3D computed tomography

BAM has developed several 3D tomographs for various purposes [2, 3]. A variety of objects from 200 mm diameter (with minimum resolution of 0.1 mm) (Figure 1a) down to small parts with 1 mm diameter (with minimum resolution of 1.5 µm) (Figure 1b) can be investigated using 3D computed tomography. To expand the measuring range to higher attenuating objects, a 320 kV micro-focus X-ray tube has been installed (Figure 2). For the reconstruction of the 3D image of the investigated object from the collected data the Feldkamp algorithm [4] is used.

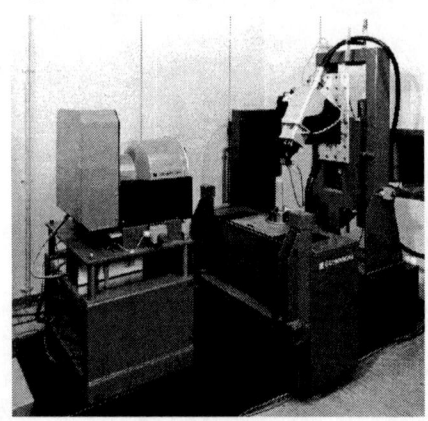

Figure 1. 3D tomograph machines: a - for objects with diameter of up to 200 mm,
b - for objects with diameter of up to 20 mm.

320kV micro focus X-ray tube

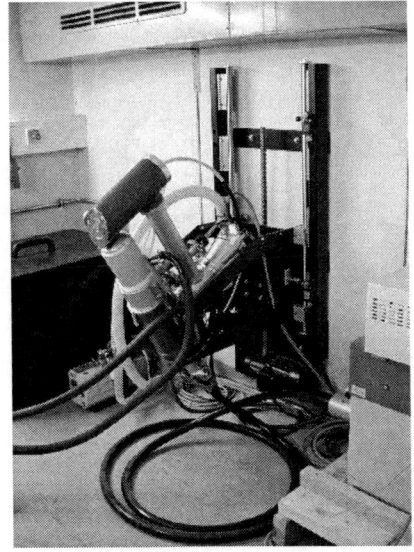

Figure 2. 320kV micro focus tube.

Most of the cellular metals are made from aluminium and thus have a high penetration depth for X-rays. But often they are of non-regular shapes with an outer cover of a different material or with the foam skin itself. This produces artefacts in tomographical images due to the high attenuation in the tested samples. In some applications foams of high attenuating materials (up to lead) could also be used. In these cases higher X-ray energy could reduce the beam hardening artefacts and exponential edge gradient effects in tomographical images.

Samples

Two types of metallic foams and one type of hollow sphere material were used in this study. Both foam samples were cuboids with 20 mm side length and 30 mm height. The hollow sphere sample was cylindrical in shape with a diameter 20mm and a height of 20mm The voxel side length in tomographical images was 40 µm.

The first sample was a porous Al alloy, known by trade name ALPORAS [6] (Figure 4a). It was produced at the Shinko Wire Company Ltd., Izusmisano, Japan by a melting route. The density of the sample was 0.45 g/cm^3.

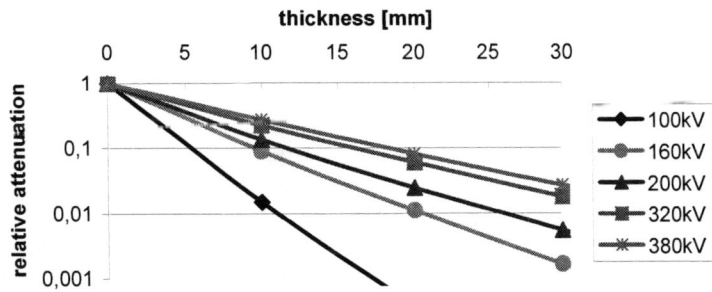

Figure 3. Relative attenuation for Fe with 2mm Cu filter.

Another sample was the material with the trade name ALULIGHT [7] (Figure 4b). This AlSi12Mg foam was produced by a powder metallurgical route at the Slovak Academy of

Science, Bratislava. The density of the sample was 0.50 g/cm³. The sample was cut from a slab with the dimensions 400 × 400 × 30 mm.

The hollow sphere material was produced by IFAM Dresden [8] (Figure 4c). An iron based solution is sprayed on small floating polystyrene foam balls in a fluidised bed. These coated balls are filled into forms, slightly pressed and sintered to any desired form.

3D image processing

For the description of the mechanical behaviour and failure mechanisms of the cellular metals several 3D software tools were developed. Relevant microstructural features of the foams are: localised mean density, pore size and pore size distribution. Starting from 3D-CT images those features can be calculated [9]. Furthermore with a 3D version of the image processing tools 'erode' and 'dilate' a separation of nodes and walls within the material can be done, the amount of material in the walls (nodes) and the mean thickness of the walls can be calculated, the density distribution of walls (nodes) can be extracted [10]. The dislocation paths and failure mechanisms can be found by fitting small parts from 3D µCT images of foams before and after stepwise strength tests [11]. Comparisons with other local properties in 3D are possible.

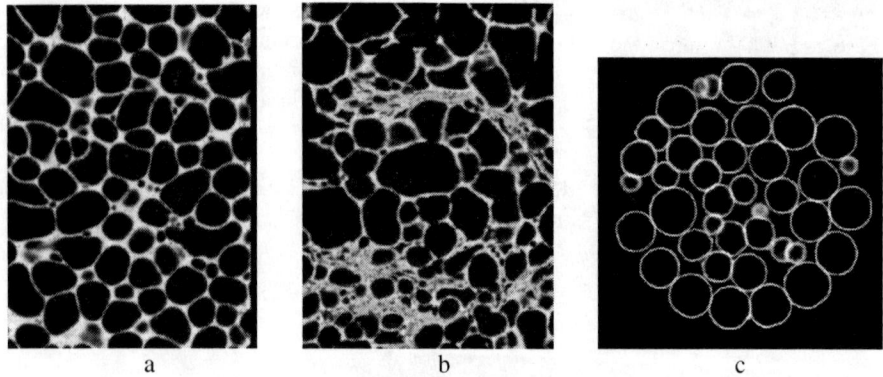

Figure 4. A vertical slice through the foam samples:
a – ALPORAS sample, b– ALULIGHT sample, c-hollow sphere material.

Density evaluation

The requirements for foams are often described using the parameter of mean density. Inside the sample this mean value can be reached by different means and may be locally dependent. A foam with small pores and thin walls can have the same overall density as a foam with large pores and with only few thick walls. Large pores included in the sample cannot be detected using simple radiography due to the overall structure of the foam.

For the density evaluation a tool to describe the homogeneity of the sample was developed, using an idea of Degischer et al. [12]. First the region has to be defined in which all points should be averaged. The averaging region is moved across the whole sample in 3D. The resulting image allows inspection of density variations within the foam. The size of the averaging region is also important. It should be set to the same size as the size of the maximum pore which is still tolerable. By now the information is still in 3D, this means, all information is not visible at one glance. Therefore a software tool, which x-rays the data using the 'mean' operation, was used. The result is a two dimensional representation of the density distribution. If there would be a larger pore inside, this would not be clearly visible by now. So all areas with a too low density over a certain region where highlighted. Highlighting of these regions before the 'x-ray' process will show up both by now: the low and high density areas. Large pores included in the sample cannot be detected using simple radiography due to the overall structure of the foam.

In Figure 5 the two dimensional representation of the density distributions is presented: in Figure 5a for ALPORAS foam and in the Figure 5b the for ALULIGHT foam. In Figure 5 low density areas (the large pores) are visible as a kind of fleck on the image. From the presented images it can be seen, that the density distribution in ALPORAS foam is more homogeneous. In both images the pores, which radius were larger than 2 mm, were highlighted. In Figure 5a we see no highlighted areas at all. This means, that in ALPORAS sample all pores had radius smaller than 2 mm. In Figure 3b it is possible to see, that density distribution in ALULIGHT sample is not homogeneous – some areas with the higher density can be detected. Also there are several pores with radius larger than 2 mm.

Pore detection and pore size calculation

In addition to localised mean density, as discussed above, the pore size and pore size distribution are also relevant. Starting from a high resolved 3D-CT image pores sizes can be calculated. For the calculation of pore size a special software tool was developed [9]. Depending on the type of foam, there are two ways to calculate the size of the pores inside.

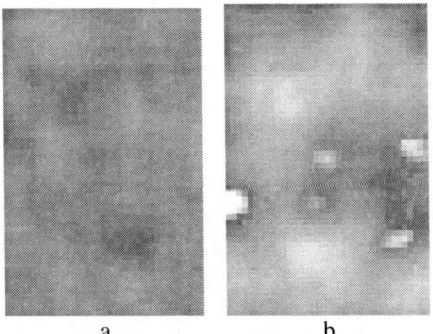

a   b

Figure 5. Two dimensional representation of density distribution:
a– ALPORAS sample, b– ALULIGHT sample.

Calculation of the pore sizes, when pores are closed

When pores are closed, first all volume outside the sample and every volume which is inside the sample, but connected to the outside, is marked. For that a search algorithm in 3D ('invader') is used [13], which expands the marked area as long as voxels in a given range of grey values, adjacent to already marked voxels, are found. Next a search algorithm is looking for non-marked voxels in a specified grey level range, which should be smaller than the first to avoid picking up voxels outside the pores. The field of search is expanded, marking voxel as long as new voxels in the grey level range adjacent to the marked one are found. Each pore is filled with a different colour (different grey level), in order to show in the image which pores are interconnected one to another. The number of voxels in each pore is counted and the

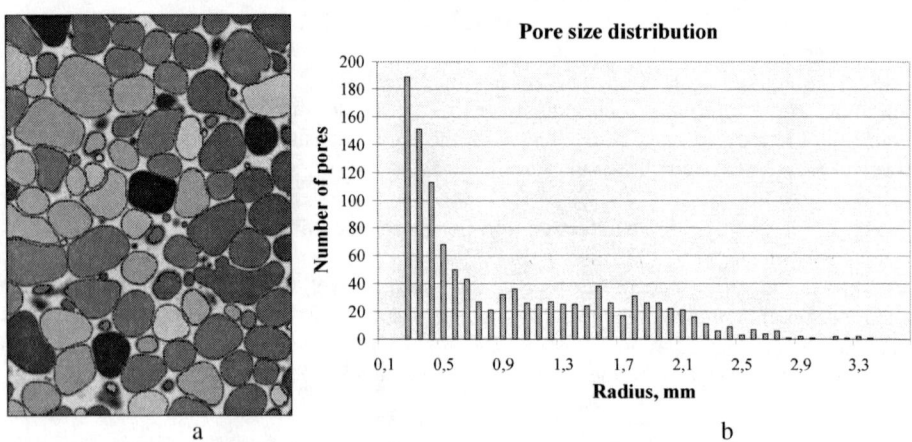

Fig. 6. Pore detection and pore size distribution:
a – example of the program output, b – pore size distribution.

coordinates are summed. When one pore is filled, the radius of an ideal sphere with same volume and the centre of gravity is calculated.

In Figure 6a an example of the program output is shown. Here the pore sizes were calculated from the image of ALPORAS sample (Figure 4a). Figure 6b shows pore size distribution for the same foam sample. The size of the smallest pore, which can be found, depends on the resolution of the image. In this case the radius of the smallest pore, which can be found, was 0.16 mm.

Calculation of the pore sizes for hollow sphere samples

The hollow spheres samples are an ideal example of closed cell foams. The calculated centre of gravity can be used as the centre of an ideal sphere, and as the radius of the spheres is also calculated, a simplified model of the hollow spheres samples can be generated. With only few parameters per cell it is an ideal input

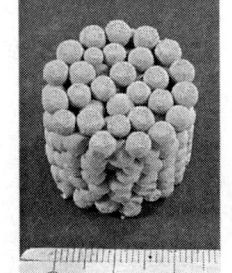

Fig. 7. The hollow spheres sample.

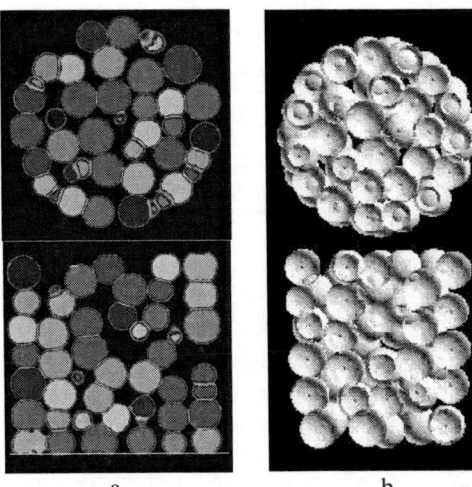

for FEM programs. Once the correlation between adjacent spheres is set, the mechanical parameters can be predicted [14]. In Figure 7 the photo of hollow spheres sample is shown.

Figure 8. Example of: a – pore detection program, b - FEM input [14].

a   b

Separation of walls and nodes

Beside the local density and the pore size (pore size distribution) an important feature of the foams is the ratio between the content of material in the walls and in the nodes. From 3DµCT data it is possible to separate walls and nodes.

For the separation of walls and nodes the image is binarised. Then it is 'eroded' in 3D until all walls disappear. The number of necessary steps depends on the measurement resolution and on the foam type and must be controlled manually. As the orientation of the walls is more or less random, the operation must be done in 3D. Then the same number of steps of '3D dilate' is performed. For the separation of walls and nodes logical operations between images have to be performed. For the separation of nodes logical AND between binarised image and 'dilated'

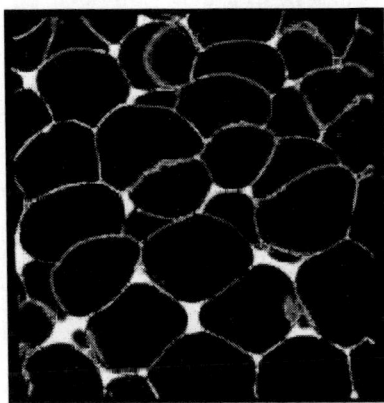

Figure 9. Separation of walls and nodes using 3D 'erode' and 'dilate'.

image has to be performed. For the separation of walls the image of the nodes is subtracted from binarised image of the foam. Figure 10 shows the image of the nodes painted over the original image in such a way that the nodes are highlighted.

Using a statistical approach the thickness and thus the total length of the walls can be calculated [15]. In the example shown, the average thickness of the walls was 20µm. In fact some areas which are not regular nodes were found. This occurs if drainage has taken place and large stiff areas are within the sample.

Internal deformation

The internal deformation of strength tested foams was studied using 3D micro tomography. The samples were tomographed before the compression and after the each step of compression. 3D images of the samples were used to find out, in which region of the foam deformation has started. To achieve this, a program for image comparison in 3D was developed. Using this program, the image of the strength tested sample is compared with the image of the sample before the strength test.

First the size of the region has to be defined, which will be compared with the region of the sample after the compression test. This region has to be as large as the size of the mean pore. The region is moved in different directions in 3D, until it fits best with respect to undeformed sample, and the shift of the parts of strength tested sample is written to an array. For all small regions of the sample we get shifts in respect to the not deformed sample in three directions: Δx, Δy and Δz.

The displacement of different parts of the sample after compression test is shown in millimetres using different grey levels which are painted over the original foam structure, in order to show the displacement of compressed foam in comparison to original foam structure. In the images presented here we show only the shift in z direction, and only in one slice of the sample, although these calculations were performed in 3D and the shifts in all directions were calculated.

The sample of ALPORAS foam was used in a strength test series with subsequent compression and CT-testing. First, the sample was axially compressed along the z direction by 5%. The CT measurements were performed before the compression of the foam, after the sample was compressed by 2%, and after 5% compression. In Figure 10 a vertical slice through the foam sample before and after the two compression steps (by 2% and by 5%) is shown. The buckling of walls is visible in the image of this foam. In Figure 11 it is shown, how much different parts of the foam moved after the compression according to their place before the compression. From presented images it is possible to see, that this foam is deforming proportionally.

As the presented images show, the program allows us to calculate shifts of the parts of the strength tested samples from the 3D tomograms on different types of the foams. This tool could be used in the future for detailed structural analysis of the foam and the visualisation and analysis of the deformation processes in the foam in 3D.

Conclusions

Studies of metal foams using a new (up to 320kV) micro focus X-ray tube were presented. The results of measurement show, that high resolution 3DµCT enables detailed structural analysis of cellular metals. Different features of cellular metals can be characterised starting from high resolution 3D X-ray tomograms:
- mean density of the foam can be calculated;
- low and high density areas can be detected;
- pore size can be calculated;
- pore size distribution can be determined;
- average thickness of the walls can be calculated;
- deformation process within the sample can be visualised.

Acknowledgments

This work was partially supported by Deutsche Forschungsgemeinschaft (DFG).

References

[1] Metal Foams and Porous Metal Structures. Editors: J. Banhart, M. F. Ashby, N. A. Fleck. Verl. MIT Publ., (1999), 416.

[2] H. Riesemeier, J. Goebbels, B. Illerhaus, Y. Onel, P. Reimers, "3-D Mikrocomputertomograph für die Werkstoffentwicklung und Bauteilprüfung", DGZfP Berichtsband 37 (1993), 280-287.

[3] B. Illerhaus, J. Goebbels, P. Reimers, H. Riesemeier, "The principle of computerized tomography and its application in the reconstruction of hidden surfaces in objects of art", 4th Int. Conf. NDT of Works of Art DGZFP, Berichtsband 45, 1 (1994), 41-49.

[4] L. A. Feldkamp, L. C. Davis, J. W. Kress, "Practical Cone-Beam Algorithm", J. Opt. Soc. Amer. A1, (1984), 612-619.

[5] MX-5 tube build by Yxlon international, Halfdangsgade 8, 2300 S Copenhagen, (john.nielsen@cph.yxlon.com, henning.wacher@cph.yxlon.com)

[6] T. Miyoshi, M. Itho, S. Akiyama, A. Kitahara, "Aluminum Foam, "ALPORAS". The Production Process, Properties and Applications", Metal Foams and Porous Metal Structures. Editors: J. Banhart, M. F. Ashby, N. A. Fleck. Verl. MIT Publ., (1999), pp. 125-132.

[7] F. Simancik. "Reproducibility of aluminium foam properties", Metal Foams and Porous Metal Structures. Editors: J. Banhart, M. F. Ashby, N. A. Fleck. Verl. MIT Publ., (1999), 235-240.

[8] H. Göhler et al., "Manufacture and properties of hollow sphere structures in sound absorption applications", Cellular metals and metal foaming technology / [Int. Conf. On cellular Metals and Metal Foaming Technology, 18th – 20th June 2001, Bremen (Germany)]. Ed.: John Banhart et al. – Bremen: Verl. MIT Publ., (2001), 391-396.

[9] E. Jasiuniene., B. Illerhaus, J. Goebbels, "3D investigation of metallic foams by Mikrotomography (µCT)", In: AIPnD (Hrsg.), Proc. 15th World Conference on Non-Destructive Testing Rom 15.-21. Oktober 2000, ID-Nr. 170; CD-ROM.

[10] B. Illerhaus, E. Jasiuniene, J. Goebbels, P. Löthman, "Investigation and image processing of cellular metals with highly resolving 3D micro-tomography (µCT)", Proc. of SPIE Vol 4503 paper 26, to be published.

[11] E. Jasiuniene, J. Goebbels, B. Illerhaus, P. Löwe, A. Kottar, 3D "Investigation of Strength Tested Metallic Foams by Micro-Tomography", Cellular metals and metal foaming technology / [Int. Conf. On cellular Metals and Metal Foaming Technology, $18^{th} - 20^{th}$ June 2001, Bremen (Germany)]. Ed.: John Banhart et al. – Bremen: Verl. MIT Publ., (2001), 251-254.

[12] H. P. Degischer, A. Kottar, "On the Non-Destructive Testing of Metal Foams", Metal Foams and Porous Metal Structures. Editors: J. Banhart, M. F. Ashby, N. A. Fleck. Verl. MIT Publ., (1999), 213-220.

[13] B. Illerhaus, J. L. Thompson, "Calculating CT data from matched geometries" DGZfP Berichtsband BB67 (CD), Computertomographie und Bildverarbeitung, Berlin, (1999).

[14] J. Gründer et al., "Modelling and simulation of the meso- and macro-mechanical properties of hollow sphere strucutres", Cellular metals and metal foaming technology / [Int. Conf. On cellular Metals and Metal Foaming Technology, $18^{th} - 20^{th}$ June 2001, Bremen (Germany)]. Ed.: John Banhart et al. – Bremen: Verl. MIT Publ., (2001), 397-402.

[15] M.R. Sené, M. Bailey, B. Illerhaus, J. Goebbels, O. Haase, A. Kulish, N. Godon, J.J. Choucan, "Characterisation of Accessible Surface Area of HLW Glass Monoliths by High Energy Accelerator Tomography and Comparison with Conventional Techniques", Project report Nuclear Science and Technology, EUR 19119 EN, (1999), 62.

SECTION V

ADDITIONAL ABSTRACTS

PERSPECTIVES ON HIGH-VOLUME AUTOMOTIVE APPLICATIONS OF LIGHTWEIGHT CELLULAR METALS AND STRUCTURES

Ray Jahn and Andrew M. Sherman
Ford Motor Company, FRL Manufacturing Systems, MD 3135
2101 Village Rd
Dearborn, MI 48124 USA

Abstract

Laminated and cellular materials have been proposed as a way of achieving higher weight reductions in automotive structures than available with high strength steels, aluminum, magnesium and polymer composites. However, it remains to be seen whether the applications in production vehicles predicted for these materials will develop. One reason is the high cost of making the materials and fabricating them into components. Also, although it is easy to show that such materials offer superior properties, e.g. specific stiffness, it is much more difficult to translate this into lighter weight in fully engineered structures. Among the difficulties are that in applications having complex loading the directional properties of these materials may be a disadvantage, the load carrying characteristics of the available joining methods do not permit full utilization of the materialís capabilities, or other aspects of the specific application compromise the materials in some way. In this paper we present a cost/benefit analysis of the use of lightweight materials for automotive weight reduction, which helps establish cost and weight target ranges for the use of cellular and laminated metals.

Processing and Properties of Lightweight Cellular Metals and Structures
Edited by Amit Ghosh, Tom Sanders and Dennis Claar
TMS, 2002

MECHANICAL BEHAVIOR OF PERIODIC HOLLOW SPHERE FOAM

Wynn S. Sanders and Lorna J. Gibson
Massachusetts Institute of Technology
Dept. of Matls. Science & Engineering
77 Massachusetts Ave., Rm. 8-135
Cambridge, MA 02139 USA

Abstract

Many metallic closed-cell foams contain manufacturing defects such as cell wall curvature and irregular cell sizes that greatly reduce their overall mechanical properties. Eliminating these defects could improve performance by as much as a factor of ten at low densities. Hollow sphere foams provide a possible solution to the drawbacks of closed-cell metal foams because they can be manufactured into relatively defect-free structures. In this work, finite element modeling was used to evaluate the mechanical behavior of periodic hollow sphere structures. The relative thickness of the spheres and the bond size between the spheres was varied to produce relative densities of 2-20 percent. Both the elastic and yield behavior of the hollow sphere structures were fully evaluated. It was determined from the simulations that the mechanical performance of hollow sphere foams is between the theoretical performance of open- and closed-cell foams. Preliminary experiments have validated the finite element simulations.

MECHANICAL BEHAVIOR OF A CLOSED-CELL ALUMINUM FOAM

Carl M. Cady, George T. Gray, Carl P. Trujillo
Los Alamos National Laboratory, MST-8, MS G755, Los Alamos, NM 87545 USA
and
Toshi Mukai
Osaka Municipal Technical Research Institute, Osaka 536-8553 Japan

Abstract

The compressive deformation behavior of a closed-cell Aluminum foam manufactured by Alporas was evaluated under static and dynamic loading conditions as a function of temperature. High strain rate tests (2000/s) were conducted using the split Hopkinson pressure bar. Quasi-static and intermediate strain rate tests were conducted on a hydraulic load frame. There appears to be little change in the flow stress behavior as a function of strain rate, but the behavior is strongly influenced by temperature. Annealed specimens will be tested to evaluate the effect of pre-existing substructure and the strain rate sensitivity of the material. Localized deformation and Stress State instability issues will be discussed in detail since the behavior over the entire range of strain rates indicates non-uniform deformation.

TAILORED COMPONENT FABRICATION USING STABILIZED ALUMINUM FOAM

Greg Mills and Scott Nichol
Cymat Corporation
1245 Aerowood Dr., Missasauga, ON L4W 1B9 Canada

Abstract

The use of stabilized aluminum foam (SAF) in component design can have numerous advantages. By taking advantage of SAFís ability to provide energy absorption, acoustic damping, lightweight, electromagnetic shielding, recyclable or non-flammable nature, components can be tailored to meet a broad range of criteria with a single material. With these material advantages in mind, Cymat is targeting the transportation industry as having the greatest need for a product such as ours. The use of SAF in automotive interiors and crush zones can provide increased passenger safety as well as producing a more fully recyclable automobile. Along with automotive opportunities, applications in architectural and military applications are also being developed. Cymat Corp is in a position to provide these benefits in a variety of forms; with our foam technology we are capable of producing a wide range of densities and thickness in foam panels and 3-D castings.

Processing and Properties of Lightweight Cellular Metals and Structures
Edited by Amit Ghosh, Tom Sanders and Dennis Claar
TMS, 2002

PROCESSING AND PROPERTIES OF STEEL FOAM MATERIALS AND STRUCTURES

T. Dennis Claar, Virgil Irick, Kenneth Kremer and Jim Adkins
Fraunhofer USA, Center for Manufacturing & Advanced Materials
501 Wyoming Rd., Newark, DE 19716 USA

Abstract

The powder metallurgy-based foaming process is capable of producing foams from a variety of metals, including aluminum, tin, lead, copper, and iron. This paper will focus on the processing and properties of steel foams for weight reduction of engineering structures. Steel foam precursors are produced by mixing iron powders with alloying elements and a foaming agent, and then consolidating the powder mixtures into dense compacts via hot isostatic pressing, hot pressing, extrusion, or hot rolling processes. The precursor material is then foamed by heating into the semi-solid state, during which evolved gas forms the closed-cell foam structure. The effects of precursor consolidation method, as well as various alloy and foaming agent additions, on steel foam microstructure will be discussed. The relationship between steel foam compressive strength and porosity level will be shown and compared with aluminum and titanium metal foams. The fabrication and characterization of steel foam sandwich panels and foam-filled tubes will be described. Potential uses of steel foams in ship structures, automotive, and other applications will be presented.

Processing and Properties of Lightweight Cellular Metals and Structures
Edited by Amit Ghosh, Tom Sanders and Dennis Claar
TMS, 2002

DEVELOPMENT OF VIBRATION DAMPING TECHNIQUES IN METALLIC-INTERMETALLIC LAMINATE (MIL) COMPOSITES

Aashish Rohatgi, Kenneth S. Vecchio
University of California, Mech. & Aeros. Eng.,
9500 Gilman Dr., EBU-2, Rm. #262, La Jolla, CA 92093-0411
and
John B. Kosmatka
University of California, Struct. Eng.,
9500 Gilman Dr., La Jolla, CA 92093-0085 USA

Abstract

Lightweight Ti-Al_3Ti metal-intermetallic laminate (MIL) composites with Lodengraf damping have been developed. Lodengraf damping was achieved by creating cylindrical cavities in the composites and filling them with low density glass spheres. For a given Ti-Al3Ti composition, the damping characteristics of the MIL composites was controlled by varying the location and size of the cavities, and by varying the size and quantity of glass spheres used to fill these cavities. These *porous* MIL composite plates were vibration tested using an acoustic excitation and a non-contacting scanning laser vibrometer to determine their modal properties (natural frequency, mode shapes and damping properties). An extra layer of ferritic stainless steel was incorporated in several *porous* Ti-Al_3Ti composites. Vibration induced stress oscillations may produce eddy current damping in such composites, in addition to the Lodengraf damping. The dependence of damping on the processing variables i.e. the size and distribution of perforations, and the size and amount of glass beads, will be highlighted. The effect of introducing porosity on the mechanical strength and toughness will also be described.

AUTHOR INDEX

Adkins, J., 3, 287
Anderson, R., 189
Ashby, M.F., 243

Balch, D.K., 251

Cady, C.M., 285
Church, B.C., 157
Cirincione, R., 189
Claar, T.D., 3, 61, 287
Clark, J.L., 137
Clyne, T.W., 15, 97
Cochran, J.K., 127, 137, 147, 157, 167, 223
Curran, D.C., 97

Davis, N.G., 177
Dunand, D.C., 177, 251

Fleck, N.A., 243
Flemmig, K., 61

Gao, T., 115
Gergely,V., 97
Ghosh, A.K., 73, 211
Gibson, L.J., 284
Goebels, J., 271
Goehler, H., 61
Gray, G.T., 285

Hayes, A.M., 223
Hebsur, M.G., 85, 233
Hurysz, K.M., 167

Illerhaus, B., 271
Irick, V., 3, 287

Jahn, R., 283
Jasinien, E., 271

Kantzos, P.T., 233
Kosmatka, J.B., 288

Kottar, A., 271
Krause, D.L., 233
Kremer, K., 3, 287
Kupp, D., 61

Laag, R., 25
Lavoie, J.A., 261
Lee, K.J., 127, 137, 167

Markaki, A.E, 15
Mazumder, J., 47
McDowell, D., 127, 223
Mills, G., 286
Mukai, T., 285
Mumm, D., 189

Nadler, J.H., 147
Nichol, S., 286
Nutt, S.R., 107, 261

Oh, R., 167
Oruganti, R.K., 211

Rajaram, S., 261
Rajner, W., 25
Rohatgi, A., 288

Sanders, Jr., T.H., 127, 137, 147, 157, 167
Sanders, W.S., 284
Shen, H., 261
Sherman, A.M., 283
Shrotriya, P., 201
Simancik, F., 25
Soboyejo, W.O., 189, 201
Sokolinsky, V.S., 261
Song, Z., 107
Stiles, E., 47
Sun, Y.Q., 115
Sypec, D.J., 35

Tong, X., 73

Trujillo, C.P., 285

Vecchio, K.S., 288
von Bremen, H.F., 261

Waag, U., 61
Westgate, S.A., 15

Whittenberger, J.D., 233

Zhou, J., 189, 201
Zhou, Y., 261
Zupan, M.,

SUBJECT INDEX

Adhesively Bonded, 15
Acceleration amplitude, 256
Aerospace, 61
Alcan, 81
Alporas, 81
Alulight, 25, 81
Al-A356, 77
Al – 5056, 42
Al – 6101, 203
Al - 6061, 6, 80
Al – 6063, 7
Al – 7075, 80, 251
Al-Si-4Cu, 6, 81
AlSi12, 26
Alumina particles, 116
Armor, 8

Beam stiffness, 15, 234
Bending strength, 234
Bending stiffness, 31
Binder, 129
Blast Protection, 135
Blast Threat, 8
Bonding, bridge-type, 193
Biomedical implants, 61
Brazing, 42
Buckling, 205

Calcium carbonate, 99
Calcium hydride, 149
Cantilever specimens, 261
Capillary induced pressure, 101
Capillary induced thinning, 101
Casting, counter-gravity, 87
Casting, gravity, 87
Casting, investment, 87
Casting, Lattice Block, 85
Cell, Open/Closed, 4, 81, 201, 211, 261
Cell, Equiaxed, 4, 81
Ceramic Particles, 116
Coefficient of thermal expansion, (CTE), 51, 145
Compaction, 39

Compression test, 7, 78, 99, 194, 225, 237, 246, 264
Contamination, 76
Controlled porosity, 61, 73
Core, column-supported, 79
Corrugated, 39
Crash Energy management, 7
Creep, 214
Creep recovery, 214
Cu-Ni, Cu-Ag alloys, 157

Decomposition, 97
Density evaluation, 140, 150, 179, 275
Designed material, 47
Design guideline, 83
Diffusion barrier, 149
Diffusion bonding, 211
Diffusivity, 149
Dihedral angle, 110
Direct Metal Deposition, 47
Direct reduction process, 127
Drainage, 121
Droplet, 73
Ductile dimples, 143
Duocell, 201
DURALCAN, 98
Dynamic exitation, 261
Dynamic peak stress, 228, 285

Egg-box, 243
Elastic modulus, 26, 78, 201, 259
Electrical resistance, 15
Electro discharge machining, 78, 202
Electroless Ni, 212
Electroslag refining, 91
Energy Absorption, 10, 226, 243, 247
ERG, 81, 202
Ethanol, 119
Evolution of foam structure, 101
Expanded Metal, 29
Extrudate velocity, 171
Extrusion, 128
Extrusion, liquid, 73

Extrusion defect, 144

Fabrication, component, 286
Face Sheets, 6
Fatigue Strength, 68
Fibers, 19, 75
Fibers, stainless steel, 16
Fiber Mat, 17, 75
Finite Element analysis, 218
Fire Resistance, 7
Flocking Process, 16
Fluid Flow, 89
Foam baking, 98
Foam expansion and shrinkage, 109
Foam, Reinforced, 28
Foam, silica, 69
Foam, syntactic, 251
Foam-filled tube, 6
Foam height, 119
Foam-Sandwich, 6
Foams, periodic, 74-77, 284
Foams, steel, 287
Foams, Stochastic, 37, 77
Foaming agent, 4, 99
Foaming rate, 107, 182
Foaming, solid state, 177
Foil-core, 75
Forging, 77, 81
Four-point Bend test, 29
Fragment Simulating Particles, 7
Fraunhofer, 4, 61, 81
Free energy, 102

Gas bubbles, 108
Gas-liquid interface, 101
Geometric hardening and weakening, 211
Gibbs surface elasticity, 122
Gibbs-Marangoni effect, 122
Heat Exchanger, 40, 157
Heat Sink, 127
High pressure argon, 185
High volume application, 283

Hollow sphere, 63
Hollow sphere, periodic, 80, 285
Homogenization method, 47
Honeycomb, 129, 211
Hot isostatic pressing, 177

Ideal solution, 151
Image Processing, 274
Inclusion, non-metallic, 91
Insert, removable, 73
Inter-cell slip, 112
Intermetallic Compounds, 61
Internal oxidation, 116
In-vitro, 68

Kinetic stability, 116

Laser Beam, 48
Laser Extensometer, 17
Lattice Block, 81, 85, 233
Lattice strain, 256
Ligament, 91, 235
Ligament stress, 239
Linear cell, 127
Liquid extrusion, 73
Liquid infiltration, 253
Liquid meniscus, 212
Liquid stream, 73
Loading, dynamic, 223
Loading, out-of-plane, 223
Loading, parallel, 215
Loading, quasistatic, 223
Loading, transverse, 215

Manufacturing, 29, 128
Manufacturing:computer-aided, 49, 75
Maraging steel, M-200, M-350, 131, 138, 141, 224
MAR-M 247, 95
Melt, thickened, 108
Melting – localized, 10
Metal Foam Composites, 28

Metal-intermetallic laminate, 288
Methylcellulose, 159, 170
Microbalance, 17
Microsphere, 255
Mg alloys, 26
Multifunctionality, 3, 40, 127
Multi-material, 47

Natural frequencies, 261
Net shape part, 6, 76
Nichrome, 38
Nickel, 81
Nickel-base alloy, IN617, IN718: 85, 127, 211, 233
Normalized density, 81
Normalized strength, 81

Osteoblast, 68
Oxide paste, 137
Oxide reduction, 152

Parametric study, 54
Partial pressure, 151
Perforated sheet, 27, 40
Pipe structure, 83
P/M process, 25
Plastic bending, 205
Plastic collapse, 204, 230
Plateau border, 101, 110
Plateau stress, 38, 81, 195, 228
Polymer particles, 115
Pore detection, 275
Pore distribution, bimodal, 76
Pore growth, 182
Pore size, 178
Porosity, controlled, 61, 73
Porosity, gradeded, 73
Porosity, interconnected, 71
Porosity, multiscale, 73
Porosity, open/closed, 69, 184
Porosity, terminal, 177
Porous coating, 69

Potential drop, 17
Powder, submicron size, 127
Powder paste, 131
Precursor, 98
Prismtic cavities, 83
PTFE, PVC, 117

Reactive sintering, 61, 71
Relative Density, 43, 244
Reticulated Open Cell, 69
Rheology, 107, 167
Rheometry, capillary, 169
Rule-of-mixtures, 81

Sandwich panel/structure, 5, 15
Scaffold, polymer, 67
Scaffold, titanium, 70
Semi-solid zone, 75
Sesquioxide, Fe-,Cr-, 147
Shear instability, 216
Shear modulus, 42
Silica, amorphous, 255
Silicon carbide particle, 101, 116
Sintered P/M material, 161
Sintering time, 189
Skin, solid, 27, 76
Slip band, 204
Solid oxide fuel cell, 127
Solidification, 75, 252
Spherical cavities, 77
Split Hopkinson apparatus, 228
Spray, 73
Square cell, 157
Stabilization, 115,
Stainless Steel-304, 7, 42,
Stainless Steel-316L, 16
Stainless Steel Assembly, 16
Steel Sheet, Brazed, 16
Stiffness, 15
Stiffness of core, 23
Stiffness-to-weight ratio, 26
Strain hardening, 214
Strain measurements, 253

Strength, anisotropic, 78, 215
Strength, isotropic, 78
Strength, through-thickness, 243
Structure, periodic, 73, 128
Structure, stochastic, 73
Strut, 203
Subelement test, 240
Suction pressure, 121
Super Invar, 127, 140
Superplasticity, 177
Surface tension, 122, 212
Surfactant, 116
Swelling, 64
Synchrotron, 253

Tertiary creep, 216
Textile Laminates, 37
Thermal conductivity, 131, 159
Theoretical strength and modulus, 81
Thermal cycle, 179
Thermochemical processing, 157
Thermodynamic Stability, 119
Thermodynamic modeling, 151
$TiAl_3$, 64
Titanium Alloys, 61, 81, 177, 189
Ti-6Al-4V powder, 191
Titanium Aluminide, 63
Titanium Hydride, 5, 98
Tomography, 272
Tool Path, 56
Topology Optimization, 49, 86
Transformation stresses, 177
Transient Liquid Phase, 40, 212
Triple point, 212
Truss, periodic, 37
Truss, tetrahedral, 43
Truss structure, 35, 83, 86
TSP, 181

Unit Cell, 58
UTS prediction, 142

Vacuum arc remelting, 91
Vibration damping, 14, 61, 288
Vibration frequencies, 261
Viscosity, 107, 116

Wall slip, 111
Wax Pattern, 88
Weldability, 15
Welding, in-situ, 75
Welding, Resistance, 21
Welding, Spot, 15
Welding zone, 75
Wetting Behavior, 115
Wetting Angle, 119
Wire-weld structure, 74
Woven Wire Mesh, 39, 211
Woven wire bonded, 214

X-ray diffraction, 253

Yield/Peak strength, 81, 214, 238, 246

Zn-4Cu, 81